Data Fitting and Uncertainty

Tilo Strutz

Data Fitting and Uncertainty

A practical introduction to weighted least squares and beyond

2nd, revised and extended Edition

Springer Vieweg

Tilo Strutz
Leipzig, Germany

ISBN 978-3-658-11455-8
DOI 10.1007/978-3-658-11456-5
Springer Wiesbaden Heidelberg New York Dordrecht London

Library of Congress Control Number: 2015957585

Springer Vieweg
Springer Fachmedien Wiesbaden GmbH is part of Springer Science+Business Media (www.springer.com)

Preface

The topic of optimisation based on the method of least squares has nothing lost of its relevance and topicality. This gives me the chance to update the content of this book. Aside from many little improvements, there are also some major additions compared to the first edition.

The set of outlier detection methods has been extended by two robust methods. The first is a modification of the z-score method replacing the mean value by the median and replacing the standard deviation by the median absolute deviation. The second method is the random sample consensus (RANSAC), which estimates the most suitable set of model parameters in the presence of a probably high portion of outliers. These two methods are also implemented and tested in the software, which is available together with this book. The software also has been extended enabling the provision of weights in the input file along with conditions and observations.

Subsection 6.4.5 has been extended by two examples showing the performance of the Levenberg-Marquardt method in application to a non-linear optimisation problem.

The chapter 'Supplemental Tools and Methods' has been complemented by a section describing the treatment of correlated data. Correlation appears, when observations or measurements tend to have errors with the same sign, e.g., if the errors are influenced by a common mechanism. In addition, the text about alternative optimisation methods has been revised and complemented with a subsection explaining the training of neural networks using backpropagation.

Besides this, Wiener filters and least-mean-square (LMS) filters are discussed as they have a strong relation to the method of least squares, and many examples have been added.

I would like to thank Dr. Rainer Feistel for the valuable input regarding the text about uncertainty.

Leipzig, July 2015 Tilo Strutz

Preface to 1st edition

The method of least squares has been developed by Carl-Friedrich Gauß in 1795 and has been published 1809 in volume two of his work on celestial mechanics *Theoria Motus Corporum Coelestium in sectionibus conicis solem ambientium.* Although it is a rather old technique, it is used still today in many practical applications and has lost nothing in its importance, but has been developed further in several directions.

The idea of this method is to determine the behaviour of systems for which just a limited number of samples, i.e. pairs of input and output, are given. If a certain model function describing the system is presumed, then the parameters of this function are searched which explain the system best in terms of minimisation of the quadratic error. In case of simple systems, applying the method of least squares reduces to, for instance, drawing an approximating curve through a scatter plot of data points.

The early fields of application included, for instance, astronomy (modelling of celestial bodies' movement) and geodesy with the purpose of alleviating measurement errors. Nowadays, this method of approximating data points is probably applied in all sciences, especially those dealing traditionally with statistics as, for instance, physics, biology, economy, or psychology, but also in more recent fields as computer vision. There is basically no difference between experimental data and computer generated data with respect to the data analysis via the method of least squares.

The principle of least squares is generally straightforward, however, its computational intricacy is heavily dependent on the character and complexity of the model function which should describe the behaviour of the system under investigation. So, from the introduction of this data-fitting method in the 19th century until now many numerical techniques for proper and elegant treatment have been developed. In addition, mathematical tools for the evaluation of the fitting results have had to be derived.

The intention of this textbook is to introduce the topic in a more comprehensible manner enabling the reader to solve his/her own specific data-fitting problem. It describes the method of least squares and the treatment of uncertainty in a recipe-like way. Emphasis is put on the unified and generic approach from which the solutions for particular applications can be derived. The book fills the gap between those explaining only very simple data-fitting techniques and those looking at the topic at very high mathematical level. The text discusses many real-life and simulated examples and is accompanied with working source code in the programming language C, which is provided on the publishers website (see 'OnlinePlus' area).

The book addresses mainly engineers and software programmers or corresponding undergraduates, which are not primarily interested in proofs and derivations of the formulae, but want to quickly become acquainted with the topic and its pitfalls in order to solve a particular fitting problem. A basic mathematical background, especially a familiarity with matrices and partial derivatives as well as some knowledge about variances and standard deviations, is beneficial. Readers, not yet familiar with the method of least squares, are recommended to start with the first chapter followed by the others in sequential order. Every chapter finishes with a list of test questions, enabling a cross-check of whether the subject has been understood.

The material is organised in two parts. The generic basics of data fitting by the Gaussian method of least squares are introduced in the first part. Several examples of linear systems as well as of nonlinear systems are discussed thoroughly in order to enable the reader to solve similar problems.
Chapter 1 explains the general idea of data fitting and defines the terminology and the notation. Chapter 2 presents all principal formulae for the application of the least-square method. In Chapter 3, methods for the estimation of weights are proposed. This is essential, for instance, when uncertainties of recorded data points are not known in advance or the data contain outliers. A clustering approach is discussed, which separates the possible outliers from the bulk of useful data points. An assessment of the results of the data fitting is described in Chapter 4.

The second part of the book is mostly dedicated to the underlying theory and also some numerical methods are discussed in more detail.
It starts with Chapter 5 which is about basic linear algebra and algorithms for matrix inversion. Chapter 6 explains the idea of least squares introducing the principle of maximum likelihood. On this basis, different techniques for the solution of linear and nonlinear fitting problems (i.e. optimisation) are discussed. In Chapter 7 some methods and tools are collected which are helpful in connection with data fitting.

The appendix contains two chapters. The first is about investigations regarding a cluster-based outlier-detection method. The second chapter describes a data-fitting software and discusses several issues of implementing the least-squares method. It concludes with a detailed performance tests of the software.

Leipzig, July 2010 Tilo Strutz

"He felt that his whole life was some kind of dream and he sometimes wondered whose it was and whether they were enjoying it." – Douglas Adams, The Hitchhiker's Guide to the Galaxy

Contents

Part I

Framework of Least-Squares Method

"Everything must be made as simple as possible. But not simpler." – Albert Einstein

Chapter 1

Introduction to Data-Fitting Problems

1.1 What is data fitting?

The classical setup for data fitting is based on an experiment or measurement. The measurements yield different results, dependent on the experimental conditions. For instance, we can think of measuring the electrical current I (*observation*) in a resistor R dependent on the given voltage U (*condition*) across this resistor.

Typically, the underlying functional relation between the outcome of the measurement and the corresponding conditions is known or assumed. This relation is described by a mathematical model, which is called *model function* throughout the book. If we expect, for instance, the resistor R to be constant, the functional connection would be $I = U/R$ according to Ohm's law. If the electrical component is of any other type, the relation between I and U could be different and another suitable model has to be used.

The goal of data fitting is to find those parameters of the model function that best describe the connection between observations and conditions.

The mathematical description is as follows.
The outcome of the measurement or experiment is called observation y. It is dependent on several conditions, described by a vector $\mathbf{x}$. The parameters of the model are combined in a vector $\mathbf{a}$ with M elements a_j, $j = 1 \ldots M$. They are unknown and have to be determined by data fitting methods. Thus, the model function is

$$y = f(\mathbf{x}|\mathbf{a}) \,. \tag{1.1}$$

In case of the aforementioned Ohm's law, the current I is equivalent to the observation y, the vector of conditions reduces to a scalar $x = U$, and $\mathbf{a} = a = R$ is the only parameter to be determined.

Sometimes more than one value is observed for a single experiment. We could, for instance, think of measuring not only the current I flowing through the resistor, but also its temperature T. However, since the corresponding model function $T = f(I)$ will be different, y is only considered as a scalar for the moment and not as a vector of observations.

It is impossible to change the experimental conditions continuously in practice. Discrete points indicated by variable i have to be selected instead. Besides this, the observations are affected by random errors (noise). That is why the problem has to be re-formulated as

$$y_i = f(\mathbf{x}_i|\mathbf{a}) + \varepsilon_i \; . \tag{1.2}$$

The vector $\mathbf{x}_i$ contains the condition for experiment i and can be regarded as a set of independent variables, whereas the y_i are considered as dependent variables. It is assumed that $\mathbf{x}_i$ is known exactly or that the error in $\mathbf{x}_i$ is at least negligibly small compared to the error in the measured observation y_i. Only $\mathbf{a}$ is unknown and has to be calculated using suitable data-fitting methods. The repetition of a certain experiment under the same condition $\mathbf{x}_i$ leads to different values of y_i, because observations are subject to the random error ε_i. This error is presumed having a mean of zero. Doing the experiment an infinite number of times would reveal the true *parent distribution* of y_i. In practice, however, the parent distribution is unknown and can only be estimated to a certain accuracy.

Since we are looking for an optimal solution for $\mathbf{a}$, data fitting is a kind of optimisation problem. Strictly speaking, it concerns minimisation, as will be shown later on. It must be emphasized that the optimum of $\mathbf{a}$ is never absolute, but rather always related to the measured data and the assumptions taken with regard to the reliability of the observations y_i or the distribution of their errors [Mey84]. When the model parameters $\mathbf{a}$ are found, the model function can be used for the interpolation and, provided that the model is not only an approximation for a limited subset of conditions, also for the extrapolation of data.

Data fitting is often referred as *data reduction*, since the number of N observations is reduced to a mere M parameters plus the selection of a suitable model function. Keeping in mind that the model does not represent all observations exactly, but instead, acts as approximation, it also can be regarded as special kind of lossy data compression [Str09a].

In some applications, the true functional connection between the experimental conditions $\mathbf{x}_i$ and the observations y_i is not known. In this case, it is a common method to approximate y for a limited range of $\mathbf{x}$ by using a superposition of special functions of higher order like polynomials, trigonometric functions, or something else. However, the decision for or against a certain model function must be made very carefully, since it will heavily influence the result of the data-fitting procedure. In particular, model functions that are not based on a true or realistic physical, chemical, or other model should not be used to extrapolate data.

1.2 Notation

Three basic types of variables are used in this book. Scalar variables are always denoted in *italics*, for instance, y, $f(x)$, a_1 etc. Vectors are expressed in small, bold-faced letters, such as $\mathbf{x} = (x_1, x_2, \dots x_N)^{\mathrm{T}}$. The exponent T means "transposed", making $\mathbf{x}$ a column vector. Please note that T is not in *italics* because it is not a variable. The same holds true for indices, which are not consecutive numbers such as i in y_i but rather simply indicate that a certain variable is related to something else, like σ_{y}^2. The letter y is not a serial number here; it signals that σ^2 is the variance in y. Finally, matrices are denoted with capital, bold-faced letters such as $\mathbf{J}$ and $\mathbf{W}$, for example.

The elements of matrices are scalars, thus the elements of matrix $\mathbf{A}$ are denoted by a_{ij} with $i = 1, \dots, N$ and $j = 1, \dots, M$ as the row and column indices, respectively. The set of integers is denoted by $\mathbb{Z}$. $||\mathbf{x}||$ is the Euclidean norm of column vector $\mathbf{x}$, i.e., $||\mathbf{x}|| = \left(\mathbf{x}^{\mathrm{T}} \cdot \mathbf{x}\right)^{1/2} = \left(\sum_i x_i^2\right)^{1/2}$. Pages 267f list most of the variables, which are used throughout the book.

Readers who are not familiar with matrix algebra are requested to read Section 5.1 first.

A signal (or sequence of values) is written as $\{x[n]\}$, while $x[n]$ is the sample (or signal value) at the position n[1]. The brackets indicate that the signal is discrete in n. Parentheses are used in case of continuous signals (e.g., $f(t)$).

1.3 Linear vs. nonlinear problems

The discussion of the linearity of a data-fitting problem is indispensable, since the kind of treatment and its complexity depend heavily on this property.

In engineering, the term "linear" is associated with things like linear time invariant (LTI) systems or linear signal transformations and indicates that an additive superposition of input signals leads to an additive superposition of the according

[1] For the sake of simplicity, the curly brackets are sometimes omitted and $x[n]$ is also interpreted as the entire signal.

output signals. In addition, a characteristic curve $y = f(x)$ is called linear if it is a straight line, which is basically the same because of

$$y_1 = f(x_1)\,, \qquad y_2 = f(x_2) \qquad \longrightarrow \qquad y_1 + y_2 = f(x_1 + x_2)\,. \tag{1.3}$$

In the field of data fitting, the terms "linear" and "nonlinear" are used with different meanings. They are not assigned to the relation between the observation y and the vector of conditions $\mathbf{x}$ of the model function $y = f(\mathbf{x}|\mathbf{a})$, but instead to the relation between y and the model parameters a_j $(j = 1 \ldots M)$! For instance, the exponential function

$$y = f(\mathbf{x}|\mathbf{a}) = a_1 \cdot \mathrm{e}^{-x} \tag{1.4}$$

is regarded as linear in $\mathbf{a}$, despite the nonlinear term e^{-x}, since the first derivative is $\partial y/\partial a_1 = \mathrm{e}^{-x}$ and no longer dependent on $\mathbf{a}$. In contrast to that

$$y = \mathrm{e}^{a_2 \cdot x} \tag{1.5}$$

is obviously nonlinear due to a_2, because $\partial y/\partial a_2 = x \cdot \mathrm{e}^{a_2 \cdot x}$ is still dependent on a_2.

Sometimes a chosen model function seems to be nonlinear at first glance, but turns to linear if treated properly. Since data fitting of nonlinear functions is in general more difficult than data fitting of linear ones, the desired model should always be inspected carefully to determine whether it is in fact nonlinear or simply *pseudo-nonlinear*.

Example (1):
The model function

$$y = b_1 + (x - b_2)^2\,, \tag{1.6}$$

for example, seems to be nonlinear, since b_2 is part of a squared term. However, the function can be transformed into

$$\begin{aligned} y &= b_1 + x^2 - 2 \cdot b_2 \cdot x + b_2^2 \\ &= b_1 + b_2^2 - 2 \cdot b_2 \cdot x + x^2 \end{aligned}$$

and at closer look it becomes apparent that the equation can be rewritten as

$$y = a_1 + a_2 \cdot x + x^2\,. \tag{1.7}$$

with

$$a_1 = b_1 + b_2^2 \quad \text{and} \quad a_2 = -2 \cdot b_2\,.$$

It turns out that the model is linear in $\mathbf{a}$!
□□□

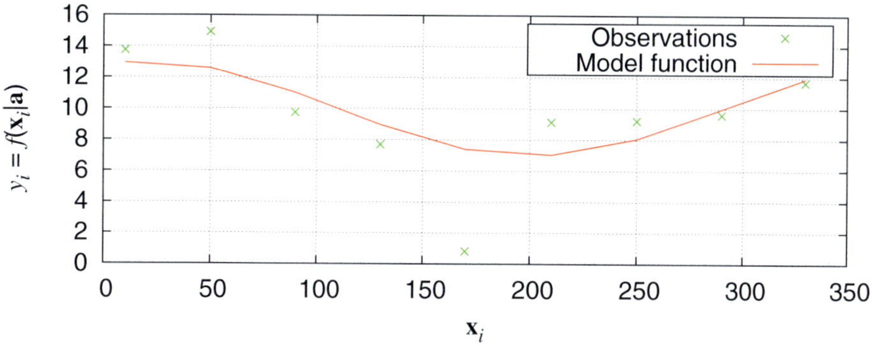

Figure 1.1: Example of a pseudo-nonlinear model function: $y = b_1 + b_2 \cdot \cos(x - b_3)$

Example (2):
Another example is the cosine function

$$y = b_1 + b_2 \cdot \cos(x - b_3) \ . \tag{1.8}$$

This model could be used to describe a circular movement, if the observation is restricted to the projection of the movement onto a screen perpendicular to the plane of movement as for instance described in [Pot06]. **Figure 1.1** shows the recording of nine observations in dependence on known rotation angles x_i.

There are three unknowns to be determined via data fitting: b_2 is the radius of the circular movement, b_3 is a phase offset dependent on the direction of the projection, and b_1 indicates the centre of the performed movement.

It is somewhat more difficult to discover a proper transformation here. Looking for a suitable trigonometric operation, we find

$$\cos(a - b) = \cos(a)\cos(b) + \sin(a)\sin(b)$$

and after its application to the initial model function (1.8)

$$y = b_1 + b_2 \cdot [\cos(x)\cos(b_3) + \sin(x)\sin(b_3)] \ .$$

Using

$$a_1 = b_1, \quad a_2 = b_2 \cdot \cos(b_3), \quad \text{and} \quad a_3 = b_2 \cdot \sin(b_3)$$

we get

$$y = f(\mathbf{x}|\mathbf{a}) = a_1 + a_2 \cdot \cos(x) + a_3 \cdot \sin(x) \ , \tag{1.9}$$

which is a linear model function with respect to **a**. After fitting of (1.9) it is straightforward to determine b_j from a_j.
□□□

Example (3):
Many applications require the fitting of points belonging to a circle, whose function is given as

$$(y_1 - a_1)^2 + (y_2 - a_2)^2 = r^2 \,, \tag{1.10}$$

with r as the radius and (a_1, a_2) as centre of the circle. The main problem here is that there are no conditions, but only observations. Therefore, this formula cannot be used directly with respect to the setup $y = f(\mathbf{x}|\mathbf{a})$. A trick is needed, and it will turn out that the problem can be converted in a linear one. Subsection 1.5.3 discusses the details.
□□□

In contrast to pseudo-nonlinear problems, real nonlinear model functions can be linearised only if alterations are permitted on the left side of the function. This is a valid operation; however, it leads also to a different optimisation problem with different results compared to the original model.

The exponential function

$$y = b_1 \cdot \exp(b_2 \cdot x) \tag{1.11}$$

is the standard example to demonstrate the linearisation of nonlinear models. In order to release b_2 from the exponent, the logarithm has to be applied to both sides of the equation

$$\begin{aligned} \ln(y) &= \ln\left[b_1 \cdot \exp(b_2 \cdot x)\right] = \ln(b_1) + b_2 \cdot x \\ &= a_1 + a_2 \cdot x \end{aligned} \tag{1.12}$$

which is a linear straight-line problem. However, as the logarithm of y is now fed into the computations instead of y itself, the errors in y are also weighted differently. The points of lower values of y influence the determination of the slope and intercept of the line now too strongly compared to higher values of y.

The data fitting is no longer optimised according to b_1 and b_2 and yields different results.

Even a variation of the exponential model

$$y = b_1 \cdot \exp(b_2 \cdot x + b_3) \tag{1.13}$$

Table 1.1: Transformations for linearisation of nonlinear functions leading to $v = a_1 + a_2 \cdot u$

Function	u	v	a_1	a_2
$y = b_1 \cdot x^{b_2}$	$\ln(x)$	$\ln(y)$	$\ln(b_1)$	b_2
$y = b_1 + \frac{b_2}{x}$	$\frac{1}{x}$	y	b_1	b_2
$y = \frac{b_2}{x + b_1}$	x	$\frac{1}{y}$	$\frac{b_1}{b_2}$	$\frac{1}{b_2}$
$y = \frac{b_1 \cdot x}{x + b_2}$	$\frac{1}{x}$	$\frac{1}{y}$	$\frac{1}{b_1}$	$\frac{b_2}{b_1}$

can be linearised. Upon closer inspection, it becomes apparent that b_1 and b_3 are indistinguishable

$$\begin{aligned}\ln(y) &= \ln\left[b_1 \cdot \exp(b_2 \cdot x + b_3)\right] \\ &= \ln(b_1) + b_2 \cdot x + b_3 \\ &= a_1 + a_2 \cdot x\end{aligned}$$

with $a_1 = \ln(b_1) + b_3$ and $a_2 = b_2$.
In fact, the ambiguity of parameters in this example might even inhibit the success of solving the data-fitting problem numerically. It must be emphasized once again that model functions should be selected carefully and simplified as much as possible.

Other functions can also be linearised in a similar manner. Setting

$$v = a_1 + a_2 \cdot u$$

as (linear) target model, the true model function can be converted using suitable substitutions for x and y. For instance, equation (1.11) was linearised by applying $v = \ln(y)$, $u = x$, $a_1 = \ln(b_1)$ and $a_2 = b_2$. **Table 1.1** provides solutions for the linearisation of some other functions (based on [Pap01]).

1.4 Example applications of linear data fitting

This Section and the next are intended to explain the broad field of data fitting applications and are not required for the comprehension of the subsequent chapters. Impatient readers may wish to skip to Chapter 2. Nevertheless, the author recommends reading these sections after the reader has familiarised himself or herself with the method of least squares. Some of the explanations therein touch on special problems that might be relevant to the reader's setup.

1.4.1 Estimation of a constant value

The simplest problem in data fitting is to determine a certain constant value, which could be the mean of some observations.

Let us assume that we want to find out how often people working in an office look out of the window during the day. We employ a statistics trainee to watch the people and let her count the occurrences of looking-out-of-the-window separately for each person. The result of a ten-day experiment watching six people each day yields the observations depicted in **Figure 1.2**a)[2]. Furthermore, we should disregard that special conditions can influence the counted numbers dramatically, for instance, if the office staff were exclusively male and the window pointed to a street with shops typically attracting young women.[3] Generally, some observations deviate from the mean stronger than others. Somebody may have spotted a beautiful bird and wants to keep track of it. Alternatively, another person could only have eyes for the trainee and does not look out of the window at all. This 'anomalous' behaviour disturbs the statistics. If we get lucky, these so-called *outliers* balance each other out. However, outliers are very often biased to one side (birds are of higher interest than the trainee is). It is therefore essential to detect these mavericks and to treat them properly, as it is discussed in Chapter 3.

The mathematical description of (1.1) simplifies to

$$y = f(\cdot|\mathbf{a}) = \text{const.} = a_1 \; . \tag{1.14}$$

No special conditions are distinguished and the outcome y of the measurement is equal to a constant value a_1 representing the entire model.

[2] The values are chosen arbitrarily and are not the result of a real experiment.
[3] The results for the summer and winter seasons will also differ dramatically in that scenario.

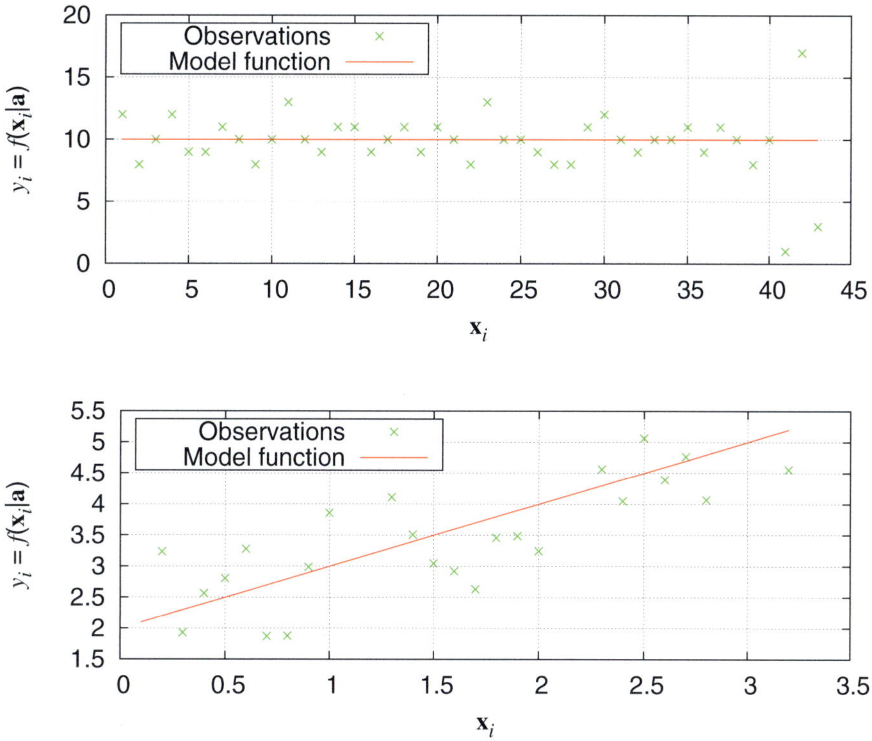

a)

b)

Figure 1.2: Example for model functions: a) determination of a constant value, b) fitting of a straight line.

1.4.2 Estimation for a straight line (linear regression)

Another very famous example of data fitting, often used in economics and medical science, for example, deals with straight lines.

The mathematical description is still very elementary

$$y = f(\mathbf{x}|\mathbf{a}) = a_1 + a_2 \cdot x \ . \tag{1.15}$$

The variable y is linearly dependent on a single condition x and two parameters (a_1, a_2) have to be determined. **Figure 1.2** b) depicts an example.

1.4.3 Polynomial function

There are scenarios where the relation between the experimental conditions and the experiment results (observations) has to be described without knowing the underlying model function. In these cases, the chosen range of conditions is approximated by a linear combination of special functions, which could be polynomials of degree $M-1$, for instance

$$y = f(\mathbf{x}|\mathbf{a}) = a_1 + a_2 \cdot x + \cdots + a_M \cdot x^{M-1} \,. \tag{1.16}$$

After the data fitting, values a_j very close to zero can indicate, for instance, that the corresponding part of the polynomial is not required to describe the examined problem. This, however, only holds true if the standard uncertainty of the observations is rather low. Additional model parameters are quickly integrated in order to fit the random component of the data. The standard uncertainty of single estimated parameters in percent (see Chapter 4) is a somewhat more reliable indication of unnecessary parts of the model function. If the random error in the observations is not too high, the uncertainties of suspicious parameters are high in comparison to other parameters.

1.4.4 Multiple linear regression

So far, the condition vector $\mathbf{x}$ had contained one single element only. We can also conceive of (M-1)-dimensional problems

$$y = f(\mathbf{x}|\mathbf{a}) = a_1 + a_2 \cdot x_1 + a_3 \cdot x_2 + \cdots + a_M \cdot x_{M-1} \,. \tag{1.17}$$

Approximation of planes

One example of multiple linear regression is the approximation of planes, which is in fact two-dimensional. The setup is

$$y = f(\mathbf{x}|\mathbf{a}) = a_1 + a_2 \cdot x_1 + a_3 \cdot x_2 \,, \tag{1.18}$$

where x_1 and x_2 might be coordinates. The parameters a_2 and a_3 are the gradients with different orientation according to the used coordinate system and a_1 is merely an offset. An example is shown in **Figure 1.3**.

Supervised classification

Typically, classification is not associated with data fitting. Nevertheless, data fitting is applicable for the distinction between objects of two classes, for instance.

According to equation (1.17) the classification is based on $M-1$ features forming the condition vector $\mathbf{x}$. The parameters a_j must be determined using a training

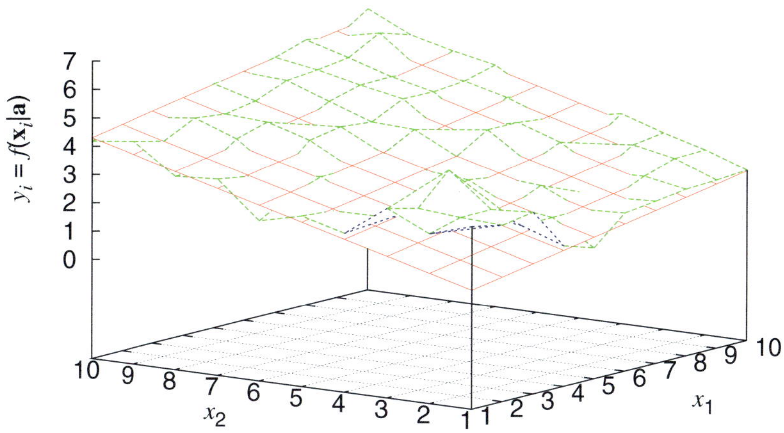

Figure 1.3: Example for a two-dimensional model function: plane approximation. The green vertices indicate the (noisy) observations y_i.

set of objects with known assignments to one of the classes. The class numbers are the observed values y_i. Once the parameters are known, the classification of objects not included in the training set becomes possible.

Practical applications typically require several independent features to distinguish the objects. As an explanation, we consider here the simplest case of classification using a single feature x. Since in supervised classification the class membership is known for the training set of objects, we can determine the distribution of the feature for each class separately (**Figure 1.4**).

Now let us assign the observation value $y_i = 1$ to the first class and the value 2 to the second class. The numbers are slightly randomised for better visualisation in **Figure 1.4**b) affecting the spread around 1 and 2. Albeit the scatter plot in Figure 1.4b) does not follow a straight line, the model function

$$y = a_1 + a_2 \cdot x$$

has to be fitted to the data.

After model parameters a_1 and a_2 are determined, new objects can be classified by using their feature value as condition x. The resulting value of y has to be compared to the threshold of 1.5: if y is larger, the object is assigned to class 2; otherwise it belongs to class 1. Even a rejection of objects could be managed, if two thresholds are applied.

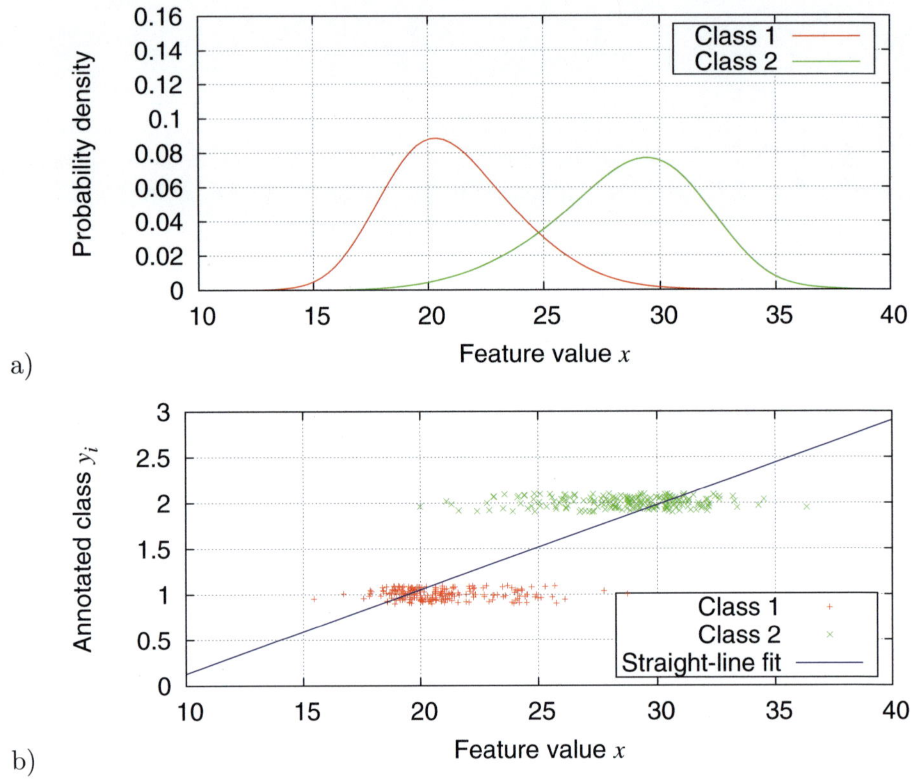

Figure 1.4: Supervised classification: a) distributions of same feature for different classes; b) assignment of class numbers; see text for details

There are only two prerequisites for a successful distinction.

First, the training set must comprise a comparable number of objects in each class. In the example of Figure 1.4b) $N_1 = N_2 = 200$ objects are used for each class. If class 1 only contained very few training objects compared to class 2, they had only weak influence on the fitting procedure and the straight line would be almost entirely captured by the second class, i.e., $a_1 \approx 2$ and $a_2 \approx 0$. This problem could be overcome simply by using different weights[4] for the classes c in order to balance their influence, i.e. $w_c = 1/N_c$.

The second prerequisite concerns the linear separability of the two classes. If, for instance, the feature values were distributed as depicted in **Figure 1.5**a), it would be impossible to make a linear decision based on only one feature. Either two or

[4] The use of weights is discussed in detail in Chapter 3.

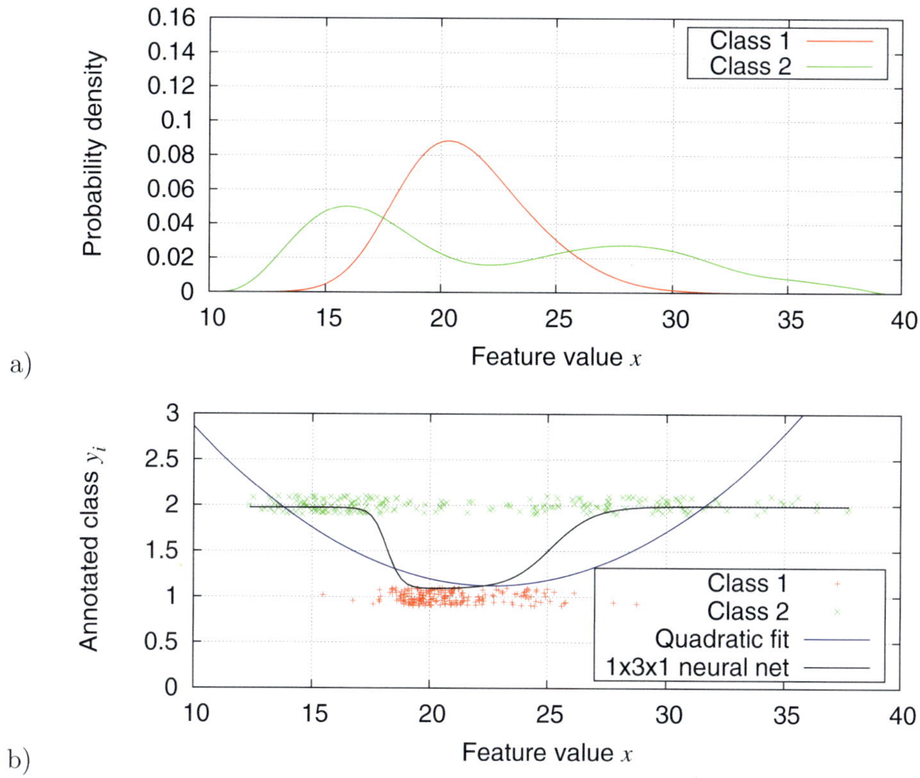

Figure 1.5: a) Class distributions that cannot be linearly separated by a single feature; b) Discrimination using a parabola vs. a neural network

more features have to be used or a nonlinear[5] approach must be applied. However, the selection of a proper model function has to be made carefully, when using the least-squares method for fitting. A parabola does not result in good distinction between these two classes, as can be seen in **Figure 1.5**b). However, applying a least-squares-based neural network (see Subsection 1.5.4) with one input and three hidden neurons (10 parameters in total) yields excellent distinction.

Affine transformation

Another example for multiple linear regression is the modelling of affine transformations of coordinates, for instance (**Figure 1.6**). For two-dimensional problems,

[5] Either nonlinear with respect to x or a nonlinear model function (see Section 1.3).

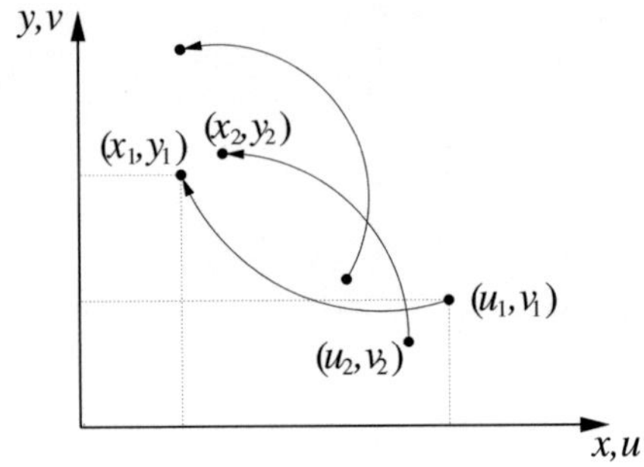

Figure 1.6: Affine transformation of coordinates

this can be described via

$$\begin{pmatrix} x \\ y \end{pmatrix} = \begin{pmatrix} a_1 \\ a_2 \end{pmatrix} + \begin{pmatrix} a_3 & a_5 \\ a_4 & a_6 \end{pmatrix} \cdot \begin{pmatrix} u \\ v \end{pmatrix} \tag{1.19}$$

or in matrix notation

$$\mathbf{x} = \mathbf{A} \cdot \mathbf{u} = \begin{pmatrix} x \\ y \end{pmatrix} = \begin{pmatrix} a_1 & a_3 & a_5 \\ a_2 & a_4 & a_6 \end{pmatrix} \cdot \begin{pmatrix} 1 \\ u \\ v \end{pmatrix} . \tag{1.20}$$

Given a set of pairs of coordinates $(u_i; v_i)$ and $(x_i; y_i)$, the parameters of the geometrical transformation have to be determined. The vectors $\mathbf{u}$ and $\mathbf{x}$ are the coordinates of points before and after the transformation, respectively. The vector $(a_1\ a_2)^{\mathrm{T}}$ introduces a translational displacement to the points, whereas $a_3 \ldots a_6$ are responsible for more complex changes, like rotation, scaling and sheering. Based on given corresponding points $\mathbf{u}_i$ and $\mathbf{x}_i$, the coefficients $a_1\ \ldots a_6$ have to be estimated. The components of $\mathbf{x}$ in equation (1.20) can be calculated independently of each other using

$$\begin{aligned} x &= a_1 + a_3 \cdot u + a_6 \cdot v \\ y &= a_2 + a_4 \cdot u + a_6 \cdot v \ , \end{aligned} \tag{1.21}$$

which is in fact the same fitting problem as the plane approximation (eq. 1.18). Only the interpretation of the data is different.

In the case of a translation combined with a pure rotation with an angle of θ, $\mathbf{A}$ would be

$$\mathbf{A} = \begin{pmatrix} a1 & \cos(\theta) & -\sin(\theta) \\ a_2 & \sin(\theta) & \cos(\theta) \end{pmatrix} .$$

However, since there are now constraints on the parameter determination, namely

$$a_3 = a_6 = \cos(\theta) \qquad \text{and} \qquad a_4 = -a_5 = \sin(\theta) \; ,$$

the two equations in (1.21) are therefore dependent on each other and the linear affine transformation has changed into a nonlinear model.

Linear prediction

A special application of data fitting is the prediction of time series (or other sequences of data, such as in image-compression algorithms [Str09a]). The experimental conditions are specified here by the time-series samples $x[\cdot]$ that have already be seen, i.e., which lie in the past. The linear prediction problem can be expressed with

$$\hat{x}[n] = f(\mathbf{x}_n|\mathbf{a}) = \mathbf{x}_n^{\mathrm{T}} \cdot \mathbf{a} \; , \tag{1.22}$$

if $\mathbf{x}$ and $\mathbf{a}$ are column vectors. $\hat{x}[n]$ is the predicted value of $x[n]$. Taking only three samples from the past for prediction, i.e. $\mathbf{x}_n^{\mathrm{T}} = (x[n-1] \; x[n-2] \; x[n-3])$, the model function is

$$\hat{x}[n] = f(\mathbf{x}_n|\mathbf{a}) = a_1 \cdot x[n-1] + a_2 \cdot x[n-2] + a_3 \cdot x[n-3] \; . \tag{1.23}$$

For optimal prediction, the signal should have a mean of zero. This can be achieved by subtracting the mean of the signal $x[n]$ from all samples beforehand.

The idea behind the procedure of prediction is to forecast the current sample value $x[n]$ dependent on its M predecessors $x[n-1] \dots x[n-M]$. Before the predictor can be applied, the model-parameters vector $\mathbf{a}$ has to be determined using a training set of known combinations of $x[n]$ and its predecessors. Please note that the mean value of the prediction error $e[n] = x[n] - \hat{x}[n]$ is assumed to be zero in general. If, however, the prediction is performed in dependence on different contexts (i.e., several vectors of parameters are used, one for each context), the mean of the prediction might deviate from zero and the correction of $\hat{x}[n]$ by an additional parameter a_0 in equation (1.23), i.e. $\hat{x}_c[n] = \hat{x}[n] + a_0$, should be considered. This is called bias cancellation [ISO00].

The scenario of predicting single pixel values in images is depicted in **Figure 1.7**. The setup is based on the presumption that the single pixel positions are scanned starting with the top-left corner, row-by-row downwards to the bottom-right corner of the image. Let us assume that the scanning process has reached the position marked with 'X'. All pixels above the thick line have been already processed and 'X' is the next value $x[n,m]$ to be predicted. A template defines which of the neighbouring pixels are involved in the computation of the estimate

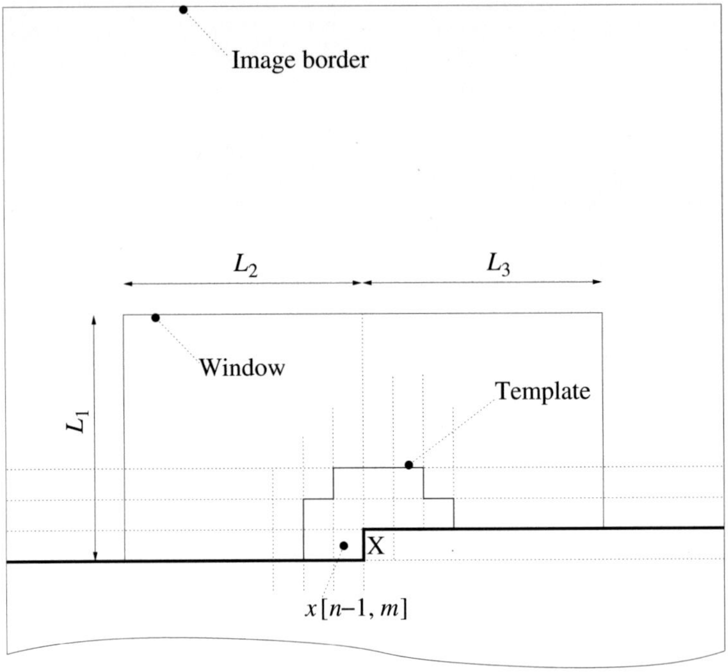

Figure 1.7: Environment for the prediction of single pixel values

$\hat{x}[n,m]$. That means these values act as conditions with respect to the model function (1.1). Given the template in Figure 1.7, the model function is

$$\begin{aligned}\hat{x}[n,m] = {} & a_1 \cdot x[n-1,m] + a_2 \cdot x[n,m-1] + \\ & a_3 \cdot x[n-1,m-1] + a_4 \cdot x[n+1,m-1] + \\ & a_5 \cdot x[n-2,m] + a_6 \cdot x[n,m-2] + \\ & a_7 \cdot x[n-2,m-1] + a_8 \cdot x[n-1,m-2] + \\ & a_9 \cdot x[n+1,m-2] + a_{10} \cdot x[n+2,m-1] \, .\end{aligned}$$

The setup further presumes that the characteristics of the image signal do not change very much within the vicinity of the current position. Thus the model parameters a_j can be determined based on the relations between the pixel values in the window defined by L_1, L_2, and L_3 in Figure 1.7. This window contains all pixel positions from which the pixel values are taken as observations. The conditions arise from the values at the corresponding template positions. In total, there are $L_1 \times L_2 + (L_1 - 1) \times L_3$ observations available for the determination of the model parameters $a_1 \dots a_{10}$, which are used afterwards for the computation of the estimate $\hat{x}[n,m]$.

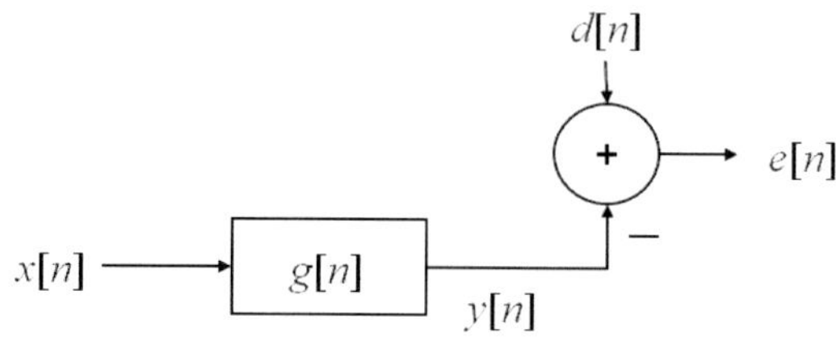

Figure 1.8: Set-up or the optimisation of the impulse response $g[n]$. See text for details.

If the template (or the window) is not located completely within the image area, an exception handling is required.

1.4.5 Wiener filter

Wiener[6] filters are filters that are optimised for a specific task. The discrete filter problem is depicted in **Figure 1.8**. Two signals are given, a reference signal $d[n]$ and another signal $x[n]$, which is to be filtered in such a manner that the filter output $y[n] = x[n]*g[n]$ is as close as possible to the reference. If the energy of the difference signal $e[n] = d[n] - y[n]$ is subject to minimisation, the problem relates to the method of least squares. The values of the reference signal correspond to the observations $d[n] \to y_i$, the signal values $x[n] \to x_{i,j}$ are the conditions, the coefficients of the impulse response are the parameters $g[n] \to a_j$, and

$$\begin{aligned} x[n] * g[n] &= \sum_{k=0}^{M-1} g[k] \cdot x[n-k] \\ &= f(\mathbf{x}|\mathbf{a}) = a_1 \cdot x_1 + a_2 \cdot x_2 + \cdots a_M \cdot x_M \end{aligned}$$

is the model function.

1.5 Selected nonlinear data-fitting problems

1.5.1 Exponential functions

Subsection 3.5.7 operates with an exponential function of type

$$y = a_1 \cdot \exp(a_2 \cdot x) \tag{1.24}$$

[6] Named after the American mathematician Norbert Wiener (1894 – 1964).

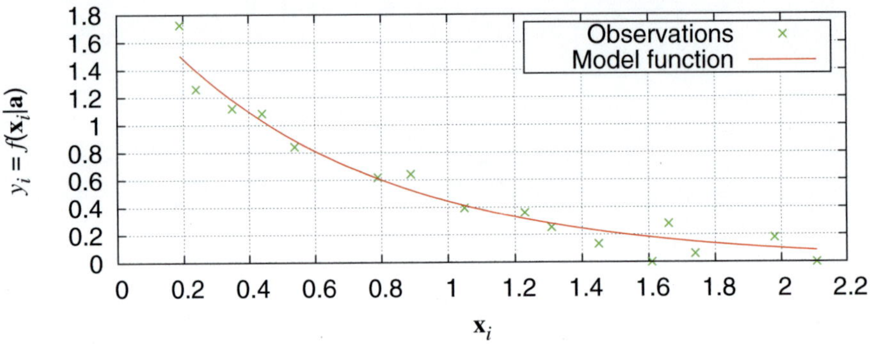

a)

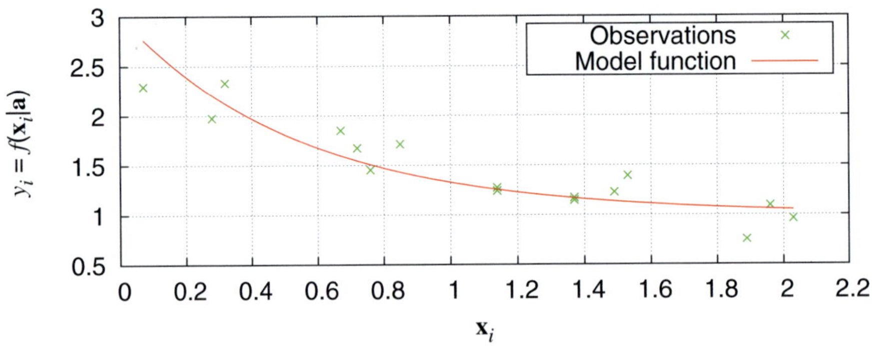

b)

Figure 1.9: Examples of exponential functions: a) $y = a_1 \cdot \exp(a_2 \cdot x)$; b) $y = a_1 + a_2 \cdot \exp(a_3 \cdot x)$

in order to study the effects of linearisation (see Section 1.3) for the selected case of $a_1 = 2, a_2 = -1.5$ (**Figure 1.9**a). The conditions of the test-data set used in this example are sampled in a non-equidistant manner. Nevertheless, it is assumed that the conditions x_i are exactly known and only the observations y_i are uncertain according to the statements in Section 1.1. There are techniques that can also deal with uncertain conditions. However this goes beyond the scope of this book and is discussed only briefly in Section 7.7.

A simple variation of equation (1.24)

$$y = a_1 + a_2 \cdot \exp(a_3 \cdot x) \tag{1.25}$$

eludes from linearisation and methods for nonlinear data fitting have to be used. **Figure 1.9**b) shows an example with $a_1 = 1.0$, $a_2 = 2.0$, and $a_3 = -1.8$.

1.5.2 Composite Gaussian bell functions

The fitting of distributions is another nonlinear problem, which can be modelled, for instance, via superposition of two Gaussian bells with different amplitudes c, mean values μ, and variances σ^2

$$y = c_1 \cdot \exp\left[-\frac{(x-\mu_1)^2}{2\cdot\sigma_1^2}\right] + c_2 \cdot \exp\left[-\frac{(x-\mu_2)^2}{2\cdot\sigma_2^2}\right] .$$

In the language of data fitting, this can be written as

$$y = a_1 \cdot \exp\left[a_2 \cdot (x-a_3)^2\right] + a_4 \cdot \exp\left[a_5 \cdot (x-a_6)^2\right] \qquad a_2, a_5 < 0 .$$

Figure 1.10 shows two examples of superposition: additive and subtractive. Despite the very different model parameters $a_1 \dots a_6$, the modelled functions look very similar. This leads to the conclusion that the fitting of those data sets is ambiguous.

1.5.3 Circle function

The function of a circle is given by

$$(y_1 - a_1)^2 + (y_2 - a_2)^2 = r^2 , \tag{1.26}$$

with r as radius of the circle centred at (a_1, a_2) (**Figure 1.11**). It is obvious that there are only observations (y_1 and y_2), but no conditions. The radius is just another parameter, say $r = a_3$. In order to fit this function into the framework of model functions a trick is required. We exchange conditions and observations and rewrite (1.26) as

$$y = 0 = (x_1 - a_1)^2 + (x_2 - a_2)^2 - a_3^2 . \tag{1.27}$$

Consequently, the observations are in principle correct, whereas the conditions are erroneous. The evaluation of the fitting results becomes more complicated (e.g. see [Kas76]).

[Coo93] and others have shown that it is also possible to linearise the model function (1.27) in the manner that was discussed in Subsection 1.3. Let us rewrite equation (1.27) as

$$\begin{aligned} 0 &= (x_1^2 - 2\cdot x_1 \cdot a_1 + a_1^2) + (x_2^2 - 2\cdot x_2\cdot a_2 + a_2^2) - a_3^2 . \\ &= x_1^2 + x_2^2 - 2\cdot x_1\cdot a_1 - 2\cdot x_2 \cdot a_2 + a_1^2 + a_2^2 - a_3^2 . \end{aligned} \tag{1.28}$$

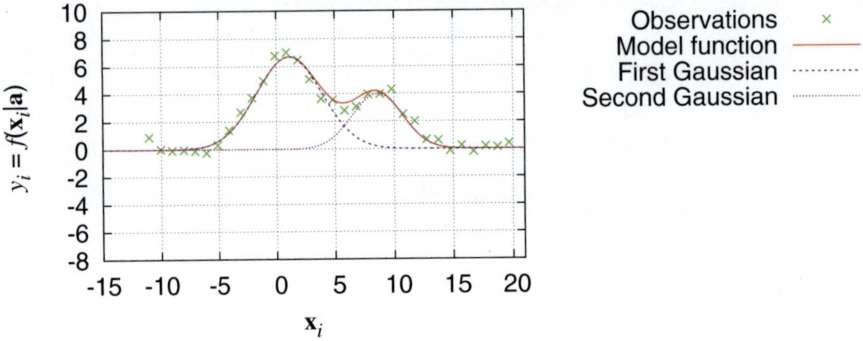

a)

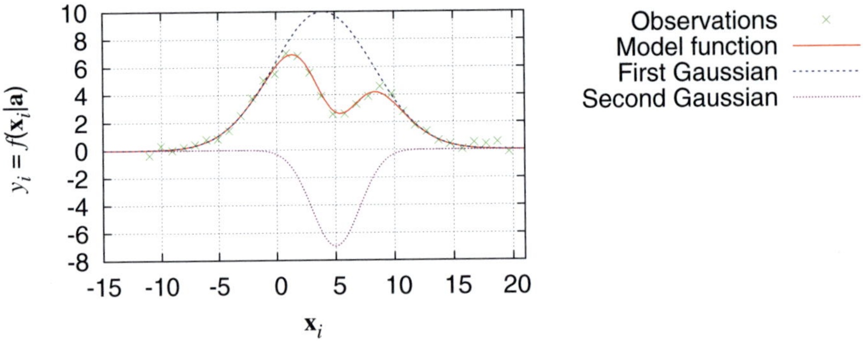

b)

Figure 1.10: Example of composite Gaussian functions; a) sum of two positive Gaussian bells with $a_1 = 6.7$, $a_2 = -0.06378$, $a_3 = 1.1$, $a_4 = 4$, $a_5 = -0.11125$, $a_6 = 8.6$ and $\sigma_\varepsilon^2 = 0.25$; b) difference of two Gaussian bells $a_1 = 10$, $a_2 = -0.27778$, $a_3 = 4$, $a_4 = -7$, $a_5 = -0.125$, $a_6 = 5$ and $\sigma_\varepsilon^2 = 0.25$

Using $b_1 = 2 \cdot a_1$, $b_2 = 2 \cdot a_2$, and $b_3 = a_1^2 + a_2^2 - a_3^2$ results in

$$-(x_1^2 + x_2^2) = -b_1 \cdot x_1 - b_2 \cdot x_2 + b_3 \ , \tag{1.29}$$

which is linear in $\mathbf{b} = (b_1 \ \ b_2 \ \ b_3)^{\mathrm{T}}$.

Please remember that x_1 and x_2 are observations, which now appear on both sides of the equation. The fitting of a linear model function is not applicable if all observations (values on the left side) are equal to zero as in equation (1.28), hence the term $(x_1^2 + x_2^2)$ must be moved to the left side of the model function.

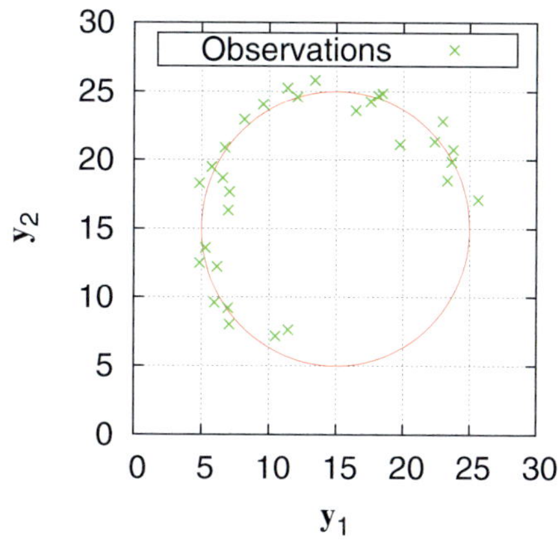

Figure 1.11: Observed points of a circular object

After fitting with this linearised version, the values of the original variables a_1, a_2, and a_3 can then be recovered by

$$a_1 = 0.5 \cdot b_1 \qquad a_2 = 0.5 \cdot b_2 \qquad a_3 = \sqrt{a_1^2 + a_2^2 - b_3}\,.$$

1.5.4 Neural networks

For a long time, artificial neural networks were regarded as a completely new method of signal processing. Because of the variety of different possible topologies of networks and kinds of signal propagation, this belief is true to a certain extend. The mostly used type of *feed-forward networks* (**Figure 1.12**), however, is nothing other than a weighted superposition of typically nonlinear functions and can be easily converted into a data-fitting task. The network in Figure 1.12 has nine parameters in total, four weights α_{lk} propagating the input values x_l to the hidden neurons, three offset parameters α_{0k} and β_0 introducing bias into the summation indicated by Σ, and two parameters β_k weighting the intermediate values h_k. The indices l and k refer to input units and hidden neurons, respectively. The intermediate values result from the weighted summation of the input values followed by a limitation of the dynamic range, which can be achieved for instance using the sigmoid function

$$h_k(t_k) = \frac{1}{1 + \mathrm{e}^{-t_k}} \tag{1.30}$$

with the intermediate value

$$t_k = \sum_l \alpha_{lk} \cdot x_l = \boldsymbol{\alpha}^{\mathrm{T}} \cdot \mathbf{x}\,. \tag{1.31}$$

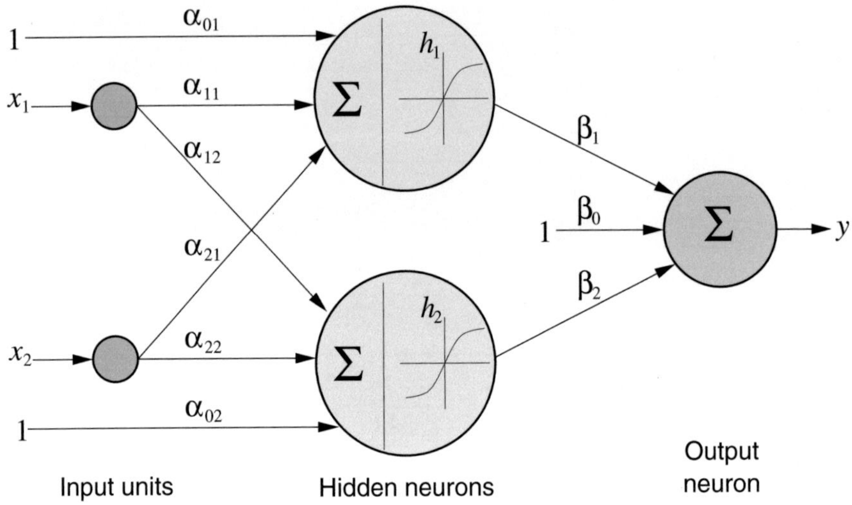

Figure 1.12: Feed-forward 2-2-1 neural network with two input units, one hidden layer consisting of two neurons, and a single output neuron

The output is merely the weighted sum of the intermediate values

$$y = \sum_k \beta_k \cdot h_k = \boldsymbol{\beta}^{\mathrm{T}} \cdot \mathbf{h} \,. \tag{1.32}$$

In order to simplify the computations, the sigmoid function (eq. 1.30) can be replaced by the hyperbolic-tangent function

$$h_k(t_k) = \tanh(t_k) = \frac{2}{1 + \mathrm{e}^{-2\cdot t_k}} - 1 \,, \tag{1.33}$$

by linear transformations of the input

$$0.5 \cdot t_k \quad \mapsto \quad t_k$$

and of the output

$$2 \cdot h_k(t_k) - 1 \quad \mapsto \quad h_k(t_k)$$

(**Figure 1.13**). A neural network using (1.33) is equivalent to a network using (1.30), but having different values for the resulting weights [Bis95].

Putting the equations (1.31), (1.32), and (1.33) together and replacing the α_{lk} and β_k by a_j, a nonlinear model function can be derived for the example given in Figure 1.12

$$\begin{aligned} y = f(\mathbf{x}_n|\mathbf{a}) = &\tanh(a_1x_1 + a_2x_2 + a_3) \cdot a_4 + \\ &\tanh(a_5x_1 + a_6x_2 + a_7) \cdot a_8 + a_9 \,. \end{aligned} \tag{1.34}$$

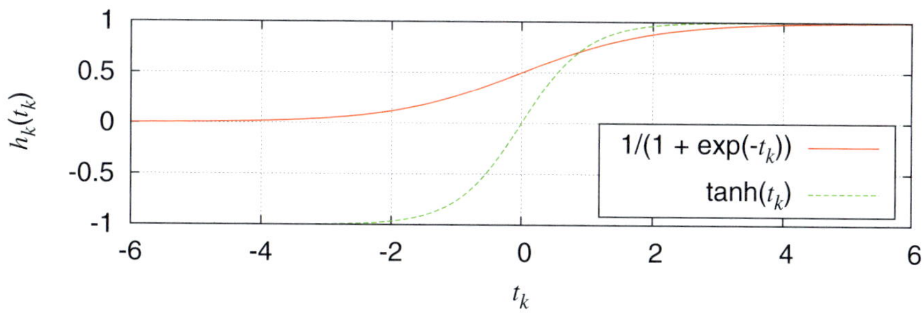

Figure 1.13: Comparison of sigmoid and hyperbolic-tangent function

In summary: the training of the neural network can be performed simply using a data-fitting procedure. Combined with a suitable weighting scheme (see Chapter 3), outliers in the training set can also be treated properly. The more neurons are involved, however, the more complex the model function will become. The alternative optimisation method is to present single pairs of $\mathbf{x}_i$ and y_i followed by weights adjustment using backpropagation of errors through the single layers as described in Subsection 7.1.4.

1.6 Test questions

1.1 What is the difference between conditions and observations?

1.2 The least-squares method deals with model functions. Which values have to be determined?

1.3 What are linear and nonlinear problems in the sense of the least-squares method? Give two examples each.

1.4 What does "linearisation of a model function" mean?

1.5 Give an example application (i.e. model function) with more than one observation per measurement.

1.6 Which values are expected to be error-free in least-squares calculations: observations, parameters, or conditions?

1.7 Itemise three practical applications that could make use of data fitting.

Chapter 2

Estimation of Model Parameters by the Method of Least Squares

2.1 What are "Least Squares"?

It is well known that the determination of N unknown variables requires a set of N linearly independent equations. If the number of equations is lower than the number of unknowns, the problem is said to be under-determined. If there are more linearly independent equations than unknown variables, it is over-determined.

The least-squares method deals with the latter case. Since experimental observations y_i are typically affected by errors expressed by ε_i

$$y_i = f(\mathbf{x}_i|\mathbf{a}) + \varepsilon_i \ , \tag{2.1}$$

it is beneficial to have as many equations (i.e., pairs of y_i and $\mathbf{x}_i$) available as possible in order to get better and more reliable statistics[1]. That means the parameters $\mathbf{a}$ can be estimated more accurately.

Suppose we are given a set of observations y_i with corresponding conditions $\mathbf{x}_i$, under which the observations have been made. The goal of data fitting using the least-squares method is to minimise the squared residual error between the observations and the values calculated via the model function

$$\chi^2 = \sum_i w_i \cdot [y_i - f(\mathbf{x}_i|\mathbf{a})]^2 \longrightarrow \text{Min} \tag{2.2}$$

Based on the maximum-likelihood principle, which is discussed in more detail in Chapter 6, the fitting problem corresponds to a minimisation problem. The value of χ^2 is dependent on the selection of a hopefully appropriate model function $f(\mathbf{x}_i|\mathbf{a})$ and on random fluctuations in the measured values of variable y_i. In

[1] It is assumed that the mean of ε is equal to zero.

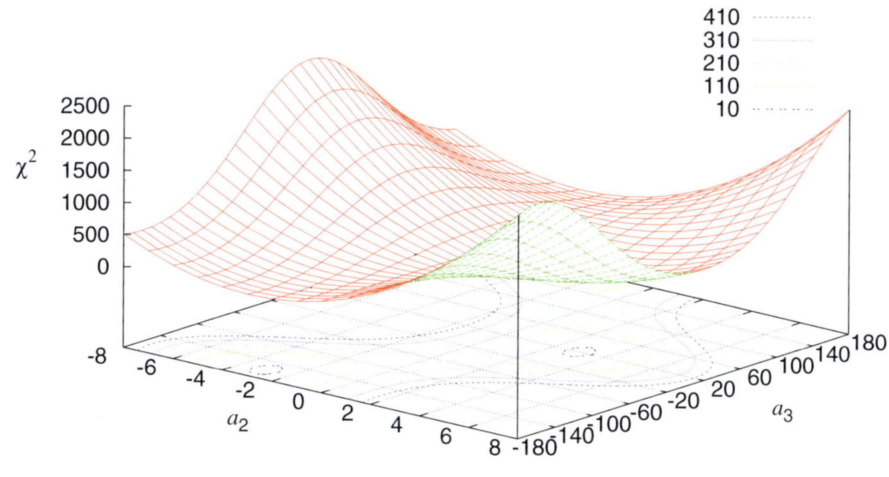

Figure 2.1: Error surface of $\chi^2 = \sum_i [y_i - a_2 \cdot \cos(x_i - a_3)]^2$ with a global minimum at $(a_2, a_3) = (3, 20)$

addition to the already introduced variables, equation (2.2) contains weights w_i. Each weight should express the reliability of the observation to which it is assigned. The more uncertain an outcome y_i of the experiment, the lower w_i should be. For example, the weight has to be close to or equal to zero when a particular observation shows indication of being an outlier. However, in a large number of cases, this reliability is not known in advance, which means the weights have to be estimated (see Chapter 3).

When observations are weighted, the procedure is primarily called *generalised least squares* or *weighted least squares* in contrast to *ordinary least squares* with $w_i = w = \text{const}$.

The influence of a chosen set of parameters $\mathbf{a}$ on χ^2 can be visualised using what are called error surfaces. The procedure for minimising χ^2 has to find the global minimum in that surface. **Figure 2.1** shows an example of the cosine model function $f(\mathbf{x}|\mathbf{a}) = a_2 \cdot \cos(x_i - a_3)$ with a global minimum at the position $(a_2, a_3) = (3, 20)$. Since the cosine is a periodic function, it is evident that this is not the only minimum. Minima can be found at $(3, 20 + k \cdot 360)$ and $(-3, 200 + k \cdot 360)$ with $k \in \mathbb{Z}$. Another example is depicted in **Figure 2.2**. When more than two parameters are involved, we speak of hyper-surfaces.

For linear problems, the position of the global minimum (i.e., the desired parameter vector $\mathbf{a}$) can be calculated directly. Simple problems, such as fitting a constant or a straight line, can easily be treated analytically by setting the par-

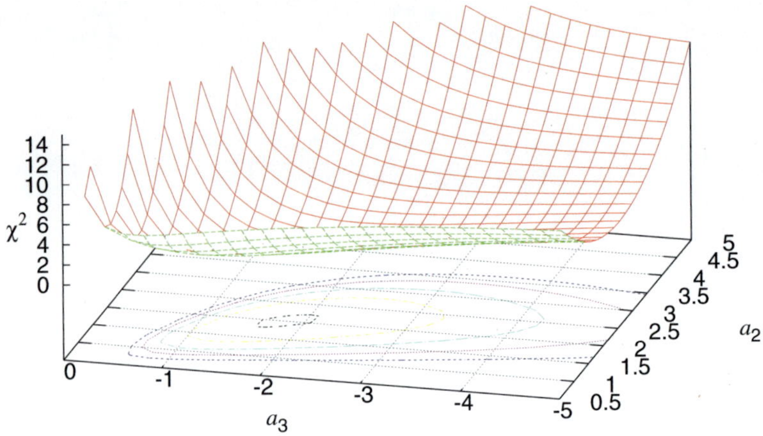

Figure 2.2: Error surface of $\chi^2 = \sum_i [y_i - a_2 \cdot \exp(a_3 \cdot x_i)]^2$ with a global minimum at $(a_2, a_3) = (2, -1.8)$

tial derivatives $\partial\chi^2/\partial a_j$ of equation (2.2) to zero and solving the corresponding equation system by hand. Section 7.4 explains one example in detail.

Nonlinear model functions require the application of methods walking from a starting point gradually towards the global minimum. It is obvious that an initial estimation of parameters **a** is needed and should be as close as possible to that minimum.

The theoretical basis of least squares is explained in more detail in Chapter 6. The following text presents the final formulae to use.

2.2 A general algorithm for minimisation problems

This section presents a general solution for the fitting of nonlinear model functions to data without any proofs or derivations in order to provide the reader with basic information required for their own software developments and implementations. Linear problems can be regarded as a subset simplifying the least-squares procedure in different terms, as will be discussed in the following.

Solving minimisation problems via the least-squares method requires a so-called *Jacobian matrix*[2]. It is of size N (number of observations) times M (number of

[2] named after the German mathematician Karl Gustav Jacob Jacobi (1804-1851)

parameters) and contains all partial derivatives of each equation $y_i = f(\mathbf{x}_i|\mathbf{a}) + \varepsilon_i$ with respect to the observation condition $\mathbf{x}_i$

$$\mathbf{J} = \begin{pmatrix} \frac{\partial f(\mathbf{x}_1|\mathbf{a})}{\partial a_1} & \frac{\partial f(\mathbf{x}_1|\mathbf{a})}{\partial a_2} & \cdots & \frac{\partial f(\mathbf{x}_1|\mathbf{a})}{\partial a_M} \\ \frac{\partial f(\mathbf{x}_2|\mathbf{a})}{\partial a_1} & \frac{\partial f(\mathbf{x}_2|\mathbf{a})}{\partial a_2} & \cdots & \frac{\partial f(\mathbf{x}_2|\mathbf{a})}{\partial a_M} \\ \vdots & \vdots & \ddots & \vdots \\ \frac{\partial f(\mathbf{x}_N|\mathbf{a})}{\partial a_1} & \frac{\partial f(\mathbf{x}_N|\mathbf{a})}{\partial a_2} & \cdots & \frac{\partial f(\mathbf{x}_N|\mathbf{a})}{\partial a_M} \end{pmatrix} . \tag{2.3}$$

In addition, a vector containing all residuals has to be prepared

$$\mathbf{r} = \begin{pmatrix} y_1 - f(\mathbf{x}_1|\mathbf{a}) \\ y_2 - f(\mathbf{x}_2|\mathbf{a}) \\ y_3 - f(\mathbf{x}_3|\mathbf{a}) \\ \vdots \\ y_N - f(\mathbf{x}_N|\mathbf{a}) \end{pmatrix} . \tag{2.4}$$

A residual is expressed by the difference between the observation y_i and the value resulting from the model function using the current vector of parameters

$$\mathbf{a} = (a_1 \; a_2 \; \cdots \; a_j \; \cdots \; a_M)^{\mathrm{T}} .$$

The parameter vector **a** has to be filled with initial estimates at the beginning. A diagonal matrix

$$\mathbf{W} = \begin{pmatrix} w_1 & 0 & 0 & \cdots & 0 \\ 0 & w_2 & 0 & \cdots & 0 \\ 0 & 0 & \ddots & \cdots & \vdots \\ \vdots & \vdots & \vdots & \ddots & 0 \\ 0 & 0 & \cdots & 0 & w_N \end{pmatrix} \tag{2.5}$$

is the third (optional) component for solving the minimisation problem. It contains one weight for each of the N observations and expresses how important or reliable the corresponding observation is for the determination of the model parameters.[3] If all observations have the same impact, this matrix can be ignored by setting all w_i equal to one. Otherwise, proper weights have to be chosen as will be discussed in Chapter 3.

[3] Setting **W** to a diagonal matrix implicates the assumption of independent observations. If there are correlations between single measurements, then the off-diagonal elements may also be non-zero.

Putting everything together, we have to compute[4]

$$\mathbf{\Delta a} = \left(\mathbf{J}^{\mathrm{T}} \cdot \mathbf{W} \cdot \mathbf{J}\right)^{-1} \cdot \mathbf{J}^{\mathrm{T}} \cdot \mathbf{W} \cdot \mathbf{r}\,. \tag{2.6}$$

The result is a vector of parameter adjustments

$$\mathbf{\Delta a} = (\Delta a_1\ \Delta a_2\ \cdots\ \Delta a_j\ \cdots\ \Delta a_M)^{\mathrm{T}}$$

and the new vector of parameters is given by

$$\mathbf{a} \longleftarrow \mathbf{a} + \mathbf{\Delta a}\,. \tag{2.7}$$

The alternating computation of (2.6) and (2.7) has to be performed iteratively until the adjustments are close to zero. The stop criterion could be, for instance,

$$|\Delta a_j| < \epsilon \qquad \forall\, j \tag{2.8}$$

with ϵ close to the CPU precision. In addition, a safeguard against an infinite loop has to be implemented, restricting the number of iterations to a maximum value. The description above is the general procedure for solving least-squares problems of nonlinear functions. Section 2.4 will describe how the computation can be simplified for linear model functions.

The term $\mathbf{J}^{\mathrm{T}}\mathbf{W}\mathbf{J}$ in equation (2.6) is also called *normal matrix* $\mathbf{N}$. It is quadratic with $M \times M$ elements according to the number of parameters of $f(\mathbf{x}|\mathbf{a})$. Substituting $f(\mathbf{x}_i|\mathbf{a})$ by f_i this matrix is

$$\mathbf{N} = \begin{pmatrix} \sum\limits_{i=1}^{N} w_i \left(\frac{\partial f_i}{\partial a_1}\right)^2 & \sum\limits_{i=1}^{N} w_i \left(\frac{\partial f_i}{\partial a_1}\frac{\partial f_i}{\partial a_2}\right) & \cdots & \sum\limits_{i=1}^{N} w_i \left(\frac{\partial f_i}{\partial a_1}\frac{\partial f_i}{\partial a_M}\right) \\ \sum\limits_{i=1}^{N} w_i \left(\frac{\partial f_i}{\partial a_2}\frac{\partial f_i}{\partial a_1}\right) & \sum\limits_{i=1}^{N} w_i \left(\frac{\partial f_i}{\partial a_2}\right)^2 & \cdots & \sum\limits_{i=1}^{N} w_i \left(\frac{\partial f_i}{\partial a_2}\frac{\partial f_i}{\partial a_M}\right) \\ \vdots & \vdots & \ddots & \vdots \\ \sum\limits_{i=1}^{N} w_i \left(\frac{\partial f_i}{\partial a_M}\frac{\partial f_i}{\partial a_1}\right) & \sum\limits_{i=1}^{N} w_i \left(\frac{\partial f_i}{\partial a_M}\frac{\partial f_i}{\partial a_2}\right) & \cdots & \sum\limits_{i=1}^{N} w_i \left(\frac{\partial f_i}{\partial a_M}\right)^2 \end{pmatrix}. \tag{2.9}$$

$\mathbf{N}$ is obviously a symmetric matrix with respect to the main diagonal.

[4] In theory, the second derivatives of the function to be minimised also have to be taken into account for nonlinear models. This is discussed more in detail in Section 6.4. However, there are justified reasons to ignore them.

2.3 Pitfalls

Although the given solution for nonlinear least-square problems looks very convincing, attention has to be paid to some details.

One of the main problems is the assignment of suitable starting parameters $\mathbf{a}$. Obviously, the initial values should already be as close as possible to the global minimum, (i) to avoid getting trapped in a local minimum and (ii) to ensure fast convergence of the iterative procedure. This can be achieved by using either an appropriate functional connection between parameters and observations, i.e., making some assumption leading to a rough guess of $\mathbf{a}$, or by a grid search in the parameter space for promising coordinates.

For instance, given a trigonometric model function $y_i = a_1 + \sin(x_i - a_2)$, the mean value a_1 could be simply estimated by averaging all observed values y_i, under the presumption that the observations are almost equally distributed over the full range of oscillation. In contrast, a grid search requires some knowledge about the range of the parameters a_j. For each a_j a certain number of points (e.g., equidistant) in that range are selected leading to a finite set of different parameter vectors $\mathbf{a}$. The vector resulting in the smallest χ^2 is then used as starting point $\mathbf{a}_{\text{start}}$ for the fitting procedure. As an alternative to an equidistant grid, randomly generated sets of seed points in the parameter space could also be considered. If possible, the estimation based on some a priori knowledge is preferred. The grid-search method is tricky for complex model functions with many parameters and should only be applied when necessary.

The initialisation of parameters is especially crucial when there are lots of local minima on the error surface of the model function. Under certain conditions, the minimisation process will converge to undesirable values or not at all. Getting stuck in a local minimum or in a flat region of the error surface is very aggravating. In case of an unsatisfactory fit, the randomised re-initialisation of the parameters can be considered; the randomisation of the resulting parameter vector $\mathbf{a}_{\text{result}}$ with deviations of $\sigma_j \geq |a_{j,\text{start}} - a_{j,\text{result}}|$ might also be an option. The fitting must be repeated until the remaining error falls below a given threshold or the maximum number of allowed iterations is reached.

The range of the parameters is another matter to be considered. If the limits of the parameters are known, they should also be supervised in order to avoid convergence to futilely large or infinitesimally small values. The most elegant method is to fake the gradients of the error hyper-surface in such a manner that the critical regions do not attract the optimisation procedure. An example is provided in Subsection 2.6.8.

The third potential problem concerns the matrix inversion involved. Very large values in $\mathbf{N} = \mathbf{J}^{\mathrm{T}}\mathbf{W}\mathbf{J}$ become very small values in $\mathbf{N}^{-1}$. In some applications, it might be required to scale the values in order to combine the data at a comparable level of magnitude. Otherwise, the limited precision of numbers in computers will lead to unpredictable effects.

Further details about the minimisation procedure can be found in Chapter 6 as well as in Section B.7.

2.4 Simplifications for linear model functions

In case of linear model functions, the computation according to (2.6) and (2.7) can be simplified a great deal. The main advantage of this is that iterative computations are not necessary. Equation (2.6) leads automatically to the final parameter values

$$\mathbf{a} = \left(\mathbf{J}^{\mathrm{T}} \cdot \mathbf{W} \cdot \mathbf{J}\right)^{-1} \cdot \mathbf{J}^{\mathrm{T}} \cdot \mathbf{W} \cdot \mathbf{r} \,. \tag{2.10}$$

Therefore, (i) the initial values a_j can be set to zero and (ii) as a result of doing so, $f(\mathbf{x}_i|\mathbf{a})$ is equal to zero and $\mathbf{r}$ reduces to

$$\mathbf{r} = \mathbf{y} = \begin{pmatrix} y_1 \\ y_2 \\ \vdots \\ y_N \end{pmatrix} \tag{2.11}$$

All the pitfalls discussed in the previous section vanish when a linear model function is used instead of a nonlinear one. Therefore, the model function should always be selected carefully. As already mentioned in Section 1.5, it is sometimes possible to convert (pseudo-)nonlinear functions into linear ones.

If the model function is linear with respect to all parameters a_j, it can be expressed by

$$y_i = f(\mathbf{x}_i|\mathbf{a}) = a_1 \cdot f_1(\mathbf{x}_i) + a_2 \cdot f_2(\mathbf{x}_i) + \cdots + a_M \cdot f_M(\mathbf{x}_i)$$

with arbitrary functions $f_j(\mathbf{x}_i)$, which are obviously the partial derivatives of the model function $f(\mathbf{x}_i|\mathbf{a})$, i.e.,

$$f_j(\mathbf{x}_i) = \partial f(\mathbf{x}_i|\mathbf{a})/\partial a_j \,.$$

In this case, the Jacobian matrix is equal to

$$\mathbf{J} = \begin{pmatrix} f_1(\mathbf{x}_1) & f_2(\mathbf{x}_1) & \cdots & f_M(\mathbf{x}_1) \\ f_1(\mathbf{x}_2) & f_2(\mathbf{x}_2) & \cdots & f_M(\mathbf{x}_2) \\ \vdots & \vdots & \cdots & \vdots \\ f_1(\mathbf{x}_N) & f_2(\mathbf{x}_N) & \cdots & f_M(\mathbf{x}_N) \end{pmatrix}$$

and the fitting problem is reduced to a system of linear equations

$$\mathbf{y} = \mathbf{J} \cdot \mathbf{a}$$

$$\begin{pmatrix} y_1 \\ y_2 \\ \vdots \\ y_N \end{pmatrix} = \begin{pmatrix} f_1(\mathbf{x}_1) & f_2(\mathbf{x}_1) & \cdots & f_M(\mathbf{x}_1) \\ f_1(\mathbf{x}_2) & f_2(\mathbf{x}_2) & \cdots & f_M(\mathbf{x}_2) \\ \vdots & \vdots & \cdots & \vdots \\ f_1(\mathbf{x}_N) & f_2(\mathbf{x}_N) & \cdots & f_M(\mathbf{x}_N) \end{pmatrix} \cdot \begin{pmatrix} a_1 \\ a_2 \\ \vdots \\ a_M \end{pmatrix}. \tag{2.12}$$

Section 6.3 will discuss this more in detail, especially the origin of equation (2.10).

The columns of the Jacobian matrix can be regarded as different basis vectors. $\mathbf{y}$ is a weighted superposition of these basis vectors with a_j as weights.

If the observations have to be weighted according to their uncertainties, both sides of the equation must be divided by the standard deviation σ_i

$$\begin{pmatrix} \frac{y_1}{\sigma_1} \\ \frac{y_2}{\sigma_2} \\ \vdots \\ \frac{y_N}{\sigma_N} \end{pmatrix} = \begin{pmatrix} \frac{f_1(\mathbf{x}_1)}{\sigma_1} & \frac{f_2(\mathbf{x}_1)}{\sigma_1} & \cdots & \frac{f_M(\mathbf{x}_1)}{\sigma_1} \\ \frac{f_1(\mathbf{x}_2)}{\sigma_2} & \frac{f_2(\mathbf{x}_2)}{\sigma_2} & \cdots & \frac{f_M(\mathbf{x}_2)}{\sigma_2} \\ \vdots & \vdots & \cdots & \vdots \\ \frac{f_1(\mathbf{x}_N)}{\sigma_N} & \frac{f_2(\mathbf{x}_N)}{\sigma_N} & \cdots & \frac{f_M(\mathbf{x}_N)}{\sigma_N} \end{pmatrix} \cdot \begin{pmatrix} a_1 \\ a_2 \\ \vdots \\ a_M \end{pmatrix} \tag{2.13}$$

$$\mathbf{y}' = \mathbf{J}' \cdot \mathbf{a} \tag{2.14}$$

$\mathbf{J}'$ is also called the *design matrix*.

2.5 Curve approximation in case of unknown model function

In the ideal case of data fitting, the functional relation between the conditions of the measurement and its outcome, i.e., the observations, is known. The aim of applying the method of least-squares approximation is determining the parameters of this functional relation. There are, however, many practical scenarios, in which this relation is unknown, i.e., no model function is available.

Example 1

The typical goal in such cases is to approximate the functional behaviour between the taken measurements based on a series of related functions, for example, as a polynomial

$$y = f(x|\mathbf{a}) = \sum_{j=1}^{M} a_j \cdot x^{j-1} = a_1 + a_2 \cdot x + a_3 \cdot x^2 + a_4 \cdot x^3 + \cdots .$$

The more elements are involved, i.e., the higher M is, the better the observations y_i ($i = 1, 2, \ldots, N$) are represented. If the number M of model parameters approaches the number N of observations ($M = N$), then the curve $f(x|\mathbf{a})$ can be drawn exactly through all the observed data points.

Figure 2.3 shows the fitting results for a signal with five samples y_i

x_i	0.1	1.9	3.0	3.8	5.2
y_i	1.0	1.3	1.0	1.3	0.1

taken under conditions x_i. The order of the polynomial used as the model function was changed from zero ($M = 1$) to four ($M = 5$). The zero-order model simply determines the average of all observations; the polynomial of order one fits a straight line and so on. It can be clearly seen that the curve is closer to the data points the higher the number of model parameters becomes. However, the property of generalisation between observed data points becomes unreliable. With $M = 5$, for example, the model function oscillates between the first two samples, which does not seem to be a sufficient approximation. If M is too high compared to N, we talk of *over-fitting*.

Using a superposition of elementary functions as a model also forbids the extrapolation of the behaviour beyond the range of available conditions, as can be concluded from Figure 2.3.

Example 2

In some applications of data fitting, there is no actual mathematical relation based on a law of nature between the data taken as conditions and the data interpreted

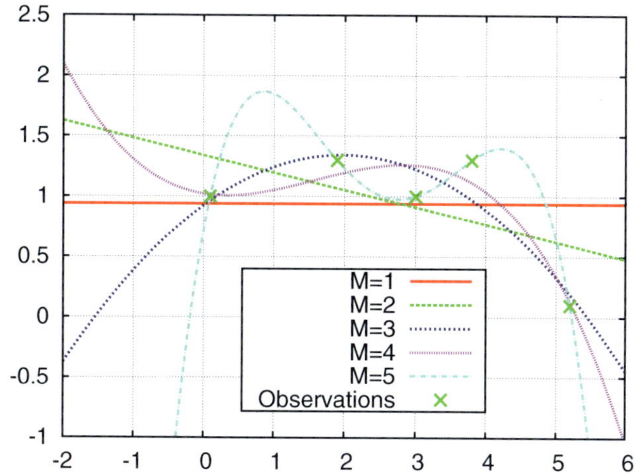

Figure 2.3: Polynomial data fitting with increasing order of model function

as observations. In economics, for example, the costs of a project often have to be estimated in the planning process. In such cases, meaningful features of products from former projects can be used as conditions and the related actual costs as observations. We can now assume that each of the features has a linear or quadratic influence on the costs. Restricting the conditions to three significant features yields, for instance,

$$\begin{aligned} y = f(x|\mathbf{a}) = a_1 + a_2 \cdot x_1 + a_3 \cdot x_1^2 + a_4 \cdot x_2 + a_5 \cdot x_2^2 + a_6 \cdot x_3 + a_6 \cdot x_3^2 + \\ a_7 \cdot x_1 \cdot x_2 + a_8 \cdot x_1 \cdot x_3 + a_9 \cdot x_2 \cdot x_3 \,. \end{aligned}$$

The estimated parameters $a_1 \ldots a_9$ can be used to determine the costs of the new project. If there is some degree of freedom, for example, if feature x_3 is not fixed yet, it can be set to a value that minimises the total costs.

Example 3

Trigonometric polynomials are also able to approximate arbitrary functions

$$y = f(x|\mathbf{a}, \phi) = \frac{a_0}{2} + \sum_{j=1}^{p} [a_j \cdot \cos(j\omega x - \varphi_j)] \tag{2.15}$$

with $\phi = \{\varphi_j\}$ and $\omega = 2 \cdot \pi / T$, where T is the assumed period of the periodic function $f(x|\mathbf{a}, \phi)$. For $p = \infty$, this polynomial corresponds to the Fourier series. Since the extrapolation outside the range of given conditions $[x_{\min} \ldots x_{\max}]$ is not recommended, this presumed periodicity does not harm the approximation of the curve. The variables a_0, a_j, φ_j, and ω are the model parameters, which have to be estimated.

2.6 Example computations

This section is devoted to some example calculations according to the model functions introduced in the first chapter, making the mode of functioning somewhat clearer. For nonlinear model functions, the initialisation of parameters and their limitation to reasonable ranges are also discussed.

2.6.1 Constant value

The model function is

$$y = f(\cdot|\mathbf{a}) = \text{const.} = a_1 \ .$$

There is a single model parameter a_1. In case of N observations y_i, the size of the Jacobian matrix $\mathbf{J}$ is $N \times 1$ and the elements are equal to $\partial f_i/\partial a_1 = 1$

$$\mathbf{J} = \begin{pmatrix} 1 \\ 1 \\ \vdots \\ 1 \end{pmatrix} .$$

Feeding $\mathbf{J}$ into equation (2.10) yields

$$a_1 = \left((1\ 1\ \cdots\ 1) \cdot \mathbf{W} \cdot \begin{pmatrix} 1 \\ 1 \\ \vdots \\ 1 \end{pmatrix} \right)^{-1} \cdot (1\ 1\ \cdots\ 1) \cdot \mathbf{W} \cdot \begin{pmatrix} y_1 \\ y_2 \\ \vdots \\ y_N \end{pmatrix}$$

$$= \left(\sum_{i=1}^{N} w_i \right)^{-1} \cdot (w_1\ w_2\ \cdots\ w_N) \cdot \begin{pmatrix} y_1 \\ y_2 \\ \vdots \\ y_N \end{pmatrix}$$

$$= \frac{\sum\limits_{i=1}^{N} (w_i \cdot y_i)}{\sum\limits_{i=1}^{N} w_i} .$$

This result is not very surprising. The constant a_1 is equal to the weighted average of all observations y_i. The denominator $\sum_{i=1}^{N} w_i$ simply normalises to the sum of all weights.

2.6.2 Straight line

Now let us increase the number of parameters from one to two for the straight-line example

$$y_i = f(\mathbf{x}_i|\mathbf{a}) = a_1 + a_2 \cdot x_i + \varepsilon_i \qquad i = 1, 2, \ldots, N \; .$$

The partial derivatives are

$$\frac{\partial f_i}{\partial a_1} = 1 \qquad \text{and} \qquad \frac{\partial f_i}{\partial a_2} = x_i$$

and the Jacobian becomes

$$\mathbf{J} = \begin{pmatrix} 1 & x_1 \\ 1 & x_2 \\ \vdots & \vdots \\ 1 & x_N \end{pmatrix} .$$

Fed into equation (2.6), we get

$$\begin{aligned}
\mathbf{a} &= \left(\begin{pmatrix} 1 & \cdots & 1 \\ x_1 & \cdots & x_N \end{pmatrix} \mathbf{W} \begin{pmatrix} 1 & x_1 \\ 1 & x_2 \\ \vdots & \vdots \\ 1 & x_N \end{pmatrix} \right)^{-1} \cdot \begin{pmatrix} 1 & \cdots & 1 \\ x_1 & \cdots & x_N \end{pmatrix} \mathbf{W} \begin{pmatrix} y_1 \\ y_2 \\ \vdots \\ y_N \end{pmatrix} \\
&= \begin{pmatrix} \sum\limits_{i=1}^{N} w_i & \sum\limits_{i=1}^{N} w_i \cdot x_i \\ \sum\limits_{i=1}^{N} w_i \cdot x_i & \sum\limits_{i=1}^{N} w_i \cdot x_i^2 \end{pmatrix}^{-1} \cdot \begin{pmatrix} \sum\limits_{i=1}^{N} w_i \cdot y_i \\ \sum\limits_{i=1}^{N} w_i \cdot x_i y_i \end{pmatrix} .
\end{aligned} \tag{2.16}$$

Applying equation (5.6) for matrix inversion results in

$$\begin{aligned}
\mathbf{a} &= \frac{1}{d} \cdot \begin{pmatrix} \sum\limits_{i=1}^{N} w_i (x_i)^2 & -\sum\limits_{i=1}^{N} w_i x_i \\ -\sum\limits_{i=1}^{N} w_i x_i & \sum\limits_{i=1}^{N} w_i \end{pmatrix} \begin{pmatrix} \sum\limits_{i=1}^{N} w_i y_i \\ \sum\limits_{i=1}^{N} w_i x_i y_i \end{pmatrix} \\
&= \frac{1}{d} \cdot \begin{pmatrix} \sum\limits_{i=1}^{N} w_i (x_i)^2 \cdot \sum\limits_{i=1}^{N} w_i y_i - \sum\limits_{i=1}^{N} w_i x_i \cdot \sum\limits_{i=1}^{N} w_i x_i y_i \\ \sum\limits_{i=1}^{N} w_i \cdot \sum\limits_{i=1}^{N} w_i x_i y_i - \sum\limits_{i=1}^{N} w_i x_i \sum\limits_{i=1}^{N} w_i y_i \end{pmatrix} .
\end{aligned}$$

with

$$d = \sum_{i=1}^{N} w_i (x_i)^2 \cdot \sum_{i=1}^{N} w_i - \left(\sum_{i=1}^{N} w_i x_i \right)^2 .$$

Without a question, the equation for **a** can also be derived manually without using the matrix approach (see Section 7.4).

2.6.3 Polynomial approximation

After reading the two preceding Subsections, the principle of the numerical solution of least-squares problems should be clear by now and we can restrict the exemplifications to the Jacobian matrix. For combinations of polynomials

$$y_i = f(\mathbf{x}_i|\mathbf{a}) = a_1 + a_2 \cdot x_i + \cdots + a_M \cdot x_i^{M-1} + \varepsilon_i \qquad i = 1, 2, \ldots, N$$

the Jacobian is

$$\mathbf{J} = \begin{pmatrix} 1 & x_1 & x_1^2 & \cdots & x_1^{M-1} \\ 1 & x_2 & x_2^2 & \cdots & x_2^{M-1} \\ \vdots & \vdots & \vdots & \cdots & \vdots \\ 1 & x_N & x_N^2 & \cdots & x_N^{M-1} \end{pmatrix} .$$

2.6.4 Plane approximation

The plane approximation is an example for multiple regression, as the condition vector is two-dimensional here, namely $\mathbf{x} = (x_1, x_2)^{\mathrm{T}}$, and the model function is

$$y_i = f(\mathbf{x}_i|\mathbf{a}) = a_1 + a_2 \cdot x_{1,i} + a_3 \cdot x_{2,i} + \varepsilon_i . \tag{2.17}$$

The Jacobian matrix must reflect that different conditions (namely $x_{1,i}$ and $x_{2,i}$) are assigned to the parameters. The gradients are

$$\frac{\partial f_i}{\partial a_1} = 1 , \qquad \frac{\partial f_i}{\partial a_2} = x_{1,i} \qquad \text{and} \qquad \frac{\partial f_i}{\partial a_3} = x_{2,i} .$$

Please note that the first index of x discriminates between the variables x_1 and x_2, while the second index is related to the different observations. The Jacobian matrix is accordingly

$$\mathbf{J} = \begin{pmatrix} 1 & x_{1,1} & x_{2,1} \\ 1 & x_{1,2} & x_{2,2} \\ \vdots & \vdots & \vdots \\ 1 & x_{1,N} & x_{2,N} \end{pmatrix} .$$

2.6.5 Linear prediction

The linear prediction of order M is described by

$$x[n] = f(\mathbf{x}_n|\mathbf{a}) = a_1 \cdot x[n-1] + a_2 \cdot x[n-2] + \cdots + a_M \cdot x[n-M] \;.$$

in which M preceding samples of a sequence of values are used to predict the current sample $x[n]$. The underlying assumption is that the sequence $\{x[n]\}$ was produced by an one-dimensional auto-regressive (AR) process of Mth order. In contrast to other model functions, the observation and the experimental conditions only differ in the time index n here.

The Jacobian matrix has to be written as

$$\mathbf{J} = \begin{pmatrix} x[m-1] & x[m-2] & \cdots & x[m-M] \\ x[m-2] & x[m-3] & \cdots & x[m-(M+1)] \\ \vdots & \vdots & \cdots & \vdots \\ x[m-N] & x[m-(1+N)] & \cdots & x[m-(M+N)] \end{pmatrix}$$

with $m, n \in \mathbb{Z}$, $m \leq n$. Parameter m defines the context of the prediction. Typically, it starts directly before the value to be predicted, i.e., $m = n$. The corresponding vector of observations is

$$\mathbf{y} = \begin{pmatrix} x[n] & x[n-1] & x[n-2] & \cdots & x[n-N-1] \end{pmatrix}^{\mathrm{T}} \;.$$

2.6.6 Cosine function

It was shown in Section 1.3 that function

$$y_i = b_1 + b_2 \cdot \cos(x_i - b_3) + \varepsilon_i \;,$$

which is nonlinear in b_3, can be expressed as

$$y_i = a_1 + a_2 \cdot \cos(x_i) + a_3 \cdot \sin(x_i) + \varepsilon_i \tag{2.18}$$

simplifying the fitting to a linear problem. The partial derivatives according to the parameters a_j are

$$\frac{\partial f_i}{\partial a_1} = 1 \;, \qquad \frac{\partial f_i}{\partial a_2} = \cos(x_i) \qquad \text{and} \qquad \frac{\partial f_i}{\partial a_3} = \sin(x_i) \;.$$

Therefore, the Jacobian matrix is simply

$$\mathbf{J} = \begin{pmatrix} 1 & \cos(x_1) & \sin(x_1) \\ 1 & \cos(x_2) & \sin(x_2) \\ \vdots & \vdots & \vdots \\ 1 & \cos(x_N) & \sin(x_N) \end{pmatrix} .$$

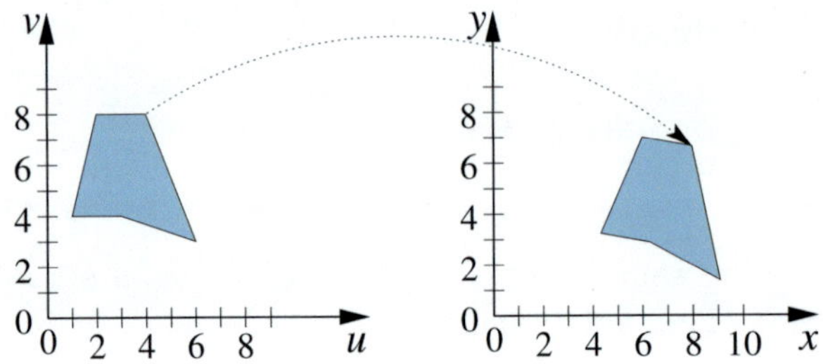

Figure 2.4: Example of translation and rotation

2.6.7 Rotation and translation of coordinates

As already stated in Subsection 1.4.4, the fitting of a rotation matrix is not trivial, since not only the conditions but also the observations appear in pairs. Let us assume that the vectors $\mathbf{u}_i = (u_i\ v_i)^{\mathrm{T}}$ are given coordinates in a first image (conditions) and the vectors $\mathbf{x}_i = (x_i\ y_i)^{\mathrm{T}}$ are the corresponding coordinates in a second image (observations) found by a matching algorithm, **Figure 2.4**.

If the second image is a translated and slightly rotated version of the first image,

$$\begin{pmatrix} x \\ y \end{pmatrix} = \begin{pmatrix} a_1 \\ a_2 \end{pmatrix} + \begin{pmatrix} \cos(a_3) & -\sin(a_3) \\ \sin(a_3) & \cos(a_3) \end{pmatrix} \cdot \begin{pmatrix} u \\ v \end{pmatrix}$$

has to be used as model function with a_3 being the rotation angle. Actually, there are two model functions, namely

$$f_{\mathrm{x}}(\mathbf{u}|\mathbf{a}) = x = a_1 + \cos(a_3) \cdot u - \sin(a_3) \cdot v \tag{2.19}$$

and

$$f_{\mathrm{y}}(\mathbf{u}|\mathbf{a}) = y = a_2 + \sin(a_3) \cdot u + \cos(a_3) \cdot v\ , \tag{2.20}$$

which are coupled via a_3. Since (2.19) and (2.20) are nonlinear functions with respect to $\mathbf{a}$, the partial derivatives

$$\begin{aligned} \frac{\partial x_i}{\partial a_1} &= 1\ , \quad & \frac{\partial x_i}{\partial a_2} &= 0 \quad \text{and} \quad & \frac{\partial x_i}{\partial a_3} &= -u_i \cdot \sin(a_3) - v_i \cdot \cos(a_3) \\ \frac{\partial y_i}{\partial a_1} &= 0\ , \quad & \frac{\partial y_i}{\partial a_2} &= 1 \quad \text{and} \quad & \frac{\partial y_i}{\partial a_3} &= u_i \cdot \cos(a_3) - v_i \cdot \sin(a_3)\ . \end{aligned}$$

are dependent on the parameters. Therefore, a suitable initialisation of $\mathbf{a}$ is required. It could be assumed, for example, that there is no rotation, but only a translational displacement

$$a_{1,\mathrm{init}} = \frac{1}{N}\sum_{i=1}^{N}(x_i - u_i) \;, \qquad a_{2,\mathrm{init}} = \frac{1}{N}\sum_{i=1}^{N}(y_i - v_i) \;, \qquad a_{3,\mathrm{init}} = 0 \;.$$

The two model functions (2.19) and (2.20) must be combined into a single framework for the fitting procedure

$$\mathbf{r} = \begin{pmatrix} x_1 - f_{\mathrm{x}}(\mathbf{u}_1|\mathbf{a}) \\ x_2 - f_{\mathrm{x}}(\mathbf{u}_2|\mathbf{a}) \\ \vdots \\ x_N - f_{\mathrm{x}}(\mathbf{u}_N|\mathbf{a}) \\ y_1 - f_{\mathrm{y}}(\mathbf{u}_1|\mathbf{a}) \\ y_2 - f_{\mathrm{y}}(\mathbf{u}_2|\mathbf{a}) \\ \vdots \\ y_N - f_{\mathrm{y}}(\mathbf{u}_N|\mathbf{a}) \end{pmatrix} \qquad \mathbf{J} = \begin{pmatrix} 1 & 0 & -u_1 \cdot \sin(a_3) - v_1 \cdot \cos(a_3) \\ 1 & 0 & -u_2 \cdot \sin(a_3) - v_2 \cdot \cos(a_3) \\ \vdots & \vdots & \vdots \\ 1 & 0 & -u_N \cdot \sin(a_3) - v_N \cdot \cos(a_3) \\ 0 & 1 & u_1 \cdot \cos(a_3) - v_1 \cdot \sin(a_3) \\ 0 & 1 & u_2 \cdot \cos(a_3) - v_2 \cdot \sin(a_3) \\ \vdots & \vdots & \vdots \\ 0 & 1 & u_N \cdot \cos(a_3) - v_N \cdot \sin(a_3) \end{pmatrix} .$$

Using this setup, the problem can be handled in the same manner as those with only one observation per condition. However, the x and y coordinates are treated separately. In the case of weighted fitting, possibly combined with the detection of outliers, this will lead to the strange effect that not the point (x, y) is assigned a weight, but x and y individually. In extreme case, one coordinate is declared as outlier and the other is not. Section 7.5 suggests an alternative treatment, which regards (x, y) and (u, v) as inseparable pairs.

2.6.8 Exponential model function

The exponential function

$$y_i = a_1 \cdot \exp(a_2 \cdot x_i) + \varepsilon_i$$

describes a nonlinear model. The partial derivatives are consequently dependent on the model parameters

$$\frac{\partial f_i}{\partial a_1} = \exp(a_2 \cdot x_i) \qquad \text{and} \qquad \frac{\partial f_i}{\partial a_2} = a_1 \cdot x_i \cdot \exp(a_2 \cdot x_i) \;. \tag{2.21}$$

The corresponding Jacobian matrix reads as

$$\mathbf{J} = \begin{pmatrix} \exp(a_2 \cdot x_1) & a_1 \cdot x_1 \cdot \exp(a_2 \cdot x_1) \\ \exp(a_2 \cdot x_2) & a_1 \cdot x_2 \cdot \exp(a_2 \cdot x_2) \\ \vdots & \vdots \\ \exp(a_2 \cdot x_N) & a_1 \cdot x_N \cdot \exp(a_2 \cdot x_N) \end{pmatrix} .$$

Facing the dependence on model parameters, it becomes apparent that an initial guess of $\mathbf{a} = (a_1, \ldots, a_M)^{\mathrm{T}}$ is required. Assuming, for instance, that the experiment analyses a decaying process, a_2 has to be negative and a_1 has presumably a magnitude similar to y_1, if x_1 is close to zero.

Using these estimates, the vector of residuals is at the beginning

$$\mathbf{r} = \begin{pmatrix} y_1 - (a_1 \cdot \exp(a_2 \cdot x_1)) \\ y_2 - (a_1 \cdot \exp(a_2 \cdot x_2)) \\ \vdots \\ y_N - (a_1 \cdot \exp(a_2 \cdot x_N)) \end{pmatrix} .$$

The application of equation (2.6) leads to a vector $\mathbf{\Delta a}$ containing the updates for

$$\mathbf{a} \longleftarrow \mathbf{a} + \mathbf{\Delta a} .$$

As long as there are distinct changes in the parameter values (see Section 2.2, eq. 2.8), the Jacobian matrix $\mathbf{J}$ and the vector of residuals $\mathbf{r}$ have to be re-computed and fed again into (2.6).

It was already mentioned above that a_2 must be negative for a decaying process. Unfortunate input data[5], however, might push a_2 towards positive values. This can be prevented by simply modifying (2.21) slightly to

$$\frac{\partial f_i}{\partial a_1} = \begin{cases} \exp(a_2 \cdot x_i) & \text{if} \quad a_2 < 0 \\ 1 & \text{otherwise} \end{cases}$$
$$\frac{\partial f_i}{\partial a_2} = \begin{cases} a_1 \cdot x_i \cdot \exp(a_2 \cdot x_i) & \text{if} \quad a_2 < 0 \\ a_1 \cdot x_i & \text{otherwise} \end{cases} .$$

Please note that the derivatives remain continuous at $a_2 = 0$.

Alternatively one should consider to set a_j to zero before computing $\frac{\partial f_i}{\partial a_j}$, if a_j is already outside the admissible range.

2.6.9 Composite Gaussian bell functions

In Subsection 1.5.2 the superposition of two Gaussian bells was discussed

$$y_i = a_1 \cdot \exp\left[a_2 \cdot (x_i - a_3)^2\right] + a_4 \cdot \exp\left[a_5 \cdot (x_i - a_6)^2\right] + \varepsilon_i ,$$

[5] especially if the model function is not appropriate

which could be used, for instance, to model the distribution of a certain variable.

Due to the nonlinear character and relative high number of parameters, the determination of the Jacobian matrix becomes a little bit more complex. The partial derivatives are

$$\frac{\partial f_i}{\partial a_1} = \exp\left[a_2 \cdot (x - a_3)^2\right] , \quad \frac{\partial f_i}{\partial a_2} = a_1 \cdot \exp\left[a_2 \cdot (x - a_3)^2\right] \cdot (x - a_3)^2 ,$$
$$\frac{\partial f_i}{\partial a_4} = \exp\left[a_5 \cdot (x - a_6)^2\right] , \quad \frac{\partial f_i}{\partial a_5} = a_3 \cdot \exp\left[a_5 \cdot (x - a_6)^2\right] \cdot (x - a_6)^2 ,$$
$$\frac{\partial f_i}{\partial a_3} = -a_1 \cdot \exp\left[a_2 \cdot (x - a_3)^2\right] \cdot a_2 \cdot 2 \cdot (x - a_3) ,$$
$$\frac{\partial f_i}{\partial a_6} = -a_4 \cdot \exp\left[a_5 \cdot (x - a_6)^2\right] \cdot a_5 \cdot 2 \cdot (x - a_6) .$$

Because of the ambiguity of additive or subtractive combination of two Gaussian bells (see subsection 1.5.2) and other effects causing an error hyper-surface with different local minima, the accurate estimation of initial parameters **a** is an extremely important task. One feasible method neglecting the ambiguity would be to fit roughly a single Gaussian at the position of the major peak of the observed values. This Gaussian curve is then subtracted from the observations. The residual undergoes the same procedure of estimating mean, amplitude, and deviation parameters, now for the second Gaussian bell. Herewith, the initial parameters are already rather close to the –hopefully– global minimum.

2.6.10 Circle function

Given the model function

$$y_i = 0 = (x_{1,i} - a_1)^2 + (x_{2,i} - a_2)^2 - a_3^2 + \varepsilon_i \tag{2.22}$$

describing a circle (see also Subsection 1.5.3), the partial derivatives with respect to the model parameters are

$$\frac{\partial f_i}{\partial a_1} = 2 \cdot (a_1 - x_{1,i}) , \quad \frac{\partial f_i}{\partial a_2} = 2 \cdot (a_2 - x_{2,i}) \quad \text{and} \quad \frac{\partial f_i}{\partial a_3} = -2 \cdot a_3 .$$

a_1 and a_2 define the circle centre whereas a_3 corresponds to its radius. Under the assumption that the points (the conditions) are fairly distributed on the circle, the centre is estimated via

$$\begin{pmatrix} a_{1,\text{init}} \\ a_{2,\text{init}} \end{pmatrix} = \begin{pmatrix} \frac{1}{N} \sum\limits_{i=1}^{N} x_{1,i} \\ \frac{1}{N} \sum\limits_{i=1}^{N} x_{2,i} \end{pmatrix} . \tag{2.23}$$

The radius estimation is a little bit more complex. It averages the distances of all points to the estimated circle centre

$$a_{3,\text{init}} = \frac{1}{N} \sum_{i=1}^{N} \sqrt{(x_{1,i} - a_{1,\text{init}})^2 + (x_{2,i} - a_{2,\text{init}})^2} \,. \tag{2.24}$$

Since the circle function (2.22) is pseudo-nonlinear, as was pointed out in Chapter 1, it would be reasonable to use instead the form

$$x_{1,i}^2 + x_{2,i}^2 = b_1 \cdot x_{1,i} + b_2 \cdot x_{2,i} - b_3 \,,$$

with

$$\frac{\partial f_i}{\partial b_1} = x_{1,i} \,, \qquad \frac{\partial f_i}{\partial b_2} = x_{2,i} \qquad \text{and} \qquad \frac{\partial f_i}{\partial b_3} = -1 \,.$$

The reconstruction of circle centre (a_1, a_2) and radius a_3 is

$$a_1 = 0.5 \cdot b_1 \qquad a_2 = 0.5 \cdot b_2 \qquad a_3 = \sqrt{a_1^2 + a_2^2 - b_3} \,.$$

It has to be mentioned that this procedure leads in general to visually pleasing fits. If, however, one seeks for a solution that minimises the sum squared geometric distances from all points to the circle, another approach has to be used. This will discussed in Section 7.7.

2.6.11 Neural networks

Depending on the number of input units and hidden neurons, neural networks are able to build rather complex nonlinear functions mapping the input (conditions $\mathbf{x}_i$) to the output (observations y_i). This makes it difficult to initialise the weights of the net (parameter vector $\mathbf{a}$) with reasonable values. Typically, they are randomly generated. The only constraint concerns the range of definition. The argument t_k of equation (1.33)

$$h_k(t_k) = \tanh(t_k)$$

should lie in the range of $-4 \ldots 4$, see also Figure 1.13 on page 25. Otherwise, small changes of the parameters would have almost no effect on the output of the neurons, since the slope of the hyperbolic tangent is too flat outside this region. Very slow convergence would be the consequence.

Figure 2.5 shows 3×32 points in 3D space, which correspond to feature vectors $\mathbf{x}$ with three features (x_1, x_2, x_3) each. The feature vectors belong to three classes 1,

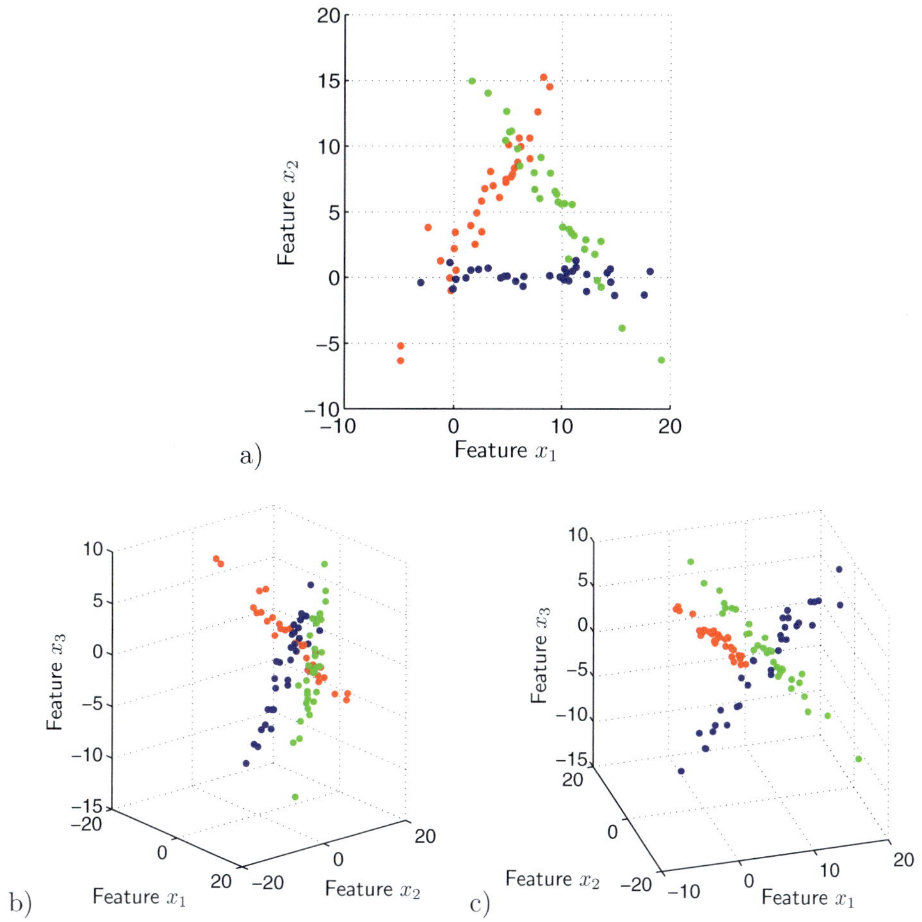

Figure 2.5: Feature vectors from three separated classes; a) view onto x_1–x_2 plane; b)+c) views illustrating the separability

2, and 3. Using a feed-forward 3-3-1 neural network (**Figure 2.6**), it is possible to correctly classify all points. It is a kind of supervised classification. The elements of the single feature vectors (conditions) are presented to the three input neurons and the correct class is expected as output (observation). The neural net combines the three feature values in such a way that the output is smaller than 1.5 for all vectors from class 1 and higher than 2.5 for all vectors from class 3 (**Figure 2.7**a).

When reducing the number of hidden neurons from 3 to 2, some vectors of class 1 and 2 are falsely assigned (**Figure 2.7**b).

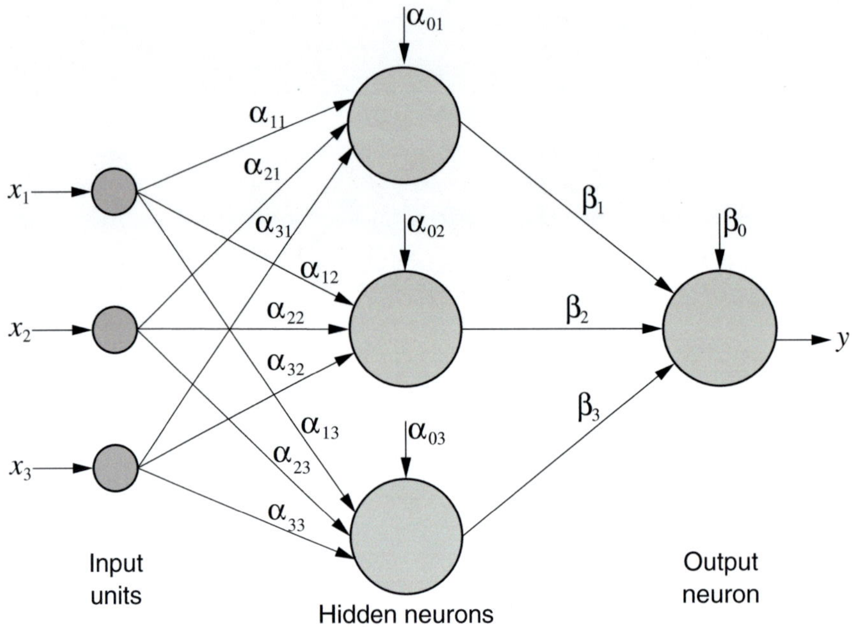

Figure 2.6: Feed-forward 3-3-1 neural network

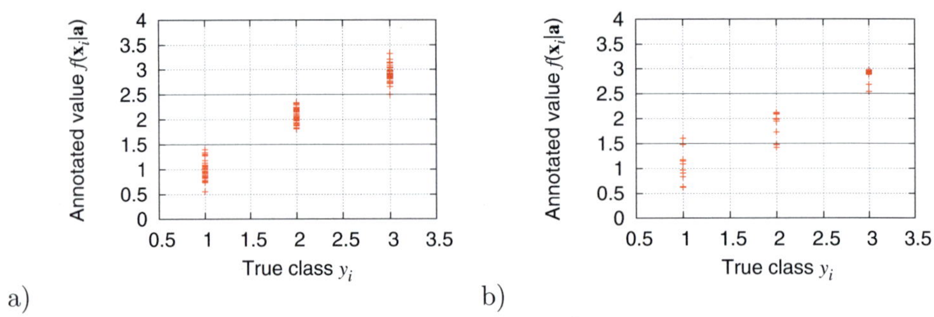

Figure 2.7: Results of classification via neural networks; a) 3-3-1 net (16 parameters); b) 3-2-1 net (11 parameters)

The major problem of neural networks lies in the error surface with multiple local minima. In addition, it is not possible to guess good starting values for the parameters. The parameters must be initialised with random values. Also a target for the minimisation criterion χ^2 should be set low in order to hopefully find the best combination of weights.

2.7 Test questions

2.1 Does the least-squares method concern minimisation or optimisation?

2.2 What is the aim of least-squares fitting?

2.3 What is an error surface? What is it dependent on?

2.4 Why do nonlinear fitting problems require an initialisation of parameters?

2.5 What is the content of the Jacobian matrix? What do its columns and rows depend on?

2.6 What is the residual vector?

2.7 Is it possible to compute the model-function parameters directly (in one pass) for linear (nonlinear) models?

2.8 Write down the fitting equations for linear (nonlinear) fitting problems.

2.9 What are the drawbacks of nonlinear model functions?

2.10 Itemise different methods for the estimation of initial parameter values.

2.11 Explain the relation between least squares and neural networks.

2.12 How must the fitting procedure be altered if there is more than one observation per condition $\mathbf{x}$?

2.13 Given the linear model function

$$f(\mathbf{x}|\mathbf{a}) = a_1 \cdot x + a_2 \cdot x^2 + a_3 \cdot \mathrm{e}^x + a_4 \cdot \sin(x) ,$$

what is the Jacobian matrix?

2.14 How many parameters have to be estimated in a 2-3-2 neural network?

Chapter 3

Weights and Outliers

3.1 What are the weights good for?

The previous chapter has already introduced an $N \times N$ matrix

$$\mathbf{W} = \begin{pmatrix} w_1 & 0 & 0 & \cdots & 0 \\ 0 & w_2 & 0 & \cdots & 0 \\ 0 & 0 & \ddots & \cdots & \vdots \\ \vdots & \vdots & \vdots & \ddots & 0 \\ 0 & 0 & \cdots & 0 & w_N \end{pmatrix}$$

containing weights w_i assigned to each of the N observation $y_i = f(\mathbf{x}_i|\mathbf{a})$. This chapter explains why it is advantageous to use these weights and how to estimate them.

The underlying idea is the following. If it is known that observation y_k has a higher uncertainty than y_l, the weight w_k should be lower than w_l. In other words: the more reliable value of y_l may influence the data-fitting calculation more than y_k.

The reliability is often dependent on the experimental conditions $\mathbf{x}_i$. **Figure 3.1** visualises data of a possible experiment for the estimation of distances by eye. It is evident that the estimation error is somehow proportional to the real distance. If the true distance is one centimetre, the error will be in the range of millimetres, whereas the estimation of hundreds of metres will result in much larger errors. The uncertainty is typically indicated by error bars (the vertical lines in Fig. 3.1).

In theory, the optimal weight is independent on the single observation y_i, but equal to

$$w_i = \frac{1}{\sigma_i^2} \,. \tag{3.1}$$

It implies the knowledge of the standard uncertainty σ_i of the parent distribution from which each single observation y_i is drawn. This uncertainty quantifies the

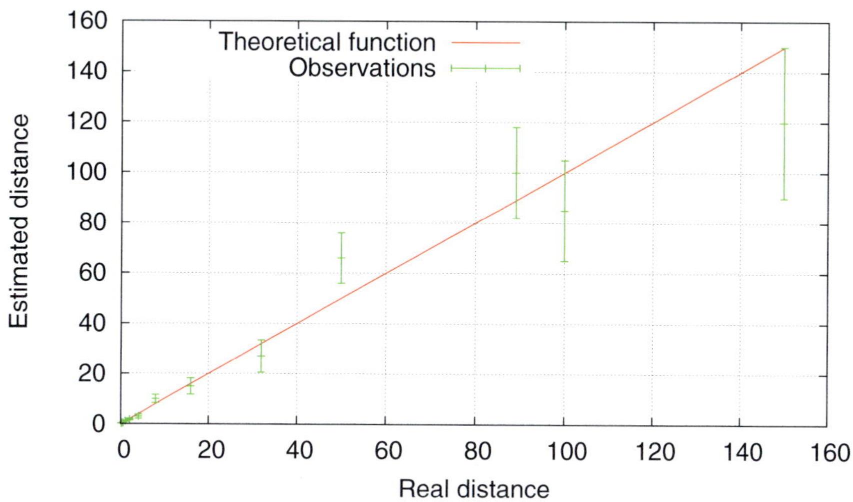

Figure 3.1: Estimation of distances: an example for uncertainties depending on the experimental condition. Uncertainties are depicted as error bars with the observed value (estimated distance) in the middle of the bars.

deviation of the measurements y_i from the true value y_t under conditions $\mathbf{x}_i$ and is, in the strict sense, the standard deviation of the error $e = y_\mathrm{t} - y_i$. In some experimental set-ups, the uncertainties of the observed values can be obtained from sources such as calibration certificates, manufactures' specification or tabulated data. Unfortunately, in very many practical cases, there is no chance neither to obtain directly σ_i nor to get knowledge about the true parent distribution[1]. The general advice is to set $\sigma_i = \sigma = 1.0$, as long as no information is available about σ_i. However, treating all observations equally can lead to poor fitting results, especially in the presence of outliers.

Instead of ignoring the weights, they should be estimated based on the available observations in order to reflect their relative importance, especially if the detection of outliers is intended (see next section). After the removal of putative outliers, it is legitimate to forget the estimated weights and to fit the model function to the reduced set of data points.

[1] In addition, the values of σ_i are based on the assumption that the errors in y_i are normally distributed, which can be different in particular scenarios.

3.2 Outliers

Many applications do not only require an estimation of weights but require, in addition, the detection of outliers. Outliers, also referred to as mavericks or contaminant observations, are data points that deviate so much from other points that they seem to be generated by a different mechanism than the 'good' observations.

When observations are subject to data analysis, at least two scenarios have to be distinguished. Either (i) outliers negatively influence the results of analysis, or (ii) the search for outliers is the main task of data analysis. In data mining, for instance, outlier detection is also regarded as the detection of novelty or anomaly. In most cases, there is a set of training values available that defines 'normality'. Security applications are examples, in which atypical behaviour by people or technical systems has to be detected. In applications with a small number of observations, however, the detection method should be able to identify outliers without prior training.

With respect to scenario (i), lively discussion can be found in the past literature on outliers as to whether to reject suspicious values or always to keep all observations. Beckman and Cook give an overview of the history of attempts to find outliers in data sets [Bec83]. Unfortunately, no established standard technique has evolved to date. A comprehensive review of the tests developed for outlier detection can be found in Barnett and Lewis [Bar94]. Outlier tests (also known as discordance tests) are mostly tailored to the statistical model generating the observations and often presume some knowledge of the number of putative outliers. Many of them can only cope with a single outlier. With the focus on machine learning and data mining, approaches to outlier detection have been surveyed by [Mar03], [Hod04], and [Zim12]. If the underlying model is not known in advance, for example, when dealing with biological or economic data, mild assumptions about the model will have to be made as a minimum.

With regard to the model, we must differentiate between (i) analytical models in the form of $\mathbf{y} = f(\mathbf{x}|\mathbf{a})$, that is, the vector of observations $\mathbf{y}$ is a function of a set of conditions $\mathbf{x}$ with certain model parameters $\mathbf{a}$, and (ii) more descriptive models, such as: "The observations must be clustered or connected". The latter type of model is often used in the field of machine learning; its parametrisation is usually possible to a certain extent. In these applications, the outlier detection is typically based on classification with a so-called rejection/novelty class [Mar03, Hod04].

For each single case of application, it must be carefully judged, whether outliers are likely to occur or whether strongly deviating observations are the normal case. The on-line "Engineering Statistics Handbook" lists the following possible reasons for outliers: [ESH]

- operator blunders,
- equipment failures,
- day-to-day effects,
- batch-to-batch differences,
- anomalous input conditions, and
- warm-up effects.

Furthermore it makes following important remark: "*The best attitude is that outliers are our 'friends'; outliers are trying to tell us something, and we should not stop until we are comfortable with the explanation for each outlier*". Everybody should therefore be keen to find outliers, since their detection and the exploration of the reason of their presence may provide new knowledge.

The enumeration above refers to real-life experiments or measurement series, and, as a minimum, must be extended by the reasons inherent in the deficiency of computational algorithms. Let us consider, for example, the tracking of objects in image sequences. A computational pattern-recognition algorithm has to analyse images taken at certain time steps. It may happen that the algorithm frequently fails to detect the desired object under certain conditions, for instance when other objects or background patterns attract the algorithm. The detection algorithm is therefore not only producing values that deviate more or less from the true object coordinates, but can also produce results that are totally wrong.

If one can expect that the number of outliers will not increase in proportion to the number of observations, the identification of outliers becomes important especially in data sets with a low number of observations. Here, the influence can typically not be neglected because the presence of outliers would strongly bias the result of data analysis.

3.3 Estimation of weights

Before starting with the determination of weights, let us recapitulate the procedure of data fitting.

Based on the analytical and general model function

$$y = f(\mathbf{x}|\mathbf{a}) \,, \tag{3.2}$$

the true outcome y of an experiment is a function of the experimental conditions $\mathbf{x}$ and certain model parameters $\mathbf{a}$. In practice, this model function is sampled with a limited number of real observations

$$y_i = f(\mathbf{x}_i|\mathbf{a}) + \varepsilon_i \qquad i = 1, 2, \ldots, N \; , \tag{3.3}$$

deviating from the true values of y by ε_i. Since the model parameters themselves have to be estimated based on the available observations y_i, only estimates $\hat{y}$ of y can be determined

$$\hat{y} = f(\mathbf{x}|\hat{\mathbf{a}}) \; . \tag{3.4}$$

$\hat{\mathbf{a}}$ is the vector of estimated model parameters. The sample number i indicates a particular estimate and its corresponding conditions

$$\hat{y}_i = f(\mathbf{x}_i|\hat{\mathbf{a}}) \; . \tag{3.5}$$

The residual[2]

$$\Delta_i = y_i - \hat{y}_i \tag{3.6}$$

expresses the deviation of each observation y_i from its estimated value $\hat{y}_i$. Please note that Δ_i is not only dependent on the fluctuations ε_i, but is also influenced by the estimation of the model parameters $\mathbf{a}$.

In regression analysis, the goal is typically to find the best model parameters $a_j \in \mathbf{a}$ with respect to a certain optimisation criterion. The presence of outliers disturbs the parameter estimation, and the deviates Δ_i are influenced as well, affecting the outlier detection in turn. Robust regression methods have therefore been developed that automatically down-weight the influence of outliers and result in more accurate model parameters [Ham86]. Many robust estimators iteratively search for optimal weights [Hol77].

In the most likely case that the uncertainty σ_i for a particular observation y_i is not known in advance, the weights have to be estimated dependent on the observations themselves. Therefore, the data fitting has to be performed first with constant weights $w_i = 1.0$ calculating model parameters $\hat{\mathbf{a}}$ which are hopefully not too far from the optimal ones. Then, the observations y_i have to be compared with the corresponding values of $\hat{y}_i = f(\mathbf{x}_i|\hat{\mathbf{a}})$ using the absolute differences

$$|\Delta_i| = |y_i - \hat{y}_i| \; .$$

Under consideration of the weighted and squared residuals of observations with respect to the modelled function

$$\chi^2 = \sum_i w_i \cdot [y_i - f(\mathbf{x}_i|\hat{\mathbf{a}})]^2 = \sum_i w_i \cdot \Delta_i^2 \; , \tag{3.7}$$

[2] The residual $(y_i - \hat{y}_i)$ is sometimes also called *deviate* to be more precisely [Sch89].

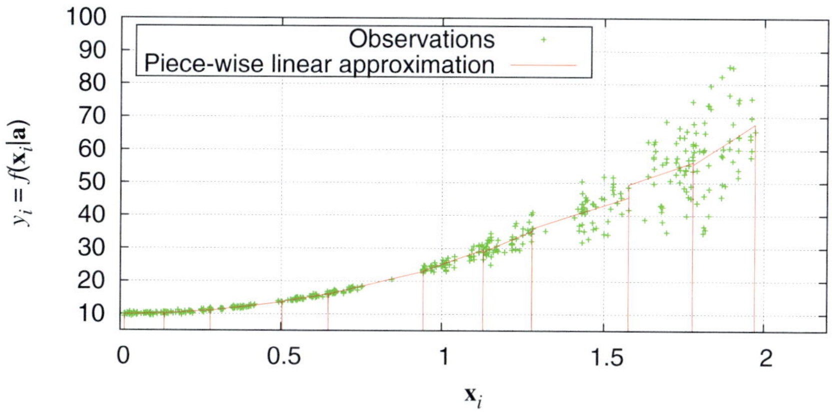

Figure 3.2: Scatter plot and piece-wise linear approximation of measured data with bins containing fifty observations each

the value of Δ_i^2 has to be mapped to a suitable weight w_i. After the determination of weights and the removal of putative outliers, the least-square fitting must be performed again.

3.3.1 Estimation by binning

In the favourable case of many measurements under the same (or at least similar) conditions, weights should be determined using estimated standard uncertainties.

If there are enough observations available for a given range of conditions, it is possible to segment ("to bin") the x axis (conditions) into K intervals. The N_k data points from each bin B_k, $k = 1, 2, \ldots K$ can be approximated, for instance, by a straight line, using the model function

$$y_i = a_1 + a_2 \cdot x_i + \varepsilon_i \qquad \forall x_i \in B_k.$$

That means, the curve is piece-wise linearly approximated before the fitting with the correct model function is carried out. **Figure 3.2** shows the start and end position of each bin (i.e. the range of conditions) and the fitted straight lines. As can be seen, the piece-wise approximation does not necessarily result in a continuous curve.

The deviation of the observations y_i from these lines indicates the uncertainty of the observations in each bin via

$$\sigma_{\mathrm{y},k} = \sqrt{\frac{1}{N_k - 2} \sum_{x_i \in B_k} \left[y_i - (\hat{a}_1 + \hat{a}_2 \cdot x_i)\right]^2} \,.$$

Obviously, the size of the bins has to be chosen carefully. It can either be fixed in size or adapted somehow dependent on the data, for example, using a constant number N_k of observations for each bin B_k. In Figure 3.2, for instance, the bins comprise fifty observations each.

Alternatively, one could use overlapping bins or a window sliding along the condition axis. At each position, the uncertainty could be determined and assigned to the observation at the window centre. In maximum, each observation would get its own uncertainty, however, at the price of additional computational costs.

3.3.2 Weights estimation using deviates

A simple method of mapping residuals (or more precise: deviates) $\Delta = (y_i - \hat{y}_i)$ to suitable weights is to take their squared values Δ_i^2 instead of σ_i^2 as weights

$$w_i = \frac{1}{\sigma_i^2} \approx \frac{1}{\Delta_i^2} \ . \tag{3.8}$$

This procedure is not correct in the sense of theoretical statistics, but under circumstances of very erroneous data a sufficient approximation. However, it turns out that a direct application of this formula can become dangerous in practice, in particular when an observation y_i is by chance equal or very close to the model function. An unjustified high weight would be assigned to this observation, compared to the others. Therefore, w_i has to be limited to a maximum value, reflecting the probable minimum squared uncertainty σ_i^2. This can be achieved by using a lower bound λ_L

$$w_i = \begin{cases} \dfrac{1}{\lambda_\mathrm{L}^2} & \text{for} \quad |\Delta_i| < \lambda_\mathrm{L} \\[2ex] \dfrac{1}{\Delta_i^2} & \quad \text{otherwise} \end{cases} \ . \tag{3.9}$$

The weights w_i obtained via (3.9) are indirectly dependent on the observations y_i. In theory, this should be avoided, since it brings bias (systematic error) into the parameter estimation. However, in very erroneous environments this possible bias can be neglected in comparison to the effect of random errors and outliers.

Normally distributed deviates

The definition of λ_L could be formulated as

$$\lambda_\mathrm{L} = \kappa_\mathrm{L} \cdot \hat{\sigma}_\mathrm{y} \ ,$$

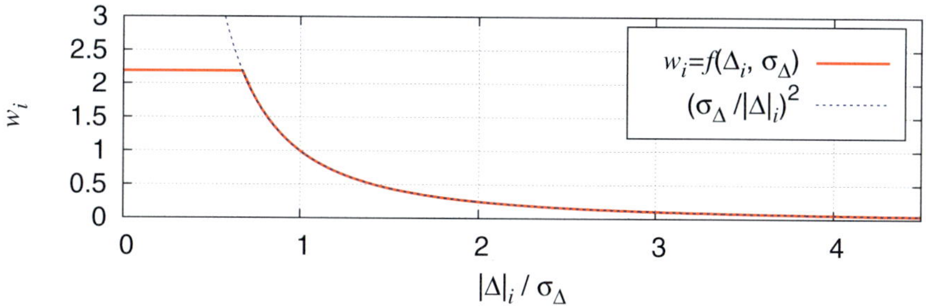

Figure 3.3: Transfer function for the mapping of deviates to suitable weights. $\kappa_L = \lambda_L / \sigma_y = 0.675$ limits the hyperbolic function (in case of normal distributed deviates).

whereas κ_L is an application-dependent factor and $\hat{\sigma}_y$ the estimated standard deviation (e.s.d.) of the observations y_i with respect to the model function

$$\hat{\sigma}_y = \sqrt{\frac{1}{N-M} \cdot \frac{\sum_{i=1}^{N} w_i \cdot [y_i - f(\mathbf{x}_i|\hat{\mathbf{a}})]^2}{\frac{1}{N}\sum_{i=1}^{N} w_i}} \,. \tag{3.10}$$

M is the number of model parameters. The sum below the fraction line normalises the weights to a mean value equal to one.

Using $\hat{\sigma}_y$ in the weighting process puts implicitly the presumption of normally distributed deviates. In addition, $\hat{\sigma}_y$ reflects only the deviation of the recorded observations, not the deviation in the parent distribution σ_i. If the number of observation N is not sufficient high, both values, $\hat{\sigma}_y$ and σ_i, can differ a lot.

Figure 3.3 depicts the weights as a function of the absolute deviates. Assuming a Gaussian distribution of deviates, the value of $\kappa_L = 0.675$ splits the observation into two halves, one getting equal weights $w_i = 1/\lambda_L^2$ and the other getting weights based on the corresponding deviate $w_i = 1/\lambda_i^2$ according to equation (3.9). Using $\kappa_L = 1.0$, for example, 68.3% of all weights would be equal.

In the beginning of Section 3.3, it was stated that the first run of data fitting must use equal weights if the uncertainty σ_i of the single observation y_i is not known. Afterwards, the estimation of the weights w_i is possible in dependence on the residuals $\Delta_i = y_i - f(\mathbf{x}|\hat{\mathbf{a}})$ and $\hat{\sigma}_y$ (equation (3.10)).

Now let us imagine that a particular data set contains observations with highly varying uncertainty. If the number N of observations is rather small ($N <<$

100), it is very likely that the data fitting yields suboptimal if not even wrong parameter values at the beginning. In addition, the computed value $\hat{\sigma}_{\text{y}}$ is high and consequently the threshold λ_{L} as well. The discrimination between most observations (with smaller absolute residuals $|\Delta_i|$) will fail, since equal weights are assigned to them. Furthermore, the derived values Δ_i do not reflect the deviation from the true model function, but describe the distances to the current modelled function $f(\mathbf{x}|\hat{\mathbf{a}})$ which can still be entirely wrong as long as the estimated parameters $\hat{\mathbf{a}}$ are far from their optimum.

Thus, the first estimate of weights is likely to be non-optimal and it is necessary to iterate the processing with alternating execution of parameter estimation by data fitting and determination of weights. The iteration stops when the changes of the weights are negligible. This procedure is also called *iteratively re-weighted least squares* and has been firstly mentioned in the context of robust regression by [Hol77].

Non-normally distributed deviates

In order to avoid the dependence of λ_{L} on $\hat{\sigma}_{\text{y}}$, one takes this deviate as threshold, which cuts the set of deviates in two halves, that is,

$$\lambda_{\text{L}} = \text{median}\left(|\Delta_i|\right), \qquad \Delta_i = y_i - f(\mathbf{x}_i|\hat{\mathbf{a}}) \tag{3.11}$$

In case of normally distributed deviates, this would approximately correspond to $\lambda_{\text{L}} = \kappa_{\text{L}} \cdot \hat{\sigma}_{\text{y}} = 0.675 \cdot \sigma_{\text{y}}$.

Taking the median element, fifty percent of all observations are assigned to equal weights. Of course, any element of another rank would differently influence the strength of weighting. However, λ_{L} should be bounded to a minimum of about $0.05 \cdot \max_i\left(|\Delta_i|\right)$ in order to prevent an unbalanced weighting if a large number of deviates is very small.

3.4 Approaches towards outlier detection

Let us assume, each observation y_i is assigned a weight w_i indicating the importance of this particular observation. In the simplest case, these weights are all equal to one. The elimination of outliers, i.e. setting their weights to zero, requires a threshold λ_{O} that is compared to the absolute deviates

$$w_i = \begin{cases} 0 & \text{for} \quad |\Delta_i| \geq \lambda_{\text{O}} \\ w_i & \qquad \text{otherwise} \end{cases} . \tag{3.12}$$

λ_O is a hard threshold and defines the range of outliers. Some application also allow soft-bounded estimations of the "degree of outlierness" [Bre00].

The determination of λ_O is the critical task in outlier detection.

Outlier detection can be applied either directly or after the estimation of weights. In general, the estimation of weights has a positive effect on outlier detection. Observations will emerge as contaminants if the weighting procedure improves the estimation of model parameters.

It is theoretically possible but not recommended to include the outlier detection into the iterations of weights estimation because this would cause problems when applying the threshold λ_O. Turning off some observations is a nonlinear operation, which could lead to an unstable iteration with slow convergence or even oscillation in the worst case. That means, iterative weighting and outlier detection must be applied consecutively.

In the following, three methods of outlier detection will be explained. The first one is quite simple and is, despite its shortcomings, widely used in many applications. The second method is more sophisticated and improves the detection results by clustering the observations in acceptable data points and contaminants, while the third method randomly excludes observations in order to find the best fitting.

3.4.1 Standardised residuals

The Method

The criterion of standardised residuals assumes that the values y to be inspected are normally distributed according to the distribution function

$$f(y) = \frac{1}{\sqrt{2\pi} \cdot \sigma_{\mathrm{y}}} \cdot \exp\left[-0.5 \cdot \left(\frac{y - \overline{y}}{\sigma_{\mathrm{y}}}\right)^2\right] \tag{3.13}$$

with a mean $\overline{y}$ and a standard deviation σ_{y}. The true value σ_{y} is not known in advance. Since the number of observations $i = 1, \ldots, N$ is limited, σ_{y} can only be estimated to a certain degree of accuracy by

$$\hat{\sigma}_{\mathrm{y}} = \sqrt{\frac{1}{N-1} \sum_{i=1}^{N} (y_i - \overline{y})^2} \, . \tag{3.14}$$

The majority of all observations drawn from a normal distribution are less distant from the mean than a certain multiple of its standard deviation. The method of standardised residuals (also called the z-score) utilises this fact to identify contaminants

$$\frac{|y_i - \overline{y}|}{\hat{\sigma}_{\mathrm{y}}} > \kappa_{\mathrm{O}} . \tag{3.15}$$

All observations leading to a standardised residual larger than κ_{O} are considered to be outliers. Many texts propose values in the range $3 \leq \kappa_{\mathrm{O}} \leq 4$. The value could also be adapted to the number of observations N according to Chauvenet's criterion [Cha71] (see Section 7.2).

In the context of data fitting via least-squares approximation, $\overline{y}$ corresponds to $f(\mathbf{x}|\hat{\mathbf{a}})$ and (3.15) turns with $\Delta_i = y_i - f(\mathbf{x}|\hat{\mathbf{a}})$ into

$$\frac{|\Delta_i|}{\hat{\sigma}_{\mathrm{y}}} = \frac{\Lambda_i}{\hat{\sigma}_{\mathrm{y}}} > \kappa_{\mathrm{O}} . \tag{3.16}$$

The threshold for outliers is therefore

$$\lambda_{\mathrm{O}} = \kappa_{\mathrm{O}} \cdot \hat{\sigma}_{\mathrm{y}} \tag{3.17}$$

with an estimated standard deviation $\hat{\sigma}_{\mathrm{y}}$ according to equation (3.10).

Median absolute deviation

The most problematic issue with z-scores is that both the mean value $\overline{y}$ and the estimated standard deviation $\hat{\sigma}_{\mathrm{y}}$ are influence by the limited number of samples and by possible bad data points. Especially the latter reason can thwart the detection of exactly these points. Using the median value $y_{\mathrm{m}} = \mathrm{median}_i(y_i)$ instead of the mean value $\overline{y}$ and the median absolute deviation $MAD = \mathrm{median}_i(y_i - y_{\mathrm{m}})$ instead of $\hat{\sigma}_{\mathrm{y}}$ allows a more robust estimation [Igl93, Rou11]. Now, equation (3.15) can be substituted by

$$\frac{|y_i - y_{\mathrm{m}}|}{1.4826 \cdot MAD} > \kappa_{\mathrm{O}} , \tag{3.18}$$

which is also called the M-score. The factor $c = 1.4826$ makes the MAD consistent with the standard deviation of a normal distribution. The reason for c is the following. Based on the normal distribution function, the range $[-\sigma_{\mathrm{y}}, +\sigma_{\mathrm{y}}]$ covers about 68.27% of all samples

$$\int_{-\sigma_{\mathrm{y}}}^{+\sigma_{\mathrm{y}}} \frac{1}{\sqrt{2\pi} \cdot \sigma_{\mathrm{y}}} \cdot \exp\left[-0.5 \cdot \left(\frac{y - \overline{y}}{\sigma_{\mathrm{y}}}\right)^2\right] \approx 0.6827 , \tag{3.19}$$

while the range $[-MAD, +MAD]$ comprises only fifty percent per definition

$$\int_{-0.6745\sigma_y}^{+0.6745\sigma_y} \frac{1}{\sqrt{2\pi} \cdot \sigma_y} \cdot \exp\left[-0.5 \cdot \left(\frac{y - \overline{y}}{\sigma_y}\right)^2\right] \approx 0.50 \,. \tag{3.20}$$

From $MAD \approx 0.6745 \cdot \sigma_y$ it follows $\sigma_y \approx 1.4826 \cdot MAD$. Please note that the estimation of σ_y via MAD still requires a sufficient number of data points.

Example:
A data set with $N = 10$ values is given: $\underline{y} = \{2\ 2\ 3\ 3\ 3\ 3\ 4\ 4\ 4\ 5\}$.
A second set $\underline{y}_o$ contains the same values plus one outlier $y_{11} = 10$.
The estimates are

set	$\overline{y}$	y_m	$\hat{\sigma}_y$	MAD
$\underline{y}$	3.300	3.000	0.949	1.000
$\underline{y}_o$	3.909	3.000	2.212	1.000

The estimators y_m and MAD are not influenced by the outlier.
□□□

With respect to equation (3.16), there is no direct solution for making the estimation of the absolute deviates Λ_i independent of possible outliers. The standard deviation σ_y, however, could be estimated via MAD. In the following text, this is referred as M-score.

Implications of the normal distribution

Although the normal distribution has its theoretical foundation, most people are not willing to accept that a measurement can deliver a result that arbitrarily deviates from the correct value. This is, however, exactly what the range of definition $-\infty \leq y \leq +\infty$ of the normal distribution is telling us. In practical cases of limited numbers N of observations y_i $(i = 1, 2, \ldots, N)$, possibly everybody will reject an observation y_i that is further apart than, say, a certain multiple of the standard deviation σ_y. Therefore, when talking about outliers, it seems to be appropriate to consider a modification of the statistical model of the observations.

Choosing the cut-off value based on the standardised residual raises another problem. Typically, it is expected that the outlier criterion will separate the cluster of 'good' observations from contaminants, especially when inspecting the plot of residuals. There should be a certain distance between the outer border of the cluster and the outliers. The standardised residual criterion, however, does not offer a separation of this kind by definition. In **Figure 3.4** a), two out of 100

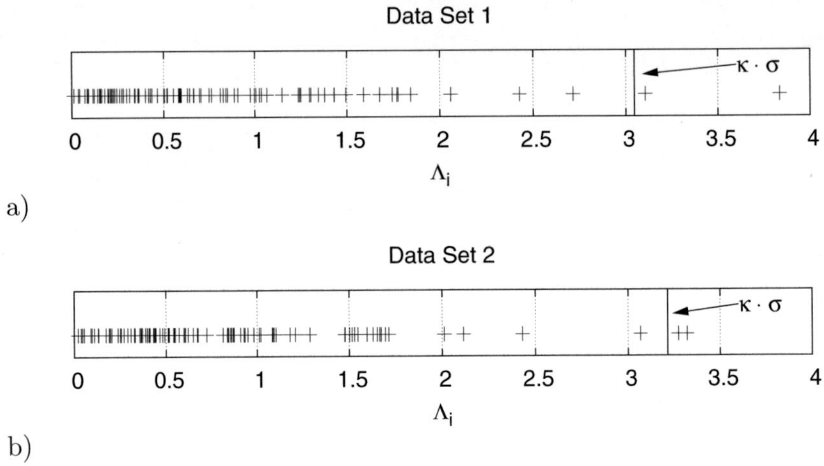

Figure 3.4: The weakness of outlier detection via standardised residuals. Example distributions of absolute deviates $\Lambda_i = |\Delta_i| = |y_i - \hat{y}_i|$ with threshold according to the $3 \cdot \hat{\sigma}_\Delta$ rule.

observations y_i drawn from a normal distribution ($\sigma_\mathrm{y} = 1.0, \overline{y} = 0$) are declared as outliers, when using $\kappa_\mathrm{O} = 3$.[3] Intuitively, one would cut off the observation at $\Lambda_i = |\Delta_i| \approx 3.8$ in maximum. The opposite is true for the example in **Figure 3.4** b). Under same conditions, two out of 100 observations are again classified as outliers. Judgement by eye would suggest three outliers, due to the large gap between 2.5 and 3.0.

The next subsection proposes an approach to outlier detection based on cluster-analysis that is independent of the estimated standard deviation $\hat{\sigma}_\mathrm{y}$ and takes into account that outliers should be apart from the bulk of 'good' observations.

3.4.2 Cluster criterion

Outlier detection based on the standardised residuals discussed above is dependent on the normal distribution of observations and the estimate of its standard deviation $\hat{\sigma}_\mathrm{y}$ (see eq. 3.14), which is rather uncertain in the case of low numbers of observations. In order to avoid this dependency, a method of outlier detection is required following another principle.

The basic idea is to find a pattern, or strictly speaking a gap, in the distribution of deviates that might point to the existence of outliers (**Figure 3.5**). Putative

[3] The estimated standard deviation is somewhat higher than 1.0 leading to a threshold of $\kappa_\mathrm{O} \cdot \sigma = 3.047$.

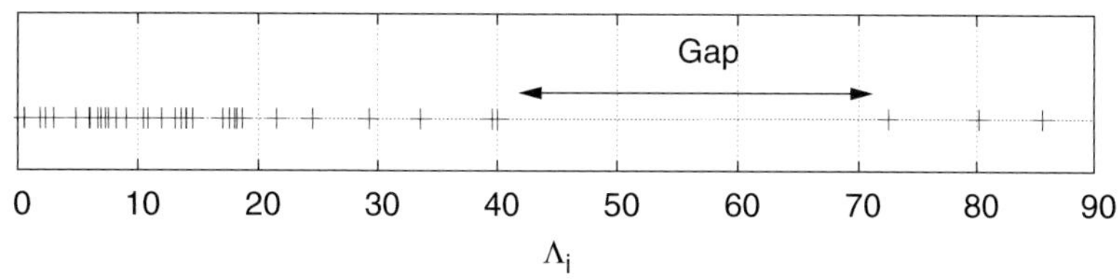

Figure 3.5: Example distribution of absolute deviates $\Lambda_i = |\Delta_i| = |y_i - \hat{y}_i|$ with a significant gap between the bulk of deviates on the left and the three suspicious points on the right.

contaminants are identified by comparing the distances between the absolute deviates $\Lambda_i = |\Delta_i| = |y_i - \hat{y}_i|$ [Str09b].

Basic approach

It is presumed that all non-outliers form a one-dimensional cluster in the sense that their corresponding absolute deviates are relatively close to each other, while the deviates of contaminant observations are more or less remote from this cluster. The sign of the deviates provides no valuable information for the process of outlier detection. In contrast, mapping all negative deviates into the range of positive numbers decreases the distances between them on average leading to fewer potential outliers, especially if they are distributed almost evenly in the negative and positive domain and their total number N is low. The example in **Figure 3.6** shows that taking the absolute values integrates exposed deviates into the set of deviates.

The algorithm for clustering the observations in good data points and putative outliers will be explained in detail in the following. Afterwards, several numerical examples will be discussed, making the procedure more transparent.

First, the absolute deviates $\Lambda = |\Delta|$ have to be sorted in ascending order

$$\cdots \leq \Lambda_j \leq \cdots \leq \Lambda_k \leq \cdots \leq \Lambda_l \leq \dots .$$

These sorted values are numbered by n

$$\Lambda_s[0] \leq \cdots \leq \Lambda_s[n] \leq \Lambda_s[n+1] \leq \dots \Lambda_s[N-1]$$

and the differences between them are calculated

$$d[n+1] = \Lambda_s[n+1] - \Lambda_s[n] .$$

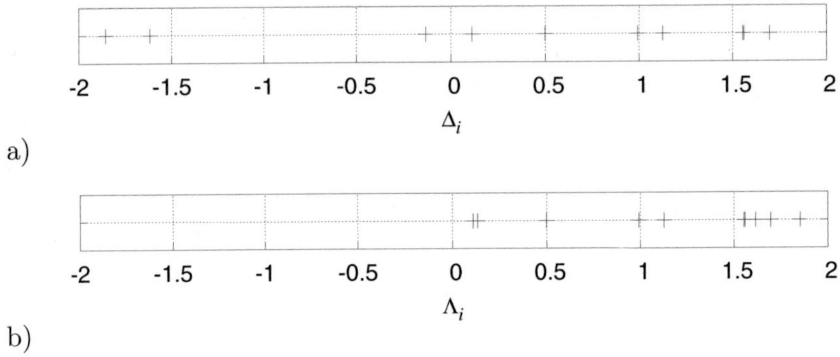

Figure 3.6: Example distribution of absolute deviates ; a) with sign, b) absolute values

In the sequel, only absolute deviates are considered and we will abandon on the attribute "absolute".

It is expected that the deviate of an outlier will show a significantly higher difference (distance) from its nearest neighbour downwards, i.e., the deviate will be more distant from the others than deviates of measurements drawn from the correct distribution.

What qualifies a distance d_b as a border (a gap) between a one-dimensional cluster of good observations and possible outliers?

1. It must be distinctly larger than a typical distance (global criterion)

$$d_\mathrm{b} \geq \kappa_1 \cdot d_\mathrm{glob} \qquad \text{with } \kappa_1 > 1\,. \tag{3.21}$$

2. It should be substantially larger than its predecessors (local criterion)

$$d_\mathrm{b} \geq \kappa_2 \cdot d_\mathrm{loc} \qquad \text{with } \kappa_2 > 1\,. \tag{3.22}$$

We define the typical distance for a certain deviate as the weighted average of distances belonging to deviates which are smaller than the deviate corresponding to the distance $d[n]$ under investigation

$$d_\mathrm{glob}[n] = \frac{1}{C_{1,n}} \cdot \sum_{j=1}^{n-1} d[n-j] \cdot v_j\,. \tag{3.23}$$

Considering only the smaller deviates avoids the influence of other potential outliers. The weighting v_j becomes weaker with increasing j, i.e. with increasing

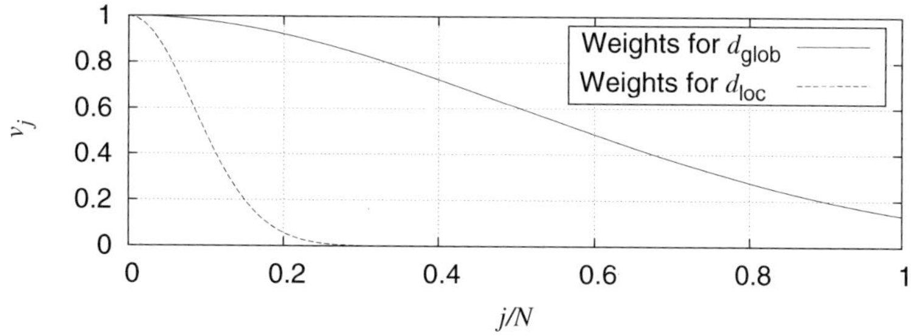

Figure 3.7: Weights v_j for the determination of $d_{\text{glob}}[n]$ and $d_{\text{loc}}[n]$ in dependence on the distance j to the difference $d[n]$ (normalised by number of observation N)

distance[4] to the point under investigation

$$v_j = \exp\left[-\frac{1}{2}\cdot\left(\frac{j}{N/2}\right)^2\right] . \tag{3.24}$$

The slope of the exponential function in eq. (3.24) depends on the number of observations N in order to make the calculation adaptive to this number (**Figure 3.7**). The constant C_1 in (3.23) normalises to the sum of weights

$$C_{1,n} = \sum_{j=1}^{n-1} v_j . \tag{3.25}$$

In principle, the second criterion (eq. 3.22) utilises the same averaging process, but with weights falling off more rapidly in order to express closeness

$$d_{\text{loc}}[n] = \frac{1}{C_{2,n}} \cdot \sum_{j=1}^{n-1} d[n-j] \cdot \exp\left[-\frac{1}{2}\cdot\left(\frac{j}{N/12}\right)^2\right] \tag{3.26}$$

with

$$C_{2,n} = \sum_{j=1}^{n-1} \exp\left[-\frac{1}{2}\cdot\left(\frac{j}{N/12}\right)^2\right] . \tag{3.27}$$

The denominators $N/2$ and $N/12$ have been determined empirically. Figure 3.7 shows that $d_{\text{glob}}[n]$ depends on all differences $d[n-j]$ with slightly decreasing

[4] Here, 'distance' refers to the index n of the sorted set of deviates.

weights, while $d_{\text{loc}}[n]$ is mainly influenced by only few neighbouring differences. Please note that both, $d_{\text{glob}}[n]$ and $d_{\text{loc}}[n]$ are different for each distance $d[n]$, reflecting the adaptive character of the approach. This property is also discussed in the examples below.

In summary, the presence of one or more outliers is indicated if there is an absolute deviate $\Lambda_{\text{s}}[n]$ showing a distance $d[n]$ from the next deviate down $\Lambda_{\text{s}}[n-1]$, which has the two properties

$$d[n] \geq \kappa_1 \cdot d_{\text{glob}}[n] \qquad \text{and} \qquad d[n] \geq \kappa_2 \cdot d_{\text{loc}}[n] \,. \tag{3.28}$$

The value of the corresponding deviate $\Lambda_{\text{s}}[n]$ is taken as the cut-off value λ_{O}. If there is more than one deviate satisfying both criteria, the one with the highest relation $d[n]/d_{\text{loc}}[n]$ is selected. If these values are identical as well, the decision is made dependent on the larger distance $d[n]$.

Using the approach described, even multiple outliers can be eliminated at once, because the observation y_{b} corresponding to the distance d_{b} fulfilling eq. (3.28) only marks the border between the two clusters of good observations and contaminants. It is evident that all observations having higher absolute deviates than y_{b} also belong to the cluster of outliers.

Parameter determination

Experiments with different kinds of data sets have shown that κ_2 can be set to a fixed value of 2, while κ_1 should be dependent on the number of observations N.

Suitable values for κ_1 have been derived from computer simulations. The idea has been to choose a value for κ_1 that will lead to a probability of about $P_{\text{o}} = 0.15$ that an outlier is detected in a data set with normally distributed deviates. For different numbers of observations N in the range of $8 \leq N \leq 2048$, 10^5 random data sets have each been investigated and the resulting plots $P_{\text{o}} = f(\kappa_1)$ have been examined. Naturally, the outlier probability P_{o} decreases with increasing values of κ_1. **Figure 3.8** shows the derived dependence on κ_1 on N. If N is a number between these points, the corresponding κ_1 can be interpolated.

Exception handling

In experiments with integer-valued observations, several deviates typically have identical values leading to many distances equal to zero. The local change in distances would be over-estimated in theses cases and the determination of $d_{\text{loc}}[n]$

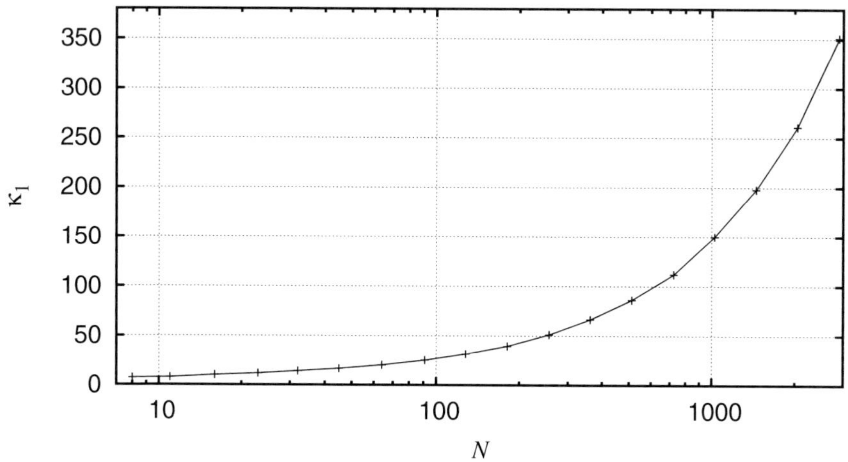

Figure 3.8: Threshold κ_1 in dependence on the Number of observations N.

is altered to

$$d_{\text{loc}}[n] = \begin{cases} d[n]/\kappa_2 & \begin{array}{l}\text{if more than the half of all} \\ \text{predecessors } d[n-j] \text{ are equal to zero}\end{array} \\ d_{\text{loc}}[n] & \text{otherwise} \end{cases} \quad . \tag{3.29}$$

The alternative value is chosen such that the local criterion is fulfilled (see eq. (3.28)), but with the lowest chance of being selected as the border between the clusters of good and bad data points.

Furthermore, the cluster-based approach described so far only separates the observations in two clusters. It does not care about the number of observations per cluster. An additional mechanism is required, which ensures that the cluster of outliers is not larger than the cluster of good observations. This can be achieved, for example, by analysing only distances $d[n]$ with $n > N/2$.

Examples

The entire procedure of outlier detection based on distances is explained in following examples.

Example 1:
The absolute deviates $\{\Lambda_i\}$ are given according to the plot in **Figure 3.9**. Inspection by eye suggests the presence of three clusters for the

Table 3.1: Intermediate values of cluster-based outlier detection, Example 1, see text for details.

n	$\Lambda_s[n]$	$d[n]$	$d_{glob}[n]$	$q[n] = \frac{d[n]}{d_{glob}[n]}$	$d_{loc}[n]$	$r[n] = \frac{d[n]}{d_{loc}[n]}$
0	1.70	0.00	0.000	0.000	0.000	0.000
1	2.00	0.30	0.000	0.000	0.000	0.000
2	2.50	0.50	0.300	1.667	0.300	1.667
3	3.10	0.60	0.402	1.492	0.464	1.294
4	3.20	0.10	0.472	0.212	0.578	0.173
5	3.70	0.50	0.373	1.341	0.196	2.553
6	4.60	0.90	0.400	2.249	0.430	2.095
7	5.10	0.50	0.500	1.000	0.816	0.613
8	10.50	5.40	0.505	10.692	0.572	9.446
9	10.70	0.20	1.335	0.150	4.451	0.045
10	18.30	7.60	1.213	6.263	1.139	6.673
11	18.40	0.10	2.231	0.045	6.235	0.016
				κ_1: 8.180		κ_2: 2.000

classification of the observations in good and bad ones.

The sorted deviates $\Lambda_s[n]$, distances $d[n]$, averaged distances $d_{glob}[n]$ and $d_{loc}[n]$, as well as the relations $q[n] = d[n]/d_{glob}[n]$ and $r[n] = d[n]/d_{loc}[n]$ for this example are listed in **Table 3.1**. The distances range from 0.10 to 7.60. Only one of them, $d[8]$, fulfils the global criterion $d[n]/d_{glob}[n] \geq \kappa_1$. Since the local criterion $d[8]/d_{loc}[8] \geq \kappa_2$ is also satisfied, a gap between two clusters has been found. In order to exclude all observations not belonging to the cluster of good points, the value of the deviate corresponding to the critical distance $d[8]$ is taken as threshold λ_O, i.e., all observations with deviates $\Lambda_i \geq \lambda_O = \Lambda_s[8] = 10.5$ are marked as outliers.

Please note that the relation $q[10] = 6.263$ is below the threshold κ_1, although the corresponding distance $d[10] = 7.6$ is higher than

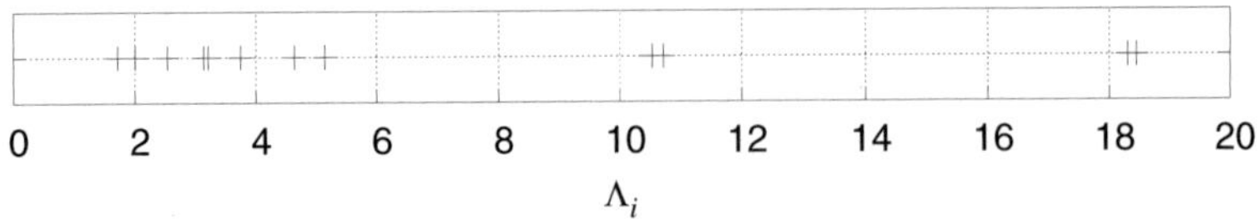

Figure 3.9: Distribution of absolute deviates in Example 1.

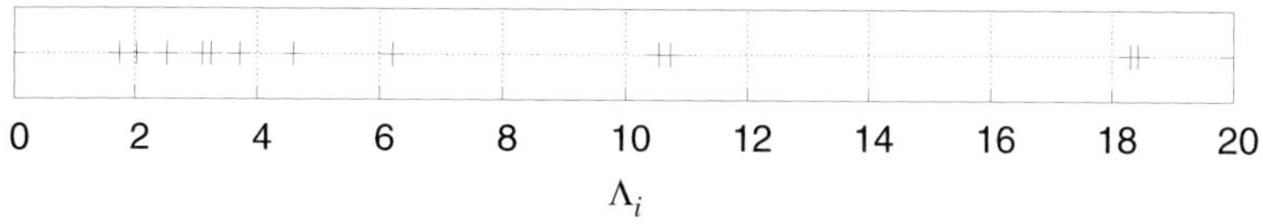

Figure 3.10: Distribution of absolute deviates in Example 2.

Table 3.2: Intermediate values of cluster-based outlier detection, Example 2, see text for details.

n	$\Lambda_s[n]$	$d[n]$	$d_{glob}[n]$	$q[n] = \frac{d[n]}{d_{glob}[n]}$	$d_{loc}[n]$	$r[n] = \frac{d[n]}{d_{loc}[n]}$
0	1.70	0.00	0.000	0.000	0.000	0.000
1	2.00	0.30	0.000	0.000	0.000	0.000
2	2.50	0.50	0.300	1.667	0.300	1.667
3	3.10	0.60	0.402	1.492	0.464	1.294
4	3.20	0.10	0.472	0.212	0.578	0.173
5	3.70	0.50	0.373	1.341	0.196	2.553
6	4.60	0.90	0.400	2.249	0.430	2.095
7	6.20	1.60	0.500	3.200	0.816	1.960
8	10.50	4.30	0.705	6.102	1.457	2.951
9	10.70	0.20	1.327	0.151	3.763	0.053
10	18.30	7.60	1.202	6.322	0.958	7.937
11	18.40	0.10	2.217	0.045	6.219	0.016
				κ_1: 8.180		κ_2: 2.000

$d[8]$. This is caused by the effect of accommodation. As one large distance has already been seen for predecessors, the new occurrence of a similar distance is not anymore an indication of contaminant observations, but only of a sparse distribution. This is an important feature of the cluster-based method. It must also be pointed out that the local condition can also be fulfilled for small distances such as, for example, $d[5] = 0.5$ if the predecessors are even smaller by chance.
□□□

Example 2:
As in any case of using thresholds, slight variation of values might cause different response of the system. Let us assume that one certain observation is somewhat more distant to the model function leading to a deviate of $\Lambda_s[7] = 6.2$ instead of 5.1 (see **Figure 3.10**). **Table 3.2** contains the resulting values. The related distance $d[8] = 4.3$ has been fallen below $\kappa_1 \cdot d_{glob} = 8.18 \cdot 0.705 = 5.7669$ which has increased itself

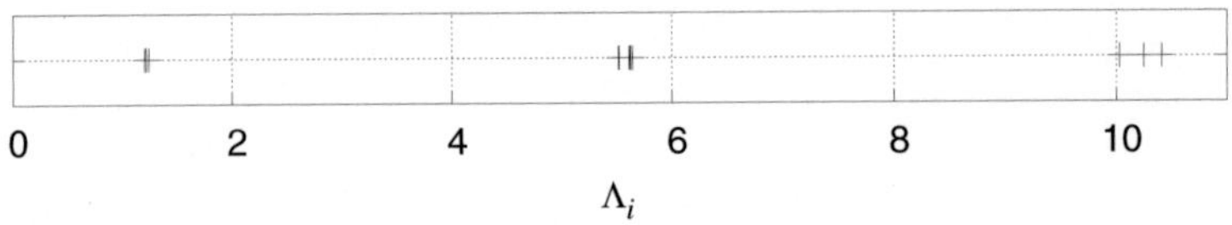

Figure 3.11: Distribution of absolute deviates in Example 3.

owing to the modified value of d_{glob}. The local criterion $r[8] = 2.951$ still signals a distinct change in the sequence of distances, but the global distance is not fulfilled anymore. The algorithm judges the distance of 4.3 as not being significant in comparison to the other distances. Since the global criterion is never fulfilled, no outliers are identified. The gaps between $\Lambda_s[7]$ and $\Lambda_s[8]$ as well as $\Lambda_s[9]$ and $\Lambda_s[10]$ are still visible in Figure 3.10, but are interpreted as sparseness of the distribution and not as indication for outliers.
□□□

Example 3:
This example reveals that the cluster-based outlier detection does not depend on the normal distribution of deviates (**Figure 3.11**). The majority of absolute residuals is in the range of $5\ldots6$, plus some distinctively smaller and higher deviates. The evaluation of the deviates results in the values listed in **Table 3.3**. There are two distances with $q[n] > \kappa_1$, $q[3] = 428.0$ and $q[9] = 9.477$.However, $d[9] = 4.38$ has the highest quotient of $r[9] = 531.891$ greater than $\kappa_2 = 2.0$. Therefore, all observations with deviates of $\Lambda_i \geq \Lambda_s[9] = 10.0$ are declared as contaminants.
□□□

Now let us consider an example in which many deviates are identical to another in pairs owing to integer-valued observations.

Example 4:
Figure 3.12 shows the absolute deviates $\{\Lambda_i\}$ used in this example. **Table 3.4** contains the intermediate values for this example. Many distances are equal to zero, because of the identical deviates. Distance $d[4] = 6.0$ exceeds the first threshold of $\kappa_1 \cdot d_{\text{glob}}[4] = 9.62 \cdot 0.207 = 1.991$ and the local criterion is satisfied as well. Since there is no other point fulfilling both criteria, the threshold λ_{O} is set to $\Lambda_s[4] = 6.8$.

Table 3.3: Intermediate values of cluster-based outlier detection, Example 3, see text for details.

n	$\Lambda_s[n]$	$d[n]$	$d_{\text{glob}}[n]$	$q[n] = \frac{d[n]}{d_{\text{glob}}[n]}$	$d_{\text{loc}}[n]$	$r[n] = \frac{d[n]}{d_{\text{loc}}[n]}$
0	1.20	0.00	0.000	0.000	0.000	0.000
1	1.21	0.01	0.000	0.000	0.000	0.000
2	1.22	0.01	0.010	1.000	0.010	1.000
3	5.50	4.28	0.010	428.000	0.010	428.000
4	5.52	0.02	1.506	0.013	3.450	0.006
5	5.60	0.08	1.130	0.071	0.785	0.102
6	5.61	0.01	0.900	0.011	0.131	0.076
7	5.61	0.00	0.719	0.000	0.025	0.000
8	5.62	0.01	0.576	0.017	0.003	3.326
9	10.00	4.38	0.462	9.477	0.008	531.891
10	10.20	0.20	1.066	0.188	3.528	0.057
11	10.40	0.20	0.963	0.208	0.948	0.211
				κ_1: 8.180		κ_2: 2.000

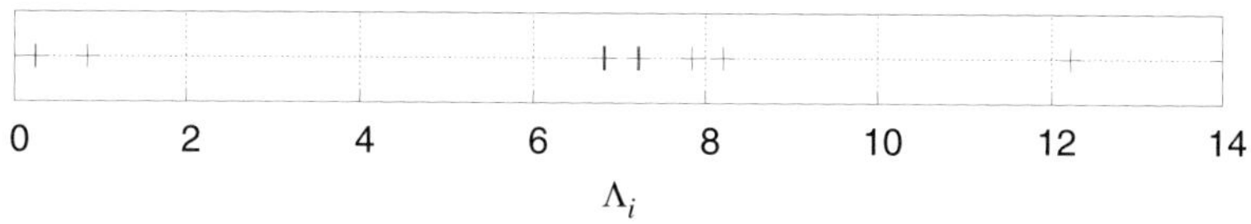

Figure 3.12: Distribution of absolute deviates in Example 4.

The result underlines the fact that the clustering of deviates works perfectly. In its application to data analysis, however, it does not seem to be appropriate to reject more that 50% of the observations. An additional mechanism is therefore needed. The threshold λ_O should only be set when the cluster of good observations contains at least a predefined percentage of the total number of observations. This could be achieved, for example, by only comparing the top 35% or less of the deviates with the thresholds κ_1 and κ_2. On the other hand, however, such a small cluster of 'good' observations could indicate an improperly chosen model $y = f(\mathbf{x}|\mathbf{a})$.
□□□

The described method of cluster-based outlier detection is not only applicable to deviates in the framework of least-squares approximation, but also suitable

Table 3.4: Intermediate values of cluster-based outlier detection, Example 4, see text for details.

n	$\Lambda_s[n]$	$d[n]$	$d_{\text{glob}}[n]$	$q[n] = \frac{d[n]}{d_{\text{glob}}[n]}$	$d_{\text{loc}}[n]$	$r[n] = \frac{d[n]}{d_{\text{loc}}[n]}$
0	0.20	0.00	0.000	0.000	0.000	0.000
1	0.20	0.00	0.000	0.000	0.000	0.000
2	0.20	0.00	0.000	0.000	0.000	0.000
3	0.80	0.60	0.000	0.000	0.000	2.000
4	6.80	6.00	0.207	29.050	0.411	2.000
5	6.80	0.00	1.742	0.000	4.242	0.000
6	6.80	0.00	1.395	0.000	1.596	0.000
7	6.80	0.00	1.149	0.000	0.319	0.000
8	6.80	0.00	0.960	0.000	0.034	0.000
9	7.20	0.40	0.808	0.495	0.002	2.000
10	7.20	0.00	0.737	0.000	0.272	0.000
11	7.20	0.00	0.624	0.000	0.104	0.000
12	7.80	0.60	0.525	1.142	0.021	2.000
13	8.20	0.40	0.513	0.779	0.411	2.000
14	12.20	4.00	0.482	8.295	0.429	2.000
				κ_1: 9.620		κ_2: 2.000

in other scenarios. The mapping of the observations y_i to absolute values Λ_i using an application-specific technique is the crucial point. The only constraint is that small values of Λ_i must correspond to good data points and large values to suspicious observations.

The advantages of the cluster-based method for outlier detection have been demonstrated in [Str09b] based on Monte Carlo simulations. The corresponding results are presented in the appendix, Chapter A.

3.4.3 Random sample consensus (RANSAC)

Basic approach

The *random sample consensus* (RANSAC) is a robust technique for estimating the most suitable set of model parameters in the presence of a probably high portion of outliers [Fis81]. It is often discussed as an alternative to the least-squares method. But in point of fact, RANSAC should be interpreted as a method of outlier detection, which assists the data fitting via least-squares approximation.

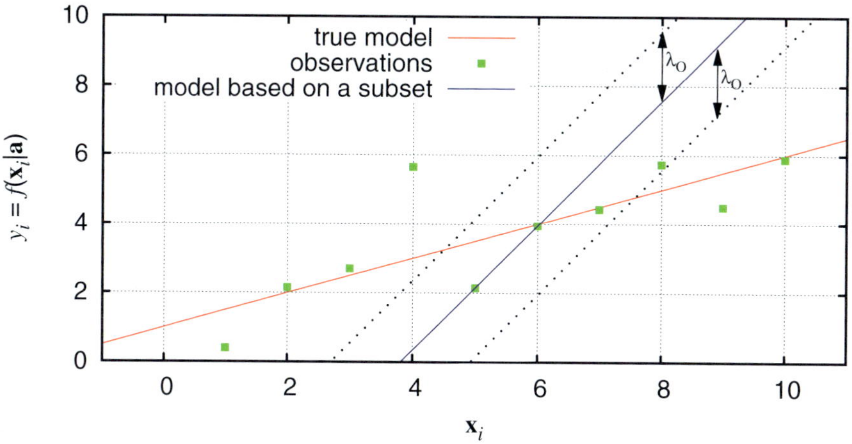

Figure 3.13: RANSAC: parameter estimation based on a subset of $M = 2$ data points ($x_i = \{5, 6\}$) in case of straight-line fitting with $y_i = a_1 + a_2 \cdot x$

The method assumes that a majority of data points supports a certain model (including a certain set of model parameters) and all other points (the outliers) are somehow distributed in the data space. In contrast to the ClubOD method, which was discussed in the previous section, the outliers must not be far away from the cluster of good points (inliers). RANSAC is mostly applied for problems in computer vision, where the portion of outliers caused by imperfections in the algorithms used for signal processing can be rather high.

Since its introduction in 1981, RANSAC has been modified and improved in several ways. An evaluation of many variants can be found in [Cho09]. RANSAC was not initially related to least-squares approximation, but, as a matter of course, can combined with the method of least squares in order to fit the remaining inliers to the model.

In the following, the basic method is explained and discussed. Some implementation details are illuminated afterwards.

Given a model function $y_i = f(\mathbf{x}_i|\mathbf{a}) + \varepsilon$, the RANSAC method generates a number of hypotheses about the model-parameter vector $\mathbf{a}$. The variable y_i is the ith observation and $\mathbf{x}_i$ the corresponding vector of conditions. The hypotheses are based on fitting randomly chosen subsets of the entire set of N data points. The subset size N_s is set equal to the number M of the model parameters. In case of fitting a straight line (**Figure 3.13**), the number of parameters is $M = 2$. Inspecting the example in the diagram, it becomes clear that using only two points leads to an exact estimate (the blue line goes through both points at $x_i = 5$

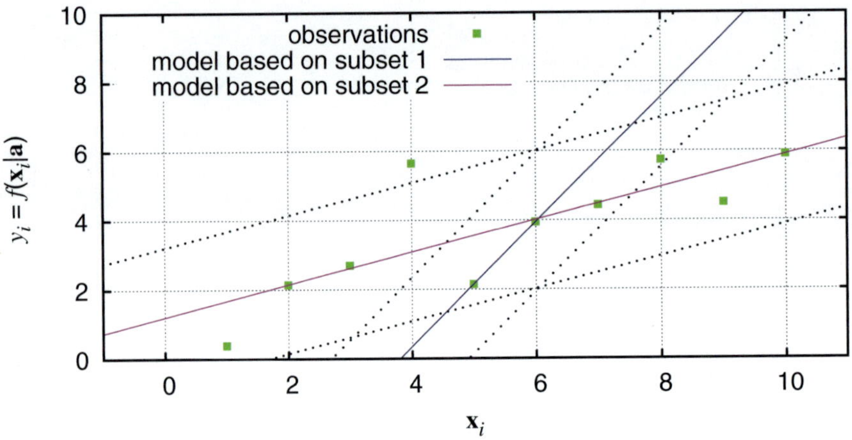

Figure 3.14: RANSAC: parameter estimation based on two different subsets $x_i \in \{5, 6\}$ (blue line) and $x_j \in \{2, 10\}$ (magenta line)

and $x_i = 6$) but at the same time the estimates of the line parameters is very uncertain. However, taking the smallest possible number of points maximises the chance that a certain subset does not contain outliers.

After the determination of the model parameters $\hat{\mathbf{a}} = (\hat{a}_j)$ $(j = 1, 2, \ldots, M)$ based on a subset of points, the algorithm checks, which of the remaining data points support this model. The deviates $\Delta_i = y_i - f(\mathbf{x}_i|\hat{\mathbf{a}})$ of all data points are compared with a threshold λ_O. In Figure 3.13, the dotted lines show the limits of $f(\mathbf{x}_i|\hat{\mathbf{a}}) \pm \lambda_\mathrm{O}$. The model with the maximum number of data points with $|\Delta_i| \le \lambda_\mathrm{O}$ is taken as the best model. The corresponding cost function is

$$c = \sum_{i=1}^{N} \begin{cases} 0 & \text{for} \quad |\Delta_i| \le \lambda_\mathrm{O} \\ 1 & \quad\quad \text{otherwise} \end{cases} \quad . \tag{3.30}$$

The set of points with $|\Delta_i| \le \lambda_\mathrm{O}$ for a certain subset is called *consensus set*, that is, the entire procedure is looking for the largest consensus set.[5] If the subset consists of the data points at $x_i = 5$ and $x_i = 6$ (blue example in **Figure 3.14**), the consensus set contains only one additional point at $x_i = 7$. Using the subset $x_j \in \{2, 10\}$ (magenta line), the slope is not as steep and the consensus set comprises nine data points, while using the same threshold $\lambda_\mathrm{O} = 1.5$.

After testing a suitable number S of different subsets, the model is fitted to all points of the largest consensus set.

[5] The equality sign in $|\Delta_i| \le \lambda_\mathrm{O}$ accounts for the artificial case that λ_O was set to zero.

The number S of randomly chosen subsets is one important parameter. In principle we could test all possible subsets. The number would be

$$S_{\text{max}} = \binom{N}{N_s} = \frac{N!}{N_s! \cdot (N - N_s)!} \,.$$

It is obvious that testing all combinations of N_s-out-of-N observations is rather time-consuming. The question is, how many subsets are sufficient. The goal is to select at least one subset that is outlier-free. Let r_{in} be the expected ratio of the number of good data points (inliers) to the total number of points N. Since this ratio is not known in advance, we have to make a worst-case estimation, for example, $r_{\text{in}} = 0.6$. If $N_s = M$ data points are selected as a minimum to estimate the M parameters of a model, r_{in}^M is the probability that all points in a certain subset are inliers and $(1 - r_{\text{in}}^M)$ is the probability that the subset contains at least one outlier.

What we are looking for is how many subsets have to be generated so that at least one subset is outlier-free with a desired probability p. Taking S subsets, the probability that the algorithm never selects a subset with only inliers is $(1 - r_{\text{in}}^M)^S$ and the desired probability is $p = 1 - (1 - r_{\text{in}}^M)^S$. The number of required trials S is therefore

$$\begin{aligned} p &= 1 - \left(1 - r_{\text{in}}^M\right)^S \\ \left(1 - r_{\text{in}}^M\right)^S &= 1 - p \\ S \cdot \log\left(1 - r_{\text{in}}^M\right) &= \log(1 - p) \\ S &= \frac{\log(1 - p)}{\log\left(1 - r_{\text{in}}^M\right)} \,. \end{aligned} \tag{3.31}$$

Please note that the number of trials does not grow with the number N of observations.

As the value of S depends on the portion r_{in} of inliers, it might be useful to adapt the number of trials as soon as a consensus set is found, that is larger than previous ones. This will significantly shorten the entire procedure.

Example:
If there are $N = 20$ observations and three model parameters (i.e. $N_s = M = 3$), there are $S_{\text{max}} = 20 \cdot 19 \cdot 18/6 = 1140$ possibilities to select N_s points out of N, while with $p = 0.9999$ and $r_{\text{in}} = 0.6$ about

$$S = \log(1 - 0.9999)/\log\left(1 - 0.6^3\right) \approx 38$$

would be sufficient. If, after some trials, a model has been found that incorporates not only 60% but, for example, 80% of the points in the consensus set, r_{in} can be set equal to 0.8 and the number of trials decreases to

$$S = \log(1 - 0.9999)/\log\left(1 - 0.8^3\right) \approx 13 \; .$$

The probability r_{in}^M, however, implies that a randomly selected data point is put back into the overall set and can again be selected for the same subset. This is not reasonable and the selection algorithm should avoid taking the same point multiple times for the same subset.[6] Hence, the probability of selecting an outlier-free subset in S trials is higher than p. Having said that, we must also note that it is possible that a certain subset can be selected multiple times causing the adverse effect.

If there is agreement in advance about the minimum size of the consensus set, the generation and testing of subsets can be stopped as soon this number is reached.

Advantages and disadvantages

RANSAC can provide robust parameter estimation even in cases when a significant number of the data points are outliers.

The estimations are based on only M points and can result in a large number of useless hypotheses also if these M points belong to the good but inaccurate data (see Figure 3.13, estimation based on the subset).

The threshold λ_{O}, which decides whether a data point belongs to the consensus set or not, is the most critical parameter. If it is too small, the consensus sets will contain too few data points for the reliable final parameter estimation. If it is too high, it could be that no consensus set will be found that is outlier-free. Either there is some prior knowledge about the typical standard deviation of the deviates Δ_i or the threshold λ_{O} must be estimated from all data, for example.

The required number S of trials, which is sufficient to find, with a certain probability, at least one outlier-free subset, does not mean at the same time that a subset has been found, that contains *all* inliers. Consequently, the number of trials should indeed be set higher than the value of S suggests.

The cost function in (3.30) is rather simple, but does not provide any information on how well the points of the consensus set fit the model. It could be that the

[6] If the subset comprises $N_{\text{s}} = M = 2$ points, for example, and one point is selected twice, the fitting of a straight line would no longer be possible.

points of the largest consensus set show a higher deviation from the corresponding model on average than points of a slightly smaller set. The modified cost function

$$c = \sum_{i=1}^{N} \begin{cases} \Delta_i^2 & \text{for} \quad |\Delta_i| \leq \lambda_\mathrm{O} \\ \lambda_\mathrm{O}^2 & \text{otherwise} \end{cases} \tag{3.32}$$

takes the single deviates into account and the chance of selecting the best hypothesis is increased.

The random approach of subset selection, however, causes non-deterministic behaviour. The final model parameters can be different for each run of RANSAC, and success, i.e. finding a suitable or even the best model, is not guaranteed. Only if all possible subsets are tested, will the method find the optimal solution with respect to the cost function in equation (3.32).

Practical implementation

Using RANSAC as a method for outlier detection within a framework of least-squares approximation means that initially, the data fitting is performed based on all data points including the outliers. Though this is not a prerequisite for RANSAC, it assists the determination of a suitable threshold λ_O. When fitting all data points including possible outliers, the model parameters are not optimal, but the deviates Δ_i indicate at least the approximate dimensions. One feasible approach for determining the threshold λ_O is to utilise the median of all absolute deviates

$$\lambda_\mathrm{O} = 3 \cdot \mathrm{median}_i(|\Delta_i|) \qquad i = 1, 2, \ldots, N \, .$$

The factor of three is empirically chosen. If the deviates are normally distributed, λ_O now corresponds to a value of about two times the standard deviation of Δ_i.[7]

In the next step, the number of trials S is determined with equation (3.31). In each round s, N_s data points are drawn from the entire set and a least-squares procedure determines the corresponding model parameters $\hat{\mathbf{a}}^{(s)}$ $(s = 1, 2, \ldots, S)$. The deviates of all data points from the hypothetical model are computed via

$$\Delta_i^{(s)} = y_i - f\left(\mathbf{x}_i | \hat{\mathbf{a}}^{(s)}\right)$$

and the costs are determined

$$c^{(s)} = \sum_{i=1}^{N} \begin{cases} \left(\Delta_i^{(s)}\right)^2 & \text{for} \quad |\Delta_i^{(s)}| \leq \lambda_\mathrm{O} \\ \lambda_\mathrm{O}^2 & \text{otherwise} \end{cases} \, .$$

[7] More about the relation of median value and standard deviation can be found in the subsection 'Median absolute deviation' on page 58.

Parameters and data points of the trial $s_{\min}$ with lowest costs

$$s_{\min} = \underset{s}{\operatorname{argmin}} \left(c^{(s)} \right)$$

are stored. The stored data points (i.e. the best consensus set) are fed into a final round of least-squares approximation yielding the model parameters. In cases of nonlinear model functions, the parameters from fitting the subset yielding the best consensus set are applied as initial guesses.

It has to be noted that, in the case of nonlinear model functions, the fitting of subsets also requires initial guesses of the model parameter. As long as no sufficient model parameters could be found (the best consensus set seen so far has, for example, less than $0.8 \cdot N$ points), the parameter values from fitting the entire data set are used. If a consensus set has been found, that is large enough, the parameters from that subset fitting are taken, resulting in the lowest costs so far.

3.5 Application of weighted data fitting and outlier detection

This section compares the least-squares fitting depending on the used weighting and outlier-detection scheme for a variety of real and simulated test data with different model functions.

The general processing pipeline is depicted in **Figure 3.15**. Each of the three blocks can be switched off independently of the others. It only makes sense to use the reset block if both of the other blocks, weights estimation and outlier removal, are enabled. When using the last processing block, the weights are only exploited for more reliable outlier detection and the final fitting is performed using equal weights. If all processing steps are bypassed, an ordinary least-squares approximation is applied. Most examples, which are discussed in the following, use weights estimation based on deviates, as has been discussed in Subsection 3.3.2. Only the last example allows for the weighting via binning (Subsection 3.3.1). The detection and removal of outliers is performed either with

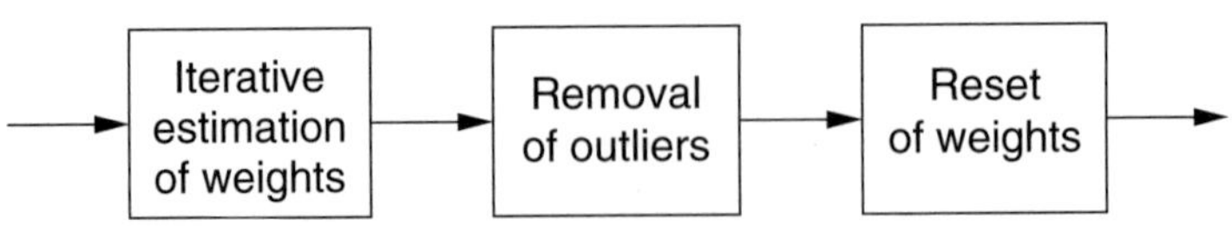

Figure 3.15: Flowchart of weighting and outlier detection

standardised residuals (z-score), median absolute deviation (M-score), cluster-based detection (ClubOD), or RANSAC (see Subsection 3.4). The thresholds for the z- and M-score methods are adaptively chosen according to Chauvenet's criterion (see Section 7.2) with $\nu_0 = 0.15$, i.e. with the assumption of 0.15 outliers per data set on average.

The single examples are discussed in several views. The pros and cons of the different detectors are enlightened. The impatient reader may jump directly to the conclusions on pages 108ff.

3.5.1 Constant value

A set of forty simulated observations y_i (normally distributed with $\overline{y_i} = 10.0, \sigma_{\mathrm{y}} = 1.340$, see Section 6.1) has been generated and extended by three outliers. It has to be fitted to the model function

$$y = f(\mathbf{x}|\mathbf{a}) = \text{const.} = a_1 \; .$$

The solution for linear model functions (see Section 2.4)

$$\mathbf{a} = \left(\mathbf{J}^{\mathrm{T}} \cdot \mathbf{W} \cdot \mathbf{J}\right)^{-1} \cdot \mathbf{J}^{\mathrm{T}} \cdot \mathbf{W} \cdot \mathbf{y}$$

simplifies to

$$a_1 = \frac{\sum\limits_{i=1}^{N} w_i \cdot y_i}{\sum\limits_{i=1}^{N} w_i} \; ,$$

as was already shown in Subsection 2.6.1. The usage of $w_i = 1.0$ for all i (i.e. equal weights) yields a mean value of 9.791 (**Table 3.5**). Without estimation of suitable weights, the z-score method declares two observations as being contaminant and the estimated parameter changes to $a_1 = 9.829$. The other methods remove three contaminants, leading to the correct value of $a_1 = 10.000$. Applying the iterated weights refinement, the result improves to $a_1 = 9.961$, if no outlier removal is involved. The number of detected outliers only increases for z-score from two to five.

The error bars in the plot of **Figure 3.16** depict the uncertainties in terms of estimated weights. Outliers have an infinite uncertainty, thus, the error bars of the outlier $(x_i, y_i) = (41, 3)$, $(42, 1)$, and $(43, 17)$ cross the entire plot. The z-score approach detects two more outliers at $(x_i, y_i) = 11, 13)$ and $(23, 13)$.

Table 3.5: Results for fitting a constant

Constant	a_1	Detected outliers
True values	10.000	3
Equal weights	9.791	0
z-score	9.829	2
M-score	**10.000**	3
ClubOD	**10.000**	3
RANSAC	**10.000**	3
Estimated weights	9.961	0
all methods	9.965	3
+ weights resct	**10.000**	3

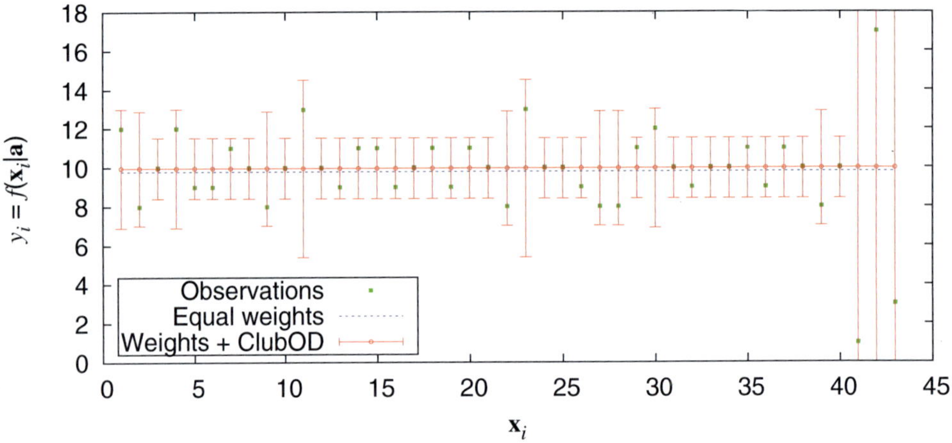

Figure 3.16: Results of fitting a constant with and without weighting of observations

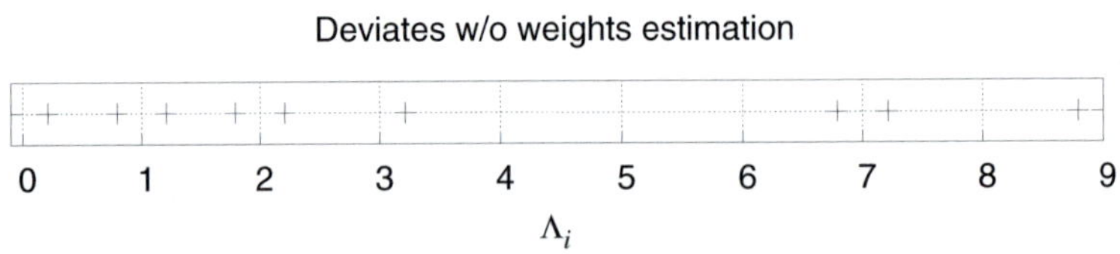

Figure 3.17: Absolute deviates according to the example of fitting a constant without weights

Table 3.6: Datasets used for straight-line fitting

i	x_i	y	$y_i^{(0)}$	$y_i^{(1)}$	$y_i^{(2)}$	$y_i^{(3)}$
1	0.1	2.1	1.704	1.704	2.113	2.113
2	0.2	2.2	2.321	3.000	2.216	4.000
3	0.3	2.3	2.208	2.208	2.345	2.345
4	0.4	2.4	2.442	2.442	2.466	2.466
5	0.5	2.5	2.577	2.577	2.581	4.000
6	0.6	2.6	2.770	2.770	2.418	2.418
7	0.7	2.7	2.493	2.493	3.076	3.076
8	0.8	2.8	2.569	2.569	2.862	2.862
9	0.9	2.9	2.888	4.000	2.342	2.342
10	1.0	3.0	3.215	3.215	1.343	1.343

The plot of deviates reveals the three distinct mavericks (**Figure 3.17**). The estimated weights bias the result towards smaller values of a_1.

In this artificial example, all methods of outlier detection result in solutions, which are identical to the true model parameter (number is boldfaced), if the weights are reset after the removal of outliers.

3.5.2 Straight line

The performance of weighted least squares applied to the fitting of straight lines shall be demonstrated using four small data sets listed in **Table 3.6**. The corresponding plots are shown in **Figure 3.18**.

The process generating the (error-free) output is

$$y = 2.0 + 1.0 \cdot x \; .$$

The first data set $y_i^{(0)}$ is a randomised version of y with normally distributed errors using a constant deviation of $\sigma_i = \sigma = 0.25$. In $y_i^{(1)}$ two additional outliers are introduced artificially at $i = 2$ and $i = 9$. For sets $y_i^{(2)}$ and $y_i^{(3)}$ the deviation starts with $\sigma_1 = 0.01$ and then doubles with increasing i up to $\sigma_{10} = 5.12$. $y_i^{(3)}$ is falsified in addition by two outliers at $i = 2$ and $i = 5$.

The result of fitting these four data sets to a straight line is shown in **Figure 3.19**. The error bars indicate the estimated uncertainty of the observation (based on the estimated weight) according to equation (3.1). If the error bar crosses the entire plot, this measurement is either assigned a very low weight or it is even regarded as an outlier.

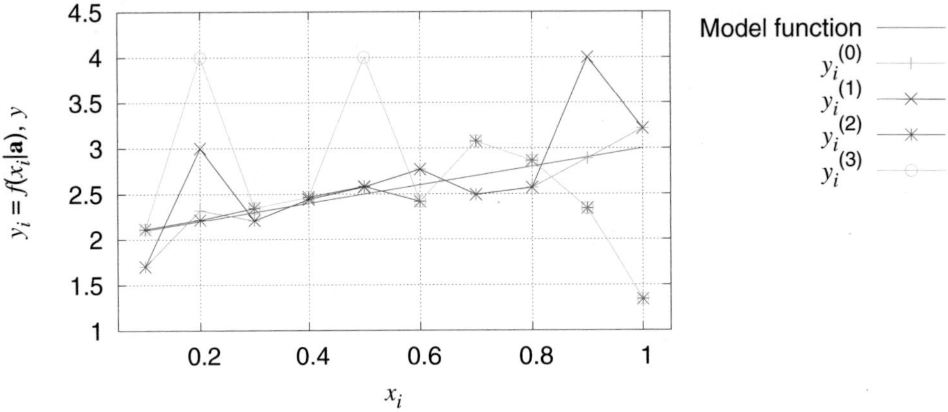

Figure 3.18: Test data for fitting of straight lines

In the presence of strong outliers, the fitted curves with application of weights are distinctly closer to the model function than those without using weights.

The estimated parameters are listed in **Tables 3.7–3.10**. For the case of $y^{(0)}$, **Figure 3.20** shows the distribution of the deviates before outlier detection with and without the estimation of weights. The plots of deviates do not indicate any presence of outliers when equal weights are used. After weights estimation, however, three observations seem to be more distant to the model function than the other explaining the results of outlier detection. Consequently, ClubOD falsely detects three outliers (Table 3.7). Also the other methods are adversely affected. Nevertheless, removing the three suspicious data points and re-running of ordinary least squares leads to results close to optimum. This first example underlines the usefulness of weights estimation. The estimated parameters are much closer to the true ones when using weighting. Even though 'z-score with weights reset' and 'RANSAC with weights reset' have the same number of detected outliers, their results are different. This is because they have only one outlier in common (at condition $x = 0.1$, while the second is different (conditions $x = 0.8$ and $x = 1.0$, respectively).

The results for $y^{(1)}$ are listed in **Table 3.8**. Only when proper weights are estimated, all four detectors are able to find the two outliers. Despite the removal of the contaminants, the estimated model parameters are not very close to the true values. This is caused by the missing influence of the correct observations at these conditions x_i. Without weighting, RANSAC is not able to find the two outliers because the determination of the decision threshold about in- and outliers is already biased by the mavericks. The plots of deviates in **Figure 3.21** show

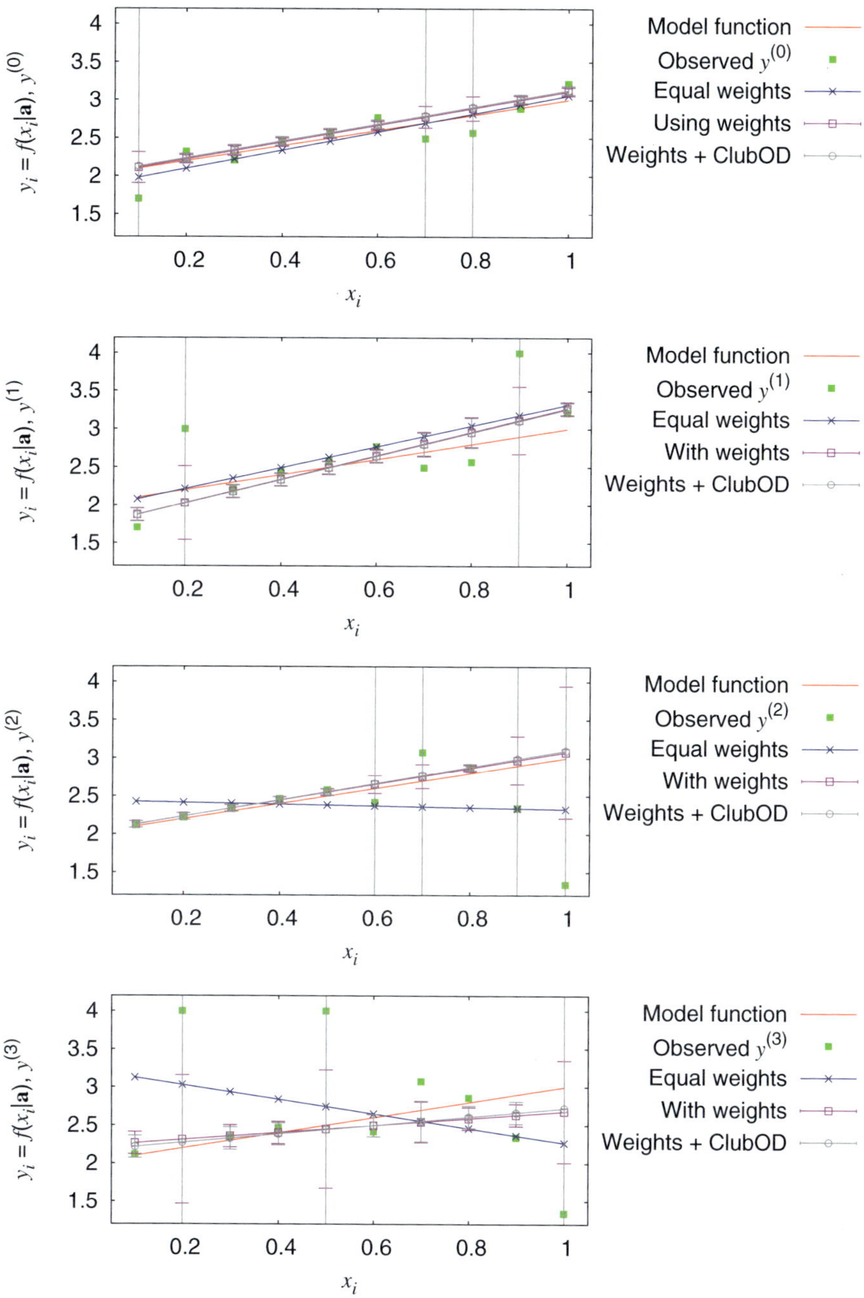

Figure 3.19: Results of fitting a straight line with and without weighting of observations and outlier detection

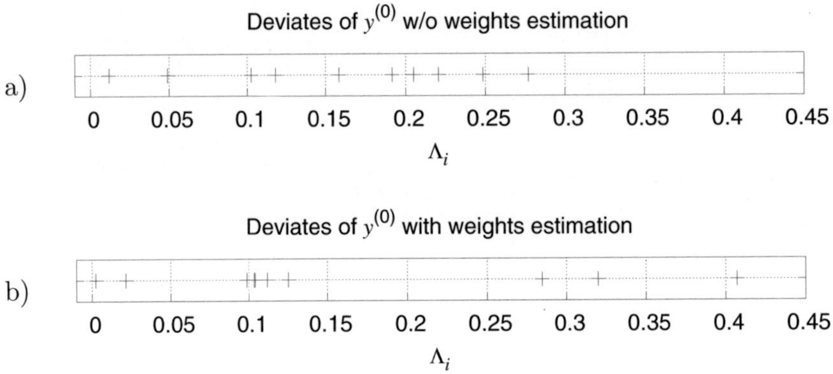

Figure 3.20: Absolute deviates according to the example of fitting a straight line $y^{(0)}$; a) without application of weights estimation, b) using the estimation of weights

Table 3.7: Results of straight-line fitting, function $y^{(0)}$

$y^{(0)}$	a_1	a_2	Detected outliers
True values	2.000	1.00	0
Equal weights	1.861	1.195	0
all methods	1.861	1.195	0
Estimated weights	**2.000**	1.111	0
Weights + z-score	2.016	1.101	2
+ weights reset	2.005	1.059	2
Weights + M-score	2.020	1.084	1
+ weights reset	2.031	**0.964**	1
Weights + ClubOD	2.016	1.112	3
+ weights reset	2.004	1.126	3
Weights + RANSAC	2.020	1.084	1
+ weights reset	2.133	0.729	2

that the two contaminants become more distant to the bulk of good observations after the estimation of weights, enabling their detection and removal.

Table 3.9 shows the results for $y^{(2)}$. Four of the 'good' observations deviate much more from the model function than others after the application of proper weights misguiding all outlier detectors (**Figure 3.22**). Fitting the remaining six observations still leads to acceptable model parameters. If the uncertainties are known, which were used in the generation process of $y^{(2)}$, they can be provided

Table 3.8: Results of straight-line fitting, function $y^{(1)}$

$y^{(1)}$	a_1	a_2	Detected outliers
True values	2.000	1.00	2
Equal weights	**1.939**	1.379	0
z-score	1.939	1.379	0
M-score	1.939	1.379	0
ClubOD	1.939	1.379	0
RANSAC	2.085	**0.916**	1
Estimated weights	1.718	1.554	0
all methods	1.712	1.546	2
+ weights reset	1.750	1.358	2

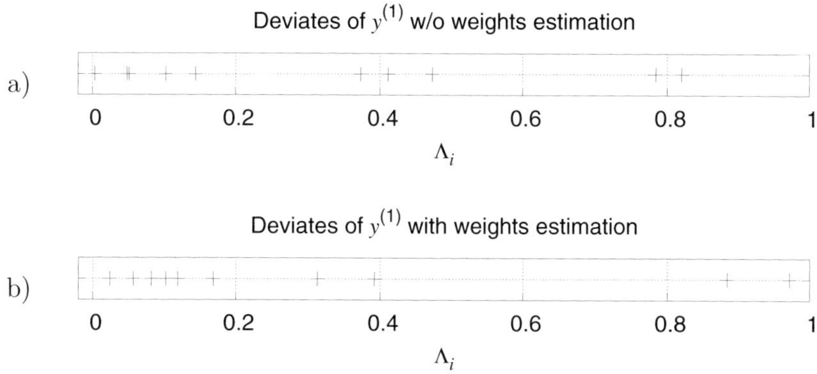

Figure 3.21: Absolute deviates according to the example of fitting a straight line $y^{(1)}$; a) without application of weights estimation, b) using the estimation of weights

Table 3.9: Results of straight-line fitting, function $y^{(2)}$

$y^{(2)}$	a_1	a_2	Detected outliers
True values	2.000	1.00	0
Equal weights	2.436	−0.109	0
z-score	2.436	−0.109	0
M-score	2.135	0.711	1
ClubOD	2.436	−0.109	0
RANSAC	1.981	1.176	2
true weights	2.001	1.110	0
Estimated weights	2.021	**1.058**	0
all methods	**2.015**	1.084	4
+ weights reset	2.015	1.084	4

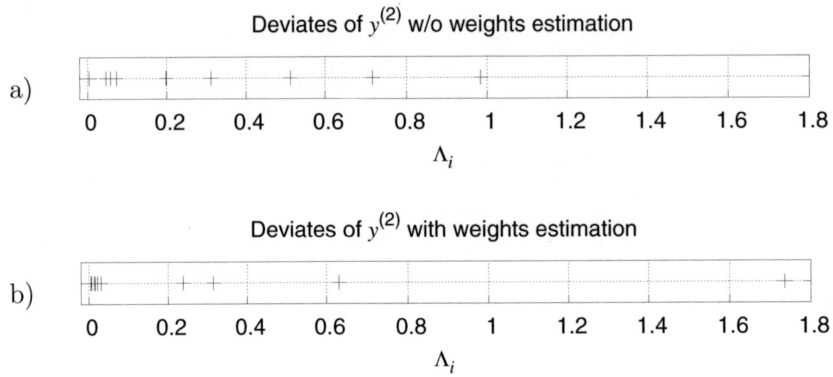

Figure 3.22: Absolute deviates according to the example of fitting a straight line $y^{(2)}$; a) without application of weights estimation, b) using the estimation of weights

Table 3.10: Results of straight-line fitting, function $y^{(3)}$

$y^{(3)}$	a_1	a_2	Detected outliers
True values	2.000	1.00	2
RANSAC	2.488	−0.196	2
other methods	3.220	−0.952	0
Estimated weights	2.220	0.460	0
all methods	2.161	0.561	3
+ weights reset	**2.150**	**0.677**	3

to the fitting process enabling an acceptable fitting result (line 'true weights' in Table 3.9).

The example of $y^{(3)}$ distinctly shows that the application of weights estimation can dramatically improve the data fitting (**Table 3.10**). The usage of equal weights results in completely wrong model parameters. As in the example of $y^{(1)}$, only weights estimation enables the proper detection of the outliers, as can be derived again from the plots of deviates (**Figure 3.23**). The best fitting is obtained in this case when we forget the weights after the removal of the contaminants.

The performance of the outlier-detection schemes is almost the same in these four examples of straight-line fitting.

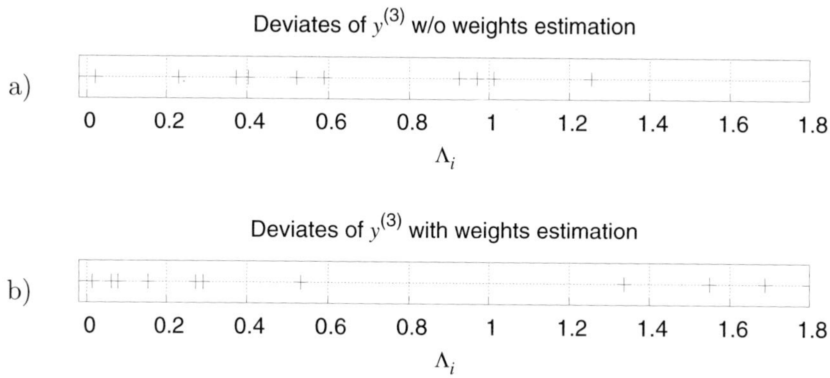

Figure 3.23: Absolute deviates according to the example of fitting a straight line $y^{(3)}$; a) without application of weights estimation, b) using the estimation of weights

3.5.3 Plane approximation

The test data for the fitting of a plane have been generated by

$$y_i = a_1 + a_2 \cdot x_{1,i} + a_3 \cdot x_{2,i} + \varepsilon_i = 0.1 + 0.2 \cdot x_{1,i} + 0.4 \cdot x_{2,i} + \varepsilon_i$$

with a randomisation of the observation using $\sigma_{\varepsilon_i} = \sigma_\varepsilon = 0.5$. In addition, an offset of 1.5 (systematic error) was introduced for the points $(x_1; x_2) = \{(1;1)\ldots(3;3)\}$, that is, the data set contains nine outliers.

The estimated model parameters and the number of detected outliers depend on the schemes used for the weights estimation (**Table 3.11**). Without using weights, only RANSAC is able to detected the outliers, due to its inherit down-weighting of points, which do not support the subset-based model. In case of weights estimation, M-score and ClubOD can find the nine outliers. This can be understood by examining the plots of deviates in **Figure 3.24**. The z-score method, however, declares twelve data points as being contaminant and RANSAC even thirteen.

The removal of the nine outliers in combination with resetting all weights to equal values yields the best approximation in this example.

In **Figure 3.25**a), it can be seen that the fitted plane (blue) is attracted by the contaminants (1;1) ... (3;3) causing high deviates in the opposite region when no weighting is used. **Figure 3.25**b) shows the improvement, if the nine outliers are removed and equal weights are used.

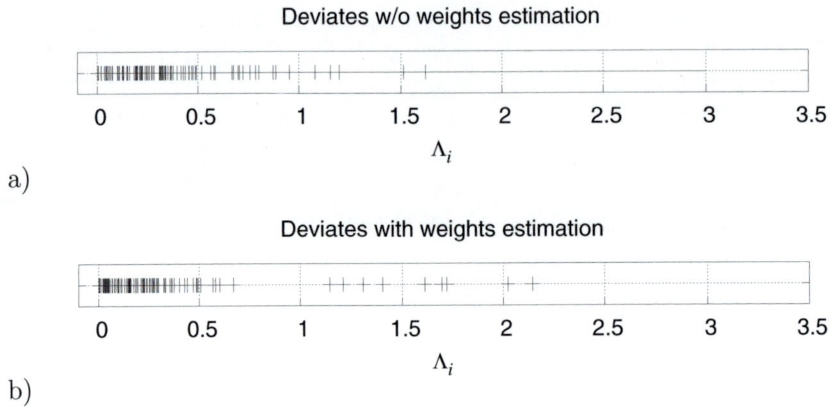

Figure 3.24: Absolute deviates according to the example of fitting a plane; a) without application of weights estimation, b) using the estimation of weights

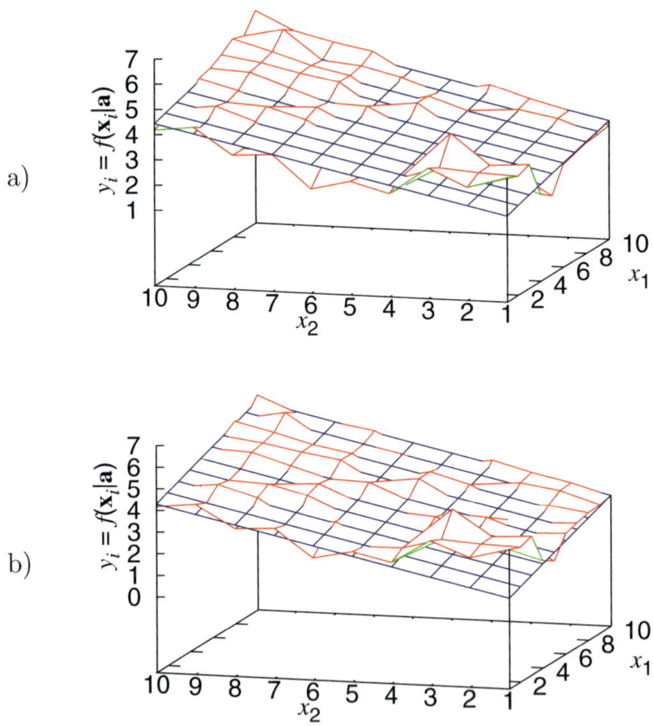

Figure 3.25: Results of fitting a plane; a) without weighting; b) after removal of all nine outliers)

Table 3.11: Results of plane fitting

Plane	a_1	a_2	a_3	Detected outliers
True values	0.1	0.2	0.4	9
Equal weights	0.954	0.142	0.334	0
z-score	0.882	0.147	0.339	1
M-score	0.762	0.157	0.349	2
ClubOD	0.954	0.142	0.334	0
RANSAC	**0.096**	**0.206**	**0.399**	9
Estimated weights	0.206	0.209	0.384	0
Weights + z-score	0.180	0.212	0.385	12
Weights + M-score	0.181	0.211	0.386	9
Weights + ClubOD	0.181	0.211	0.386	9
+ weights reset	**0.096**	**0.206**	**0.399**	9
Weights + RANSAC	0.177	0.212	0.385	13

3.5.4 Coordinates transformation

The transformation of coordinates is an example with two or more observed values per measurement (see also Subsection 1.4.4 'Affine transformation'). When dealing with images, there are two observations $(x\ y)^{\mathrm{T}}$ corresponding to two conditions $(u\ v)^{\mathrm{T}}$. The latter are coordinates in an image indicating special regions and are also called *landmarks* (see **Figure 3.26**). The task in the experiment was to find counterparts of these landmarks in a second image. The question addressed in this application is: What are the translational displacement and the angle of rotation between both images? The corresponding model function for the least-squares setup is

$$\begin{pmatrix} x \\ y \end{pmatrix} = \begin{pmatrix} a_1 \\ a_2 \end{pmatrix} + \begin{pmatrix} \cos(a_3) & -\sin(a_3) \\ \sin(a_3) & \cos(a_3) \end{pmatrix} \cdot \begin{pmatrix} u \\ v \end{pmatrix}$$

or

$$\begin{aligned} f_{\mathrm{x}}(\mathbf{u}|\mathbf{a}) =& x = a_1 + \cos(a_3)\cdot u - \sin(a_3)\cdot v \\ f_{\mathrm{y}}(\mathbf{u}|\mathbf{a}) =& y = a_2 + \sin(a_3)\cdot u + \cos(a_3)\cdot v\,. \end{aligned}$$

Since both equations are not independent of each other, but coupled via trigonometric functions of a_3, the fitting problem is nonlinear. The inspection of the landmark pattern in Figure 3.26 by eye reveals that there are obviously some mismatches, i.e. false correspondences between image regions. Indeed, nine of the 121 matched points seem to be outliers, since they show deviations of more

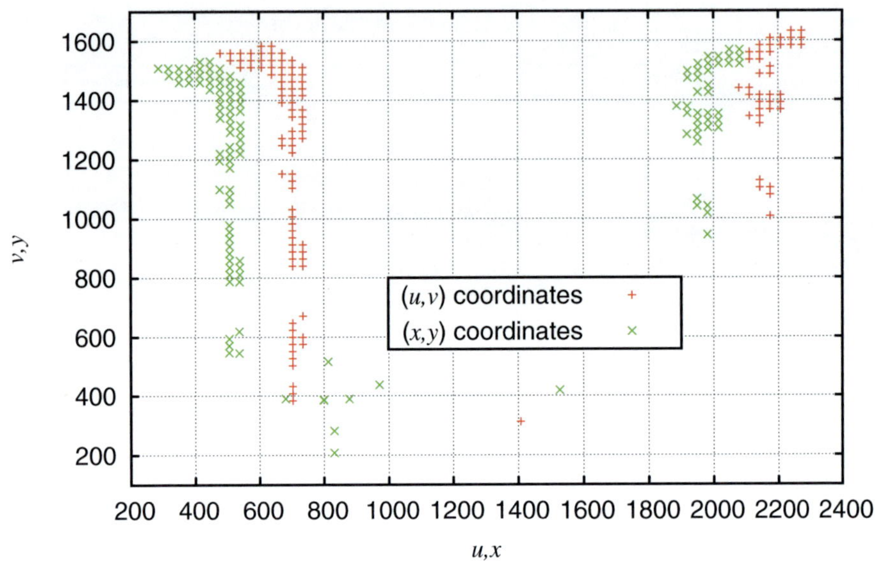

Figure 3.26: Example of two corresponding sets of coordinates with distinct outliers in the vicinity of $(x, y) = (800; 400)$; the + sign indicates coordinates in the first image (conditions), the $\times$ sign coordinates of matched regions in the second image (observations).

than two image-grid points from the modelled positions in both, x and y direction. **Table 3.12** shows the fitting results in dependence on the chosen weighting and outlier-detection schemes.

The number of outliers given in the table are counted separately for the x and y coordinates as a consequence of their independent weighting (see Subsection 2.6.7). Even though the true parameters are not known in this example, it is evident that without any weighting or outlier detection the found displacement and the rotation angle are wrong.

Based on the distribution of deviates (**Figure 3.27**), ClubOD and RANSAC are able to identify sixteen out of eighteen outliers leading to model parameters which are close to the correct ones with high probability. The z-score method finds only nine of them, M-score thirteen.

After applying the weighted least-squares procedure, the majority of all observations is very close to the model function in both coordinates (**Figure 3.28**). Only few observations (nine to be exact) deviate much stronger as can be seen in the deviates plots. ClubOD and z-score are now able to detect seventeen out-

Table 3.12: Results of fitting coordinates

Coord.transf.	a_1	a_2	a_3	Detected outliers
True values	?	?	?	18
Equal weights	−132.29	−92.08	1.668°	0
z-score	−195.64	−52.64	−0.167°	9
M-score	−198.56	−49.37	−0.294°	13
ClubOD	−202.18	−47.07	−0.396°	16
RANSAC	−202.18	−47.07	−0.396°	16
Estimated weights	−201.58	−47.57	−0.376°	0
Weights + z-score	−201.87	−47.62	−0.382°	17
Weights + M-score	−201.82	−47.72	−0.380°	18
+ weights reset	−201.82	−47.72	−0.380°	18
Weights + ClubOD	−201.87	−47.62	−0.382°	17
+ weights reset	−201.87	−47.62	−0.382°	17
Weights + RANSAC	−201.89	−47.58	−0.383°	24
+ weights reset	−201.80	−47.71	−0.380°	20

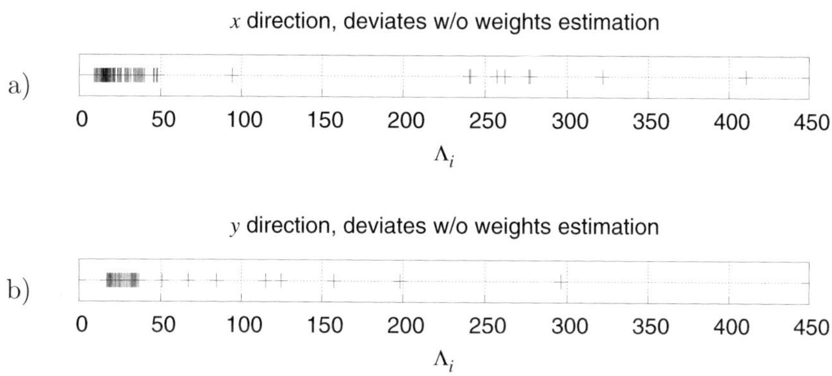

Figure 3.27: Absolute deviates according to the fitting of coordinates without weighting; a) x coordinates; b) y coordinates

liers. One out of the nine mismatches deviate only in one coordinate, while the other is rather close to the correct value. That is why the eighteenth outlier could not be detected. This underlines that it would be beneficial to treat the (multiple) coordinates of data points in a joint manner. M-score detects eighteen outliers and RANSAC finds even 24. Inspecting the deviates after application of weights estimation and RANSAC outlier detection reveals that RANSAC has removed coordinates, which deviate from the model more than 1.4 pixels. Since the observations are pixel coordinates and integer numbers by definition, each corresponding pair of coordinates that deviate more than $\sqrt{2}$ in fact can be in-

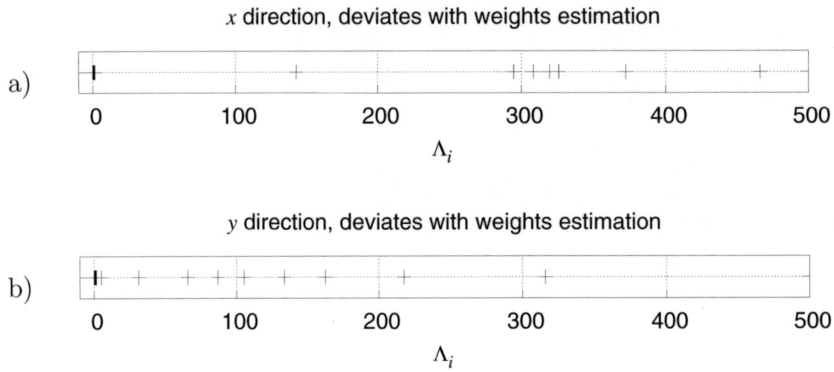

Figure 3.28: Absolute deviates according to the fitting of coordinates with weighting; a) x coordinates; b) y coordinates

terpreted as outliers. The remaining data points were assigned equal weights, because they are all very close to the model. Thus, the parameter values can be taken as the true ones. A second run of RANSAC with enabled reset of weights, however, found only 20 outliers. The varying number is a consequence of the random nature of RANSAC.

The very high accuracy of the good observation leads to the fact that they all are assigned to identical weights. Only the contaminant observations are down-weighted according to their deviates. That is why the equalisation of weights after the removal of outliers does not change the parameter values anymore (Tab.3.12).

Figure 3.29 depicts the result of the transformation of the (u, v) coordinates using the estimated translation and rotation parameters (pure outlier removal).

3.5.5 Linear prediction

In Subsection 1.4.4, the prediction of pixel values for application in image compression was introduced. The observations are identical to the pixel values themselves and it is assumed that there are no outliers. In addition, all observations are typically stemming from the same image-taking process, implicating that all values have the same uncertainty. Nevertheless, the quality of prediction can be improved when weights are incorporated. The main problem lies in the assumption that the image characteristics do not change within the window region depicted in Figure 1.7. In natural pictures, i.e. photographs, the signal properties are rather similar for neighbouring pixels, but become more and more different with higher distances. The process of model-parameter determination should take this behaviour into account. Observations close to the current position should

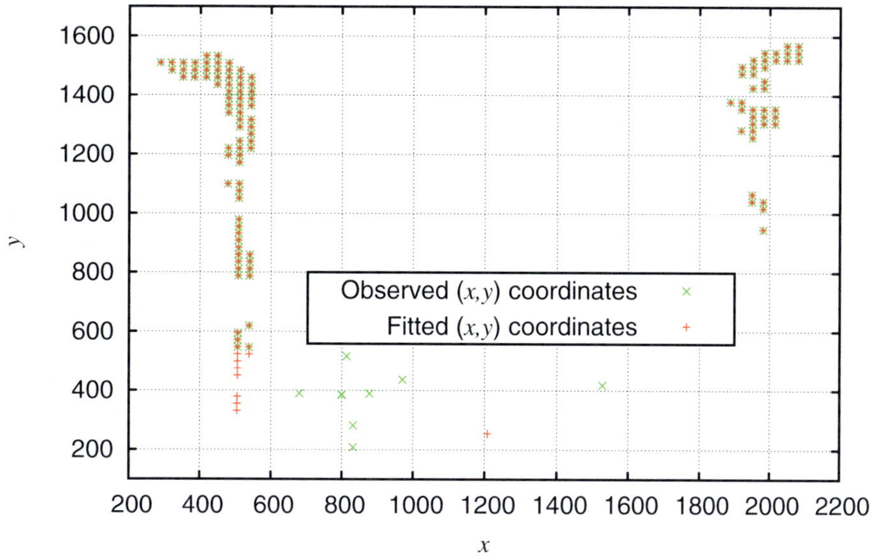

Figure 3.29: Result of the coordinates transform based on pure outlier removal

get high weights, while the weights should decrease with increasing distance. A typical relation is $w_i = 0.8^{d_i}$, if d_i is the Euclidean distance between the position of an observation $y_i = x[n-j, m-k]$ and the position of the value $x[n,m]$ to be predicted [Mey01].

3.5.6 Cosine function

The model function

$$\begin{aligned} & y = a_1 + a_2 \cdot \cos(x) + a_3 \cdot \sin(x) \\ & (\text{based on } y = b_1 + b_2 \cdot \cos(x - b_3), \text{ see Section 1.3}) \end{aligned} \tag{3.33}$$

describes a harmonic oscillation process with an oscillation frequency of $f_0 = 1/(2 \cdot \pi)$.

Example (1)

Figure 3.30 shows the plot of observations generated by computational object-detection algorithms in comparison with the true object positions. The object was performing a circular movement in the x-z plane and has been projected to

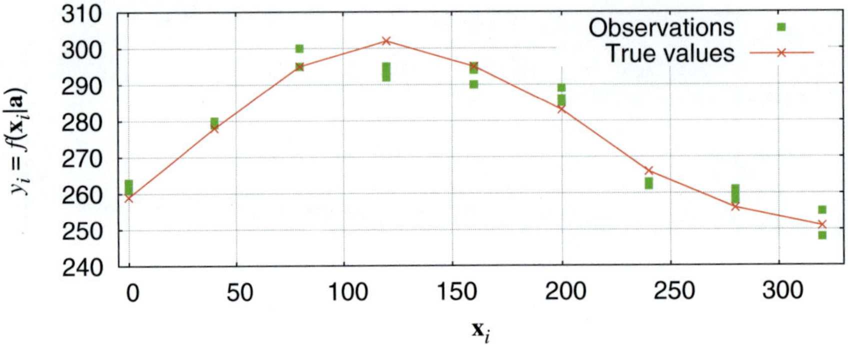

Figure 3.30: Recordings of a circular object movement. The observations y_i depend on the rotation angle x_i (in degrees)

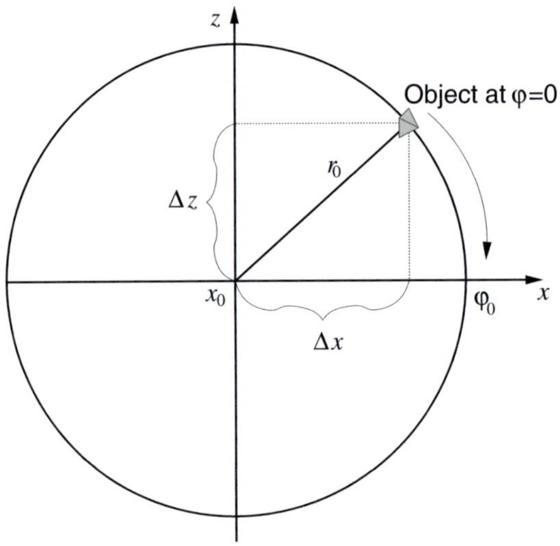

Figure 3.31: Circular movement of an object around the rotation axis (y axis) perpendicular to the x and z axes

a screen that was parallel to the x-y-plane such that the movement is mapped to a straight line on the screen. For each rotational position, three observations have been made by different detection algorithms. At zero and 320 degrees only two observations could be recorded. Please note that x, y, and z are related to the coordinate system according to **Figure 3.31** and must not be mixed with the observations $y_i = f(x_i|\mathbf{a})$ and conditions x_i.

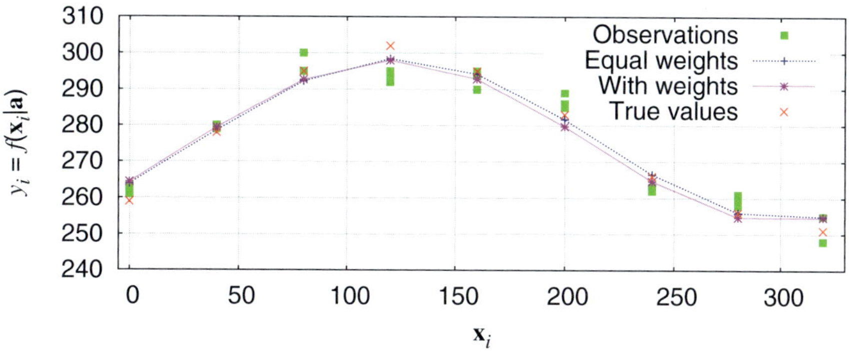

Figure 3.32: Fitting of the circular movement with and without weighting of observations

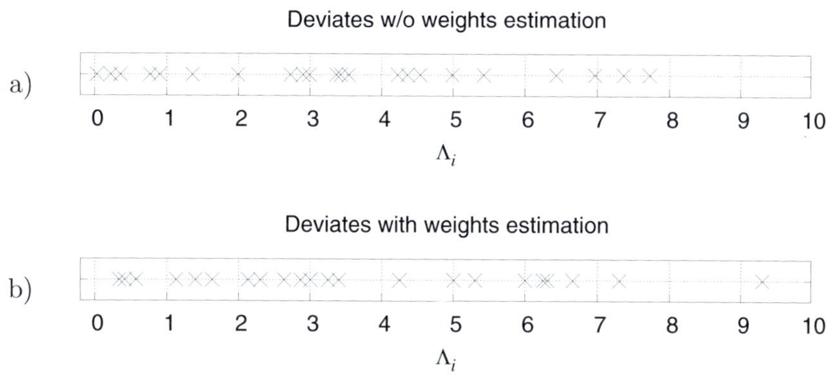

Figure 3.33: Absolute deviates according to the fitting of a cosine; a) without application of weights estimation, b) using the estimation of weights

The fitting result is almost independent of the use of weights as can be seen in **Figure 3.32**. Since the deviates are relatively homogeneous distributed (**Figure 3.33**), none of the observations was labelled as outlier. Naturally, fitting with equal weights leads to the best estimated model parameters (**Table 3.13**). Including the estimated weights, z-score and RANSAC detect one outlier each. Interestingly, it is not the same data point, as also can be deduced from the different values of estimated model parameter.

Figure 3.34 shows the object to be detected at the positions 40 degrees and 200 degrees. The images have been taken at an experimental X-ray station at the European Molecular Biology Laboratory (EMBL) in Hamburg. The object-detection algorithms have been looking for the small polygon-like object in the middle of

Table 3.13: Results of fitting circular movement (example 1)

Cosine 1	b_1	b_2	b_3	Detected outliers
True values	276.1	25.15	126.4°	0
Equal weights	**276.3**	22.19	**123.9°**	0
all methods	276.3	22.19	123.9°	0
Estimated weights	275.7	**22.32**	120.3°	0
Weights + z-score	275.6	22.30	120.1°	1
Weights + M-score	275.7	**22.32**	120.3°	0
Weights + ClubOD	275.7	**22.32**	120.3°	0
Weights + RANSAC	275.8	22.16	120.2°	1
+ weights reset	276.7	21.50	123.3°	1

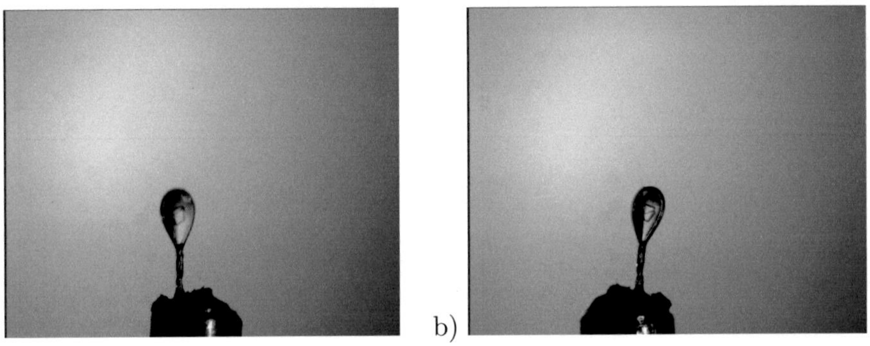

a) b)

Figure 3.34: Small protein crystal mounted in a rotating nylon loop recorded at a) 40 degrees and b) 200 degrees of rotation. The rotation axis is vertical and the horizontal movement of the crystal is observed.

the loop. This object is a protein crystal with a diameter of approximately 100 microns. It is fixed in the loop by the surface tension of the surrounding liquid. The movement of the crystal must be determined in order to centre the crystal in the X-ray beam. The beam is diffracted by the crystal lattice and the resulting diffraction pattern allows conclusion about the internal structure of the crystal and the protein molecule.

Example (2)

Figure 3.35 shows an example, where the detection of the object was much more uncertain than in the previous example. There are two reason. First, the object shows a rather elongated shape (**Figure 3.36**) making it difficult for the algorithms to agree on a certain point within this object. Secondly, even the true

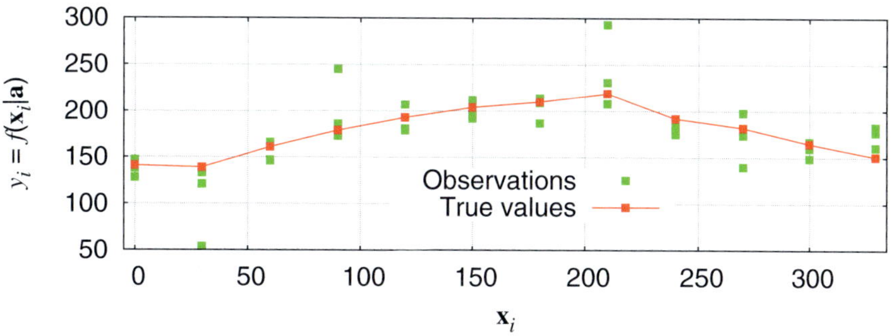

Figure 3.35: Recordings of a circular object movement with high uncertainty. The observations y_i depend on the rotation angle x_i (in degrees)

Table 3.14: Results of fitting circular movement (example 2)

Cosine 2	b_1	b_2	b_3	Detected outliers
True values	178.0	35.76	−174.9°	3
Equal weights	176.2	40.24	−174.6°	0
z-score	**178.8**	**35.58**	−178.1°	1
M-score	174.2	31.43	**−174.9**°	3
ClubOD	174.2	31.43	**−174.9**°	3
RANSAC	172.0	35.21	−178.6°	5
Estimated weights	171.7	34.81	−174.1°	0
Weights + z-score	171.5	35.05	−174.1°	6
Weights + M-score	171.7	34.64	−174.1°	3
Weights + ClubOD	171.7	34.64	−174.1°	3
+ weights reset	174.2	31.43	**−174.9**°	3
Weights + RANSAC	171.5	35.05	−174.1°	6
+ weights reset	173.1	35.20	**−174.9**°	6

data points, which were derived manually by inspecting the single images, do not lie on a smooth cosine curve. This gives rise to the suspicion that something in the experimental setup was not perfect. It could be that the stem of the nylon loop was not stiff enough and bent somewhat.

The inspection of the plot in Figure 3.35 by eye suggests at least three observations (at 30, 90, and 210 degrees) as being contaminants. Comparing the found coordinates with the positions inside the corresponding image approves that they do not lie on the crystal. Also the plot of deviates gives the same indication (**Figure 3.37**). The results after outliers removal, however, are ambiguous (**Table 3.14**). Eliminating the three outliers improves only the estimation of the phase

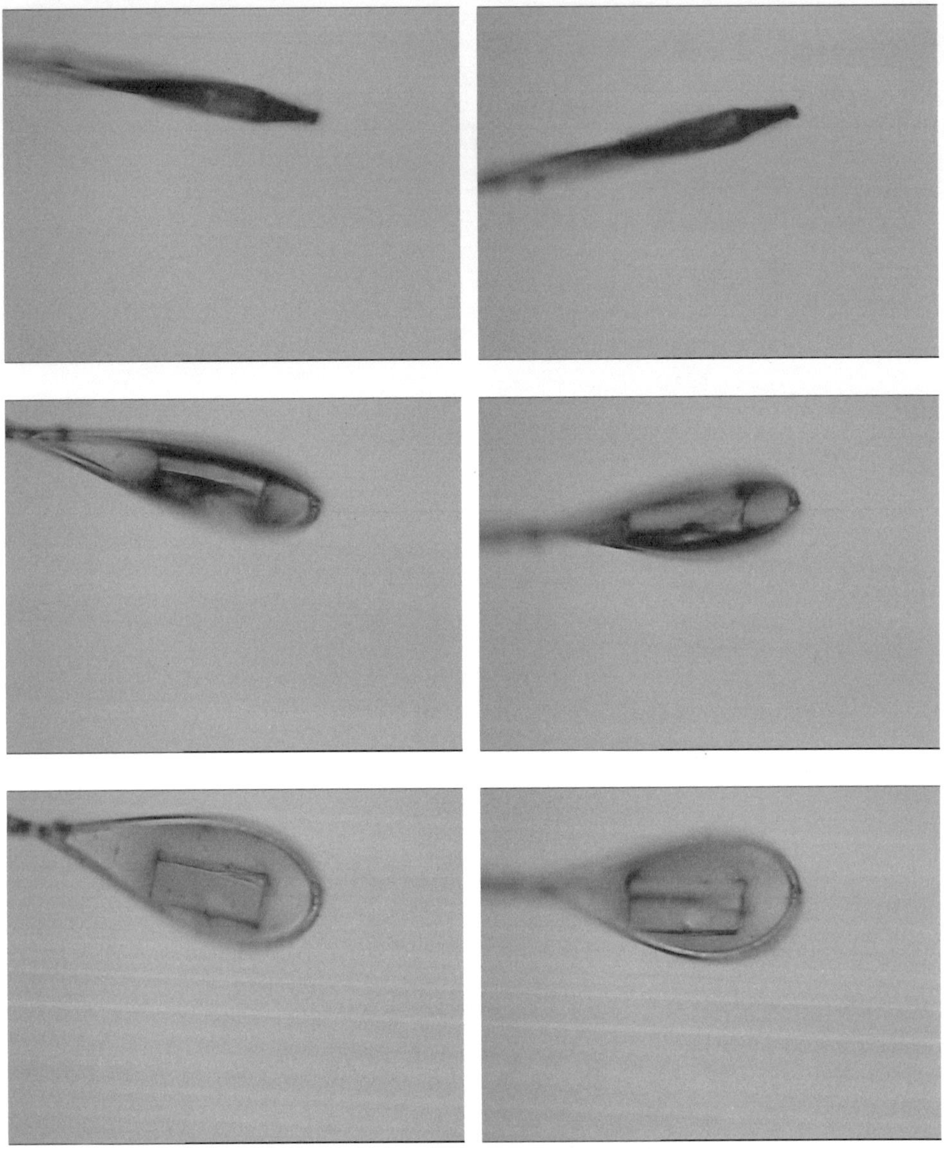

Figure 3.36: Rectangular object to be detected at different rotational positions. The rotation axis is horizontal and the vertical movement of the crystal is observed.

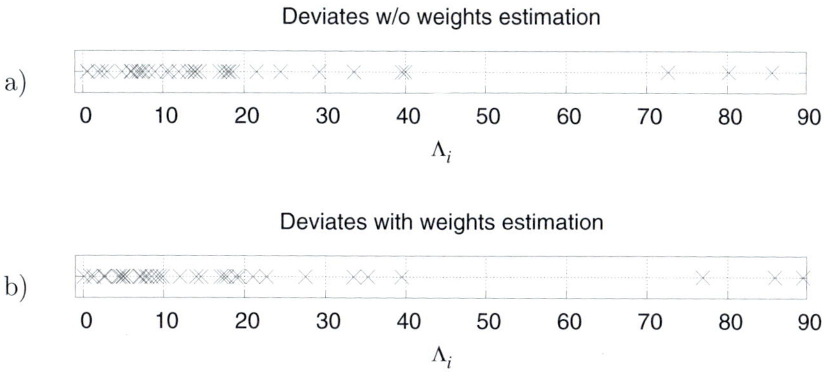

Figure 3.37: Absolute deviates according to the fitting of a cosine (example 2); a) without application of weights estimation, b) using the estimation of weights

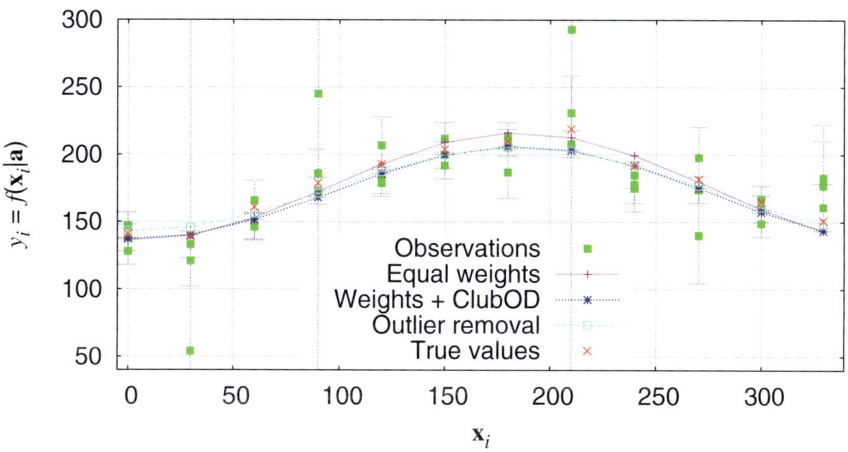

Figure 3.38: Fitting of a cosine function with and without the removal of three outliers

shift (b_3). Offset b_1 and amplitude b_2 are estimated best when only the outlier at 30 degrees is removed and the other two are kept. **Figure 3.38** shows the corresponding plots.

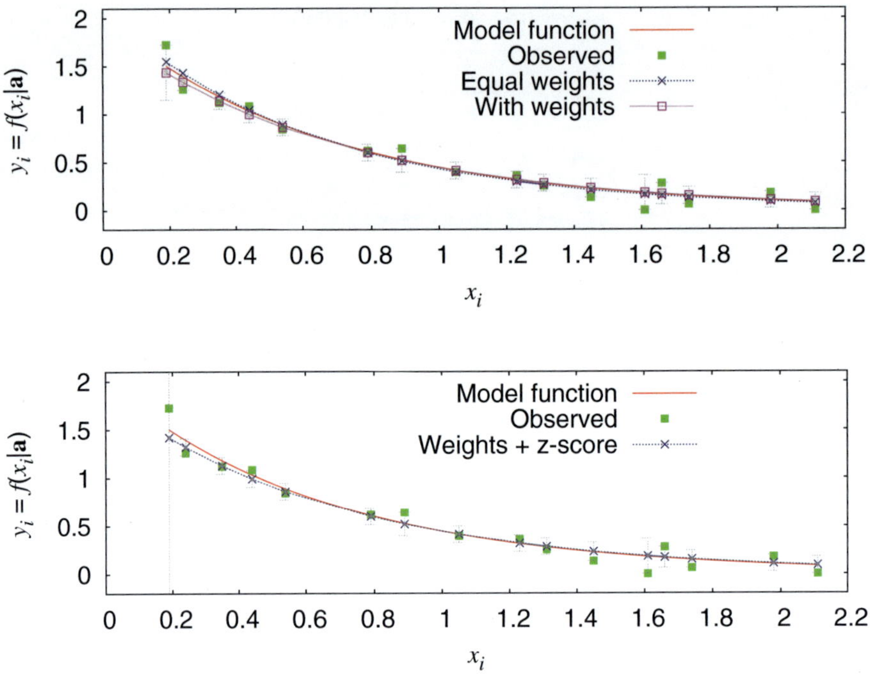

a)

b)

Figure 3.39: Results of fitting an exponential model function; a) with and without weighting of observations, b) weighting and outlier removal

3.5.7 Exponential model function

An exponential model function

$$y_i = a_2 \cdot \exp(a_3 \cdot x_i) + \varepsilon_i$$

was generated with $a_2 = 2$, $a_3 = -1.5$ and additive Gaussian noise ($\sigma_{\varepsilon_i} = \sigma_\varepsilon = 0.125$). Since this nonlinear function can be linearised according to equation (1.12), there are two approaches in estimating the parameters a_2 and a_3. First, the data fitting can be performed directly with nonlinear approximation with its drawbacks of estimation of initial parameters and iterative parameter refinement. The result is depicted in **Figure 3.39**. The resulting parameters are listed in **Table 3.15**. Automatic weighting of observations yields worse results compared to the fitting with equal weights for this example data set. Inspecting the plot of deviates (**Figure 3.40**), it becomes clear why the z-score method has declared one observation as contaminant. After weights estimation, the threshold $\lambda_\text{O} = \hat{\sigma}_\text{y} \cdot \kappa_\text{O}$

Table 3.15: Results of fitting an exponential function

Exponential	a_1	a_2	Detected outliers
True values	2.0	−1.5	0
Equal weights	**2.096**	−1.589	0
all methods	2.096	−1.589	0
Estimated weights	1.895	−1.457	0
Weights + z-score	1.866	−1.437	1
+ weights reset	1.889	**−1.461**	1
Weights + M-score	1.895	−1.457	0
Weights + ClubOD	1.895	−1.457	0
Weights + RANSAC	1.895	−1.457	0

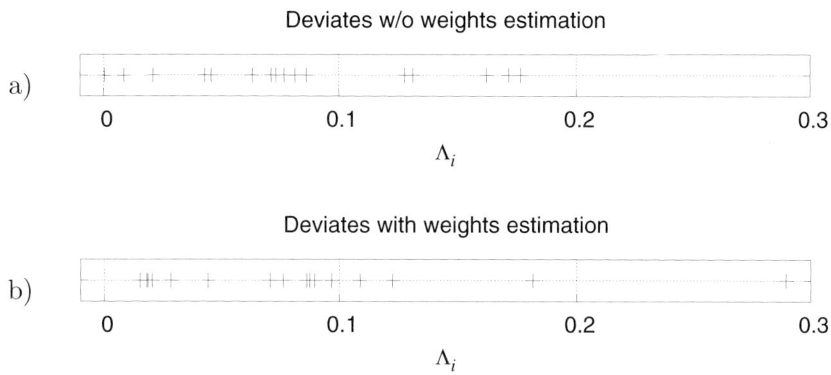

Figure 3.40: Absolute deviates according to the fitting of an exponential function; a) without application of weights estimation, b) using the estimation of weights

(eq. 3.17) is equal to 0.209, thus, one observation has a higher deviation. The other outlier-detection methods correctly keep all observations in the set of good data points.

The linearisation (see Section 1.3) simplifies the fitting distinctly. However, two observations, which are equal to zero, must be remove from the data, because a linearisation is impossible ($\log(0) = -\infty$).

As can be seen in **Table 3.16**, the parameter estimation is by chance better in this particular example compared to the nonlinear approach, also when adapted weights are used. Typically, the results after linearisation are less accurate, because it involves logarithmic weighting of the observations. The error bars in **Figure 3.41** reflect that the weights estimation is highly influenced by the lin-

Table 3.16: Results of fitting a linearised exponential function

Exp. linearised	a_1	a_2	Detected outliers
True values	2.0	−1.5	0
Equal weights	**1.996**	−1.512	(2)
z-score	1.996	−1.512	(2) + 0
M-score	1.808	−1.335	(2) + 1
ClubOD	1.996	−1.512	(2) + 0
RANSAC	1.743	−1.231	(2) + 2
Estimated weights	1.955	**−1.508**	(2) + 0
Weights + z-score	1.963	−1.515	(2) + 4
+ weights reset	**2.014**	−1.486	(2) + 4
Weights + M-score	1.948	−1.501	(2) + 1
+ weights reset	1.808	−1.335	(2) + 1
Weights + ClubOD	1.955	−1.508	(2) + 0
Weights + RANSAC	1.939	−1.493	(2) + 2
+ weights reset	1.743	−1.231	(2) + 2

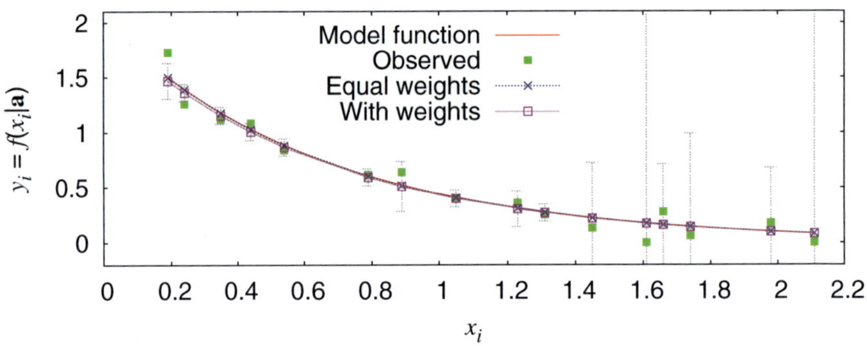

Figure 3.41: Results of fitting an exponential model function with and without weighting of observations using straight line fitting after linearisation

earisation. The errors in large values y_i are down-weighted while the errors in small values y_i are emphasized.

3.5.8 Composite Gaussian bell functions

The fitting of two superposed Gaussian bells

$$y = a_1 \cdot \exp\left[(x - a_2)^2 \cdot a_3\right] + a_4 \cdot \exp\left[(x - a_5)^2 \cdot a_6\right]$$

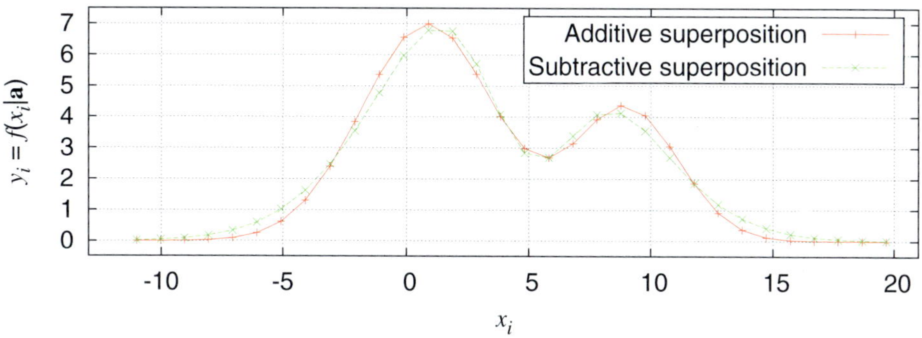

Figure 3.42: Composite Gaussian bell functions with different modes of superposition

can be ambiguous, dependent on whether the amplitudes a_1 and a_4 have the same sign or not (see also Section 1.5.2). **Figure 3.42** shows two examples with similar curves, but completely different parameters $a_1 \cdots a_6$. Both cases are discussed in the following.

Example one: additive combination

The data has been simulated with

$$y_i = 4.3 \cdot \exp\left[(x_i - 9.0)^2 \cdot (-0.11)\right] + 6.7 \cdot \exp\left[(x_i - 0.85)^2 \cdot (-0.069)\right] + \varepsilon_i$$

and $\sigma_{\varepsilon_i} = \sigma_\varepsilon = 0.25$. Dependent on the mode of weights estimation and outlier detection, the data fitting results in different parameter values (**Table 3.17**). **Figure 3.43** shows the relevant plots. ClubOD and M-score misclassify one observation each, since it deviates somewhat stronger (**Figure 3.44**). RANSAC excludes two data points. It can also be seen that the corresponding absolute deviate is not exposed anymore after the application of estimated weights. That is why it is integrated into the bulk of good data points when using ClubOD after weights estimation. The methods based on standardised residuals, however, mistake now two (M-score) or six (z-score) observations for contaminants. Surprisingly, the removal of a single outlier has almost no influence on the fitting of the additive data set. The estimated model parameters are identical up to the third decimal position compared to the fitting with equal weights in this case. The number of outliers found by RANSAC is again influence by the random selection of subsets.

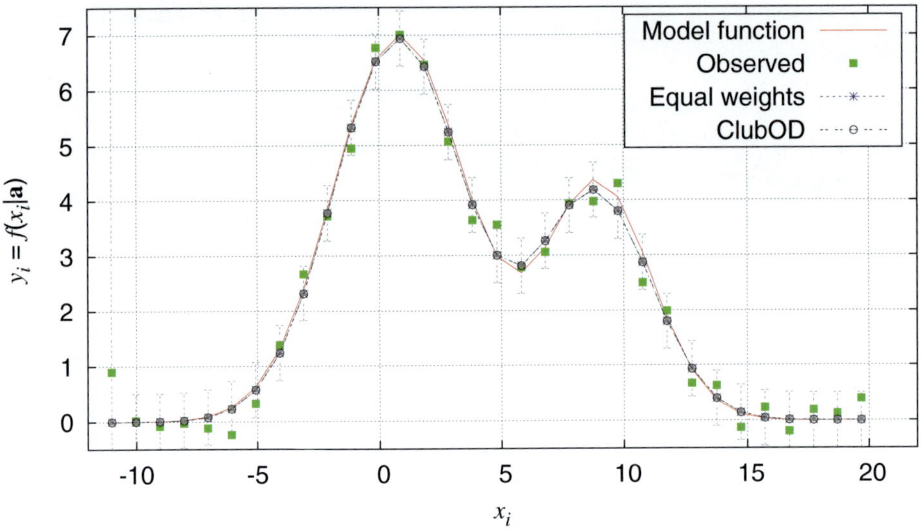

a)

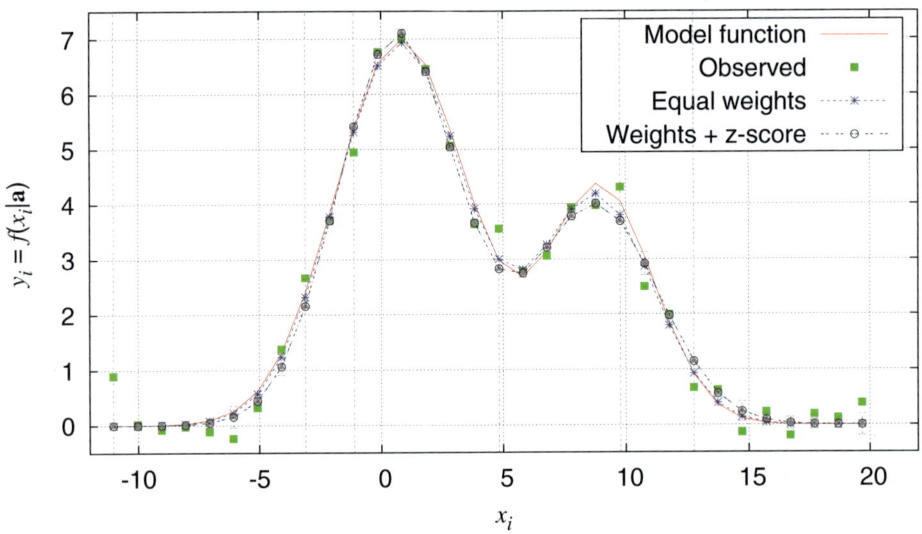

b)

Figure 3.43: Results of fitting two additively combined Gaussian bells ($a_1, a_4 > 0$); a) influence of ClubOD when using equal weights; b) outlier removal after estimation of weights

Table 3.17: Numerical results of fitting two additively combined Gaussian bells

Additive Gaussians	a_1	a_2	a_3	a_4	a_5	a_6	Outl.
True values	4.300	9.000	−0.110	6.700	0.850	−0.069	0
Equal weights	4.116	8.842	−0.097	**6.928**	**0.801**	**−0.072**	0
z-score	4.116	8.842	−0.097	**6.928**	**0.801**	**−0.072**	0
M-score	4.116	8.842	−0.097	6.929	**0.801**	**−0.072**	1
ClubOD	4.116	8.842	−0.097	6.929	**0.801**	**−0.072**	1
RANSAC	3.881	8.692	−0.087	6.932	0.754	−0.075	2
Estimated weights	3.996	8.810	−0.082	7.076	0.699	−0.082	0
Weights + z-score	3.973	8.798	−0.080	7.094	0.688	−0.083	6
Weights + M-score	4.001	8.814	−0.082	7.079	0.699	−0.082	2
+ weights reset	**4.176**	**8.865**	−0.103	6.968	0.760	−0.075	2
Weights + ClubOD	3.996	8.810	−0.082	7.076	0.699	−0.082	0
Weights + RANSAC	3.976	8.789	−0.081	7.094	0.691	−0.083	5
+ weights reset	3.920	8.707	−0.084	7.091	0.681	−0.082	6

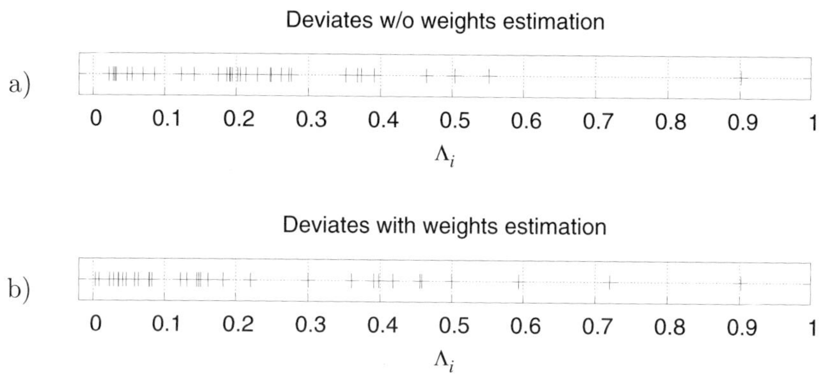

Figure 3.44: Absolute deviates according to the fitting of two additively combined Gaussians; a) without application of weights estimation, b) using the estimation of weights

Example two: subtractive combination

The data has been simulated with

$$y_i = 10 \cdot \exp\left[(x_i - 4)^2 \cdot \left(-\frac{1}{36}\right)\right] - 7 \cdot \exp\left[(x_i - 5)^2 \cdot \left(-\frac{1}{8}\right)\right] + \varepsilon_i$$

and $\sigma_{\varepsilon_i} = \sigma_\varepsilon = 0.25$.

Table 3.18: Numerical results of fitting two subtractively combined Gaussian bells

Subtractive Gaussians	a_1	a_2	a_3	a_4	a_5	a_6	Outl.
True values	10.00	4.000	0.278	−7.000	5.100	−0.125	0
Equal weights	4.283	9.077	−0.114	6.707	0.930	−0.064	0
z-score	4.283	9.077	−0.114	6.707	0.930	−0.064	0
M-score	4.283	9.077	−0.114	6.707	0.930	−0.064	0
ClubOD	4.283	9.077	−0.114	6.707	0.930	−0.064	0
RANSAC	4.214	9.099	−0.112	6.675	0.990	−0.061	1
Estimated weights	4.373	8.960	−0.124	7.012	0.892	−0.067	0
Weights + z-score	4.382	8.953	−0.125	7.026	0.890	−0.067	2
Weights + M-score	4.374	8.957	−0.123	7.028	0.886	−0.067	1
Weights + ClubOD	4.374	8.957	−0.123	7.028	0.886	−0.067	1
+ weights reset	4.299	9.019	−0.109	7.008	0.847	−0.069	1
Weights + RANSAC	4.374	8.957	−0.123	7.029	0.886	−0.067	2

In the case of Gaussian bells, which are combined via subtraction, the nonlinear fitting finds not the global but a strong local minimum owing to the misguiding initialisation of the model parameters. The procedure guessing the initial parameter values simply assumes an additive composition, i.e. $a_1 > 0$ and $a_4 > 0$, and a corresponding starting point in the parameter space is chosen. The optimisation algorithm is not able to escape this attractive minimum. **Figure 3.45** reveals that the tails of the curve and the peaks are not optimally fitted. It would be possible, in principle, to test both variants (i.e. the additive and the subtractive superposition) in the initial parameter estimation and then to compare the fitting results afterwards. **Table 3.18** lists the numerical results.

The problem of ambiguity, that is, the existence of attractive local maxima, can also occur if Gaussian bells of similar amplitude and variance are additively superposed. The resulting function would appear as a single Gaussian bell with some degradation.

3.5.9 Circle

The use of weights and the assumption of outliers do not make sense in the case of fitting a circle, since the possibly erroneous observations must be stuffed into the fitting framework as conditions, according to equation (1.27)

$$y_i = 0 = (x_{1,i} - a_1)^2 + (x_{2,i} - a_2)^2 - a_3^2 + \varepsilon_i \ ,$$

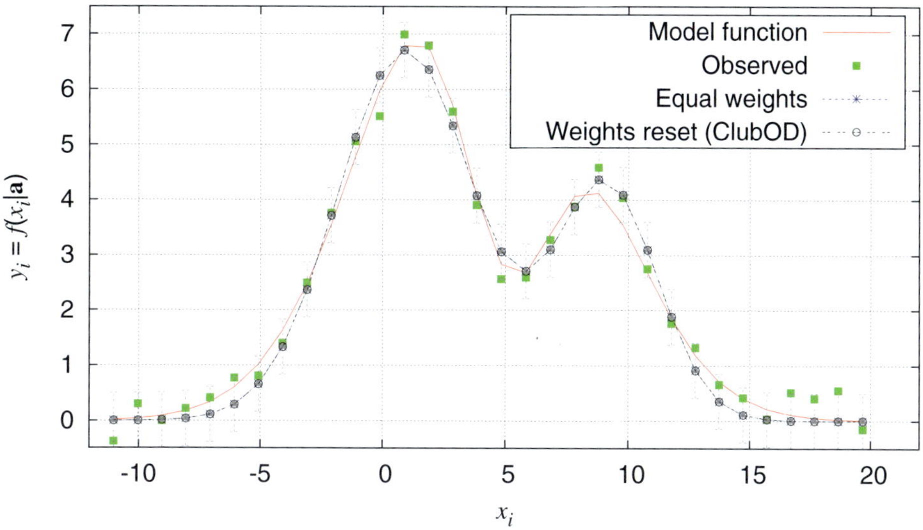

a)

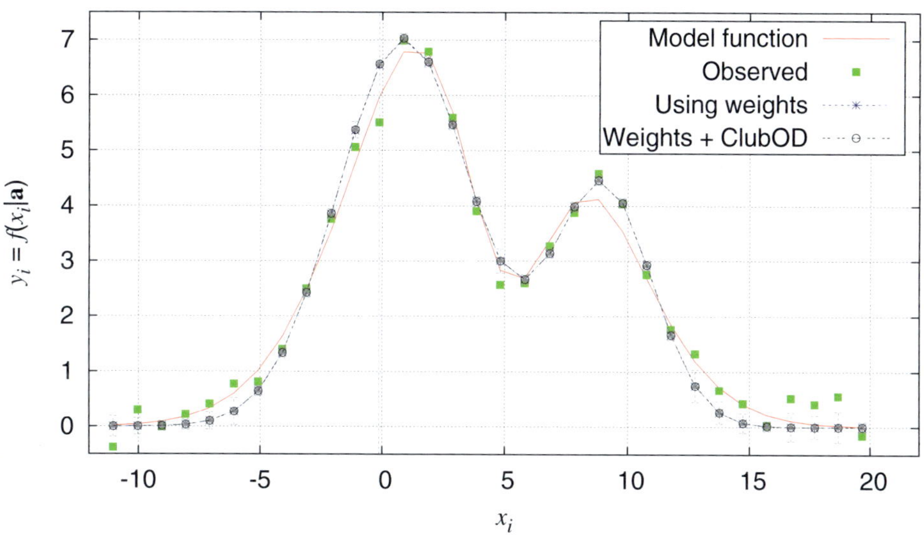

b)

Figure 3.45: Results of fitting two subtractively combined Gaussian bells ($a_1 > 0, a_4 < 0$); a) without weights; b) using weights

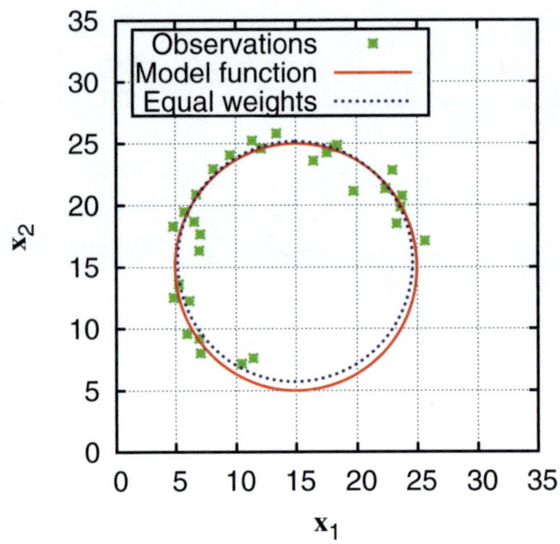

Figure 3.46: Results of fitting a circular object based on 30 observations. The true model function is $0 = (x_1 - 15)^2 + (x_2 - 15)^2 - 10^2$

whereas the 'new' observations are assumed to be correct and equal to zero (see Subsection 1.5.3). Alternatively, the linear version according to equation (1.29) could be applied.

The test data consist of thirty samples of a circle (radius of $a_3 = 10$, centre at $(a_1; a_2) = (15; 15)$) over a range of 240 degrees with randomisation of both $x_{1,i}$ and $x_{2,i}$ (normally distributed with $\sigma_{\varepsilon_i} = \sigma_\varepsilon = 1$). The least-squares approximation calculates a circle at $(a_1, a_2) = (14.9134, 15.4414)$ with a radius of $a_3 = 9.7202$ (see **Figure 3.46**).

3.5.10 Comparison of binning and deviate-based weights estimation

So far, only the weights estimator based on deviates has been investigated. This subsection uses a test signal with $N = 500$ samples enabling tests with weights estimation via binning. The chosen model function is

$$f(x|\mathbf{a}) = a_1 + a_2 \cdot x + a_3 \cdot x^2 + \varepsilon$$

with the parameters $a_1 = 10$, $a_2 = 0$ and $a_3 = 15$. The conditions x_i have been randomly chosen and the corresponding function values have been randomised (Gaussian distribution, $\mathcal{N}(0; 1/16)$, see Section 6.1) in order to simulate erroneous observations y_i. In addition to this basic noise, a random signal with increasing

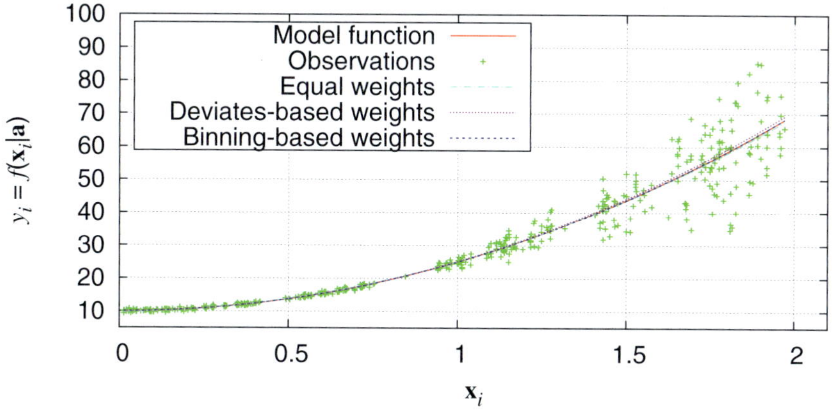

a)

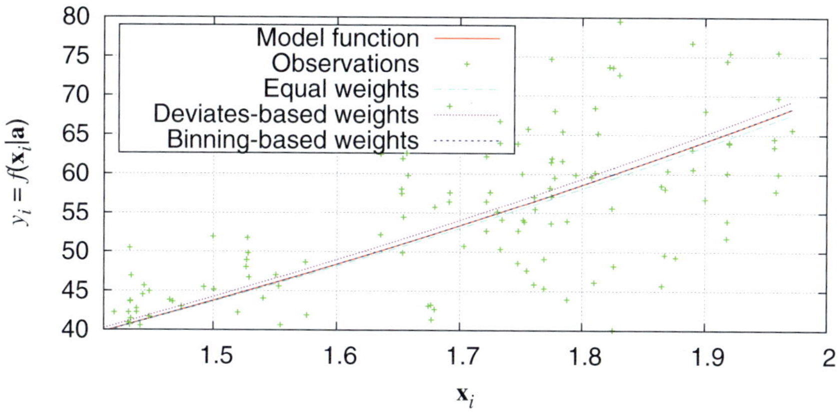

b)

Figure 3.47: Results of fitting polynomial data with and without weights. The model function is $f(\mathbf{x}|\mathbf{a}) = a_1 + a_2 \cdot x + a_3 \cdot x^2$; a) full plot; b) magnified portion

uncertainty of $\sigma_i = 0.25 + (x_i)^4$ has been added to the signal, as can be seen in **Figure 3.47**a). **Table 3.19** shows the results of the parameter estimation via least squares. The plain fitting, that is, using equal weights, shows a plot close to the correct solution in **Figure 3.47**b), however, the model-parameters estimation can be improved by proper weighting of the observations. While ClubOD never declares a data point as contaminant, the z-score approach cannot cope with the varying uncertainty at all. It relies on the estimation of the standard uncertainty of a Gaussian distribution, while the observations follow another distribution model in their entity.

Table 3.19: Numerical results of fitting a polynomial

Polynomial	a_1	a_2	a_3	Outliers
True values	10.00	0.000	15.000	0
Equal weights	9.774	1.030	14.351	0
z-score	9.671	1.446	14.124	10
M-score	9.693	1.352	14.178	114
ClubOD	9.774	1.030	14.351	0
RANSAC	10.093	−0.620	15.537	162
Estimated weights	10.007	−0.248	15.379	0
Weights + z-score	10.038	−0.416	15.508	101
Weights + M-score	10.030	−0.365	15.468	127
Weights + ClubOD	10.008	−0.256	15.385	0
Weights + RANSAC	10.057	−0.908	15.807	164
Weights via binning	**10.001**	−0.042	**15.026**	0
Binning + z-score	10.003	−0.049	14.124	95
+ weights reset	9.869	0.493	14.764	95
Binning + M-score	9.997	**0.005**	14.947	122
Binning + ClubOD	**10.001**	−0.042	**15.026**	0
Binning + RANSAC	10.013	−0.141	15.161	164

The plots of deviates are giving hints, why the results of outlier detection are so divergent (**Figure 3.48**). There is a concentration of small deviates leading to a relatively small estimated standard deviation $\hat{\sigma}_{\mathrm{y}}$. RANSAC, z-, and M-score are not able to deal with the varying standard uncertainty of the data points and cut too many of them. Especially after estimating proper weights, most of the 500 observations are relatively close to the modelled curve leading to a small threshold according to equation (3.17). Best parameter estimates are obtained when the bin-wise uncertainty of the observation is determined. The only difficulty is to assign suitable bins. In the presented example, each bin comprises observations of 50 adjacent conditions.

3.6 Conclusions

In the previous section the influence of different weighting and outlier-detection schemes on the data fitting has been investigated. Although it has not been a comprehensive test, a tendency has been become visible and some conclusions can be drawn about the usefulness of the modules depicted in Figure 3.15. The evaluation is based on the accuracy of the estimated parameter values.

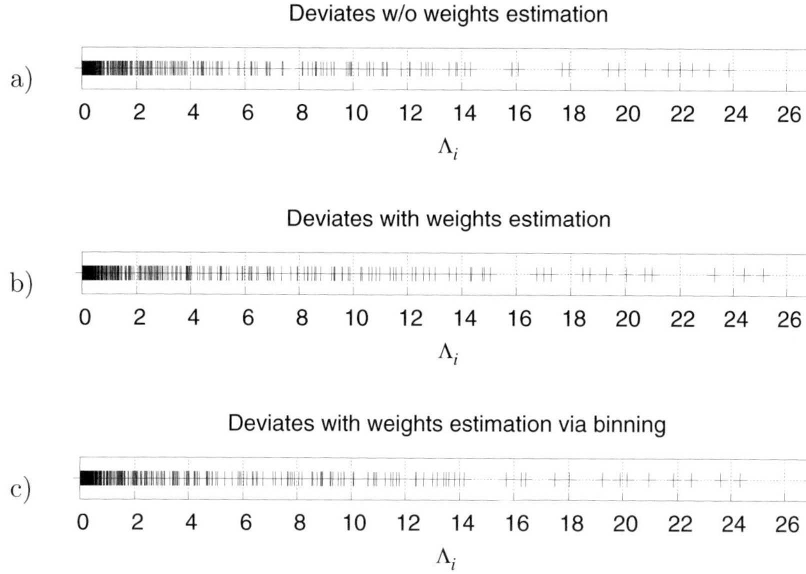

Figure 3.48: Absolute deviates according to the fitting of a polynomial function; a) without application of weights estimation, b) using the estimation of weights, c) weights estimation via binning

3.6.1 Evaluation of weighting

Equal weights versus deviates-based weights estimation

The examinations have compared the estimated model parameters in dependence on whether estimated weights have been used for data fitting or not. The application of weights estimation shows a slight advantage, especially in the case of strong outliers, but also when the observations have truly different uncertainties, as in the case of fitting the polynomial. However, the advantage is not so pronounced that the general use of weights can be recommended. This becomes different when the removal of outliers is considered, see below.

Comparison of different schemes for weights estimation

Only the example of polynomial fitting with its relative huge number of observations allows an evaluation of the correctness of the weights. **Figure 3.49** shows the estimated weights converted to estimated uncertainties of both weighting schemes in comparison with the true uncertainties used for the generation of the test data. The deviates-based approach somewhat overestimates the uncertain-

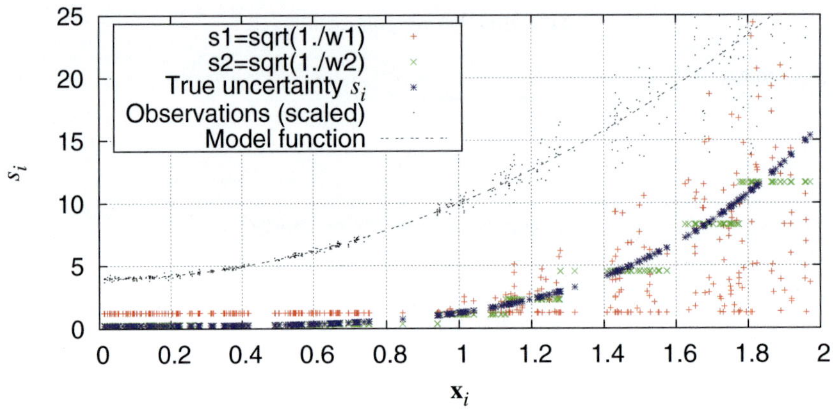

Figure 3.49: Estimated uncertainties $s1_i, s2_i$ for two different weighting schemes compared to the true uncertainty s_i. w_1 corresponds to deviates-based weighting and w_2 to weighting via binning. The observations are superposed for better comparison.

ties in the range, where identical weights have been assigned. Then it follows in principle the true uncertainties, however, the uncertainties of observations close to the fitted curve are distinctly underestimated. In contrast, the weights estimation using the binning method is able to compute uncertainties, which are very reasonable owing to the averaging effect. As already mentioned in the beginning of this section, the accuracy of uncertainties for single observations could be improved using a sliding bin or overlapping ones.

3.6.2 Comparison of outlier detectors

Table 3.20 lists the number of outliers found by the different outlier-detection methods in dependence on whether weights have been estimated or not. The last row gives a summation of the absolute differences between true and found number of outliers. The results of fitting the polynomial data were left out of the summation, because only ClubOD can cope with the varying standard uncertainties of the observations.

The comparison of the columns with and without preceding estimation of weights shows on the one hand that it is sometimes impossible to find outliers if the estimated model-function parameters are far from their optimum. On the other hand, weight estimation can lead to the detection of too many outliers as can be seen in the examples of straight-line regression of $y^{(2)}$.

Table 3.20: Summary of outlier detection for thirteen different data sets; see text for explanations. (z = z-score; M = M-score; Cl. = ClubOD; RA. = RANSAC)

Examples	True num.	w/o weigths				With weights			
		z	M	Cl.	RA.	z	M	Cl.	RA.
Constant	3	2	0	0	0	3	3	3	3
$y^{(0)}$	0	0	0	0	0	2	1	3	1
$y^{(1)}$	2	0	0	0	0	2	2	2	2
$y^{(2)}$	0	0	1	0	2	4	4	4	4
$y^{(3)}$	2	0	0	0	0	3	3	3	3
Plane	9	1	2	0	9	12	9	9	13
Coord. transf.	18	9	13	16	16	17	18	17	24
Cosine 1	0	0	0	0	0	1	0	0	1
Cosine 2	3	1	3	3	5	6	3	3	6
Exponential	0	0	0	0	0	1	0	0	0
Exp. lin.	0	0	1	0	2	4	1	0	2
Gaussian add.	0	0	1	1	2	6	2	0	5
Total difference	-	24	22	19	17	26	9	9	27
Polynomial	0	10	114	0	162	101	127	0	164
with binning	0	-	-	-	-	95	122	0	164

Naturally, RANSAC shows the best results on average if equal weights are used. Including the prior estimation of weights, the performance of RANSAC drops. Here, ClubOD and the M-score method perform best. The results of RANSAC, however, are not deterministic, as the subsets are selected randomly from the entire set of data points. Each run of RANSAC using the same data and parameters may result in different numbers of outliers. Besides, its detection power changes if the determination of the threshold λ_O is modified.

3.6.3 Usefulness of weights

Table 3.20 also reveals that the proper estimation of weights assists the outlier detection in a favourable manner for the methods M-score and ClubOD. The examination of all investigation results from Section 3.5 leads to the conclusion that it is beneficial indeed in the majority of cases to reset the estimated weights after the contaminants have been deleted from the set of observations. That means, the application of weighted least squares is mainly advantageous when it is accompanied with a suitable outlier-detection scheme. If the data set do not contain outliers or the contaminants have been successfully removed, equal weights should be used, as long as the true standard uncertainty of the single observations y_i is not known (see also page 48).

3.7 Test questions

3.1 What are the weights good for?

3.2 What are outliers?

3.3 What should be the relation between the uncertainty of an observation and its assigned weight?

3.4 Describe a method for the estimation of proper weights.

3.5 Under which circumstances should equal weights be used after the removal of outliers?

3.6 When is the estimation of weights favourable over equal weighting of observations?

3.7 Which influence do the initially estimated parameters have on the final fitting result in approaches with iteratively estimated weights?

3.8 In which cases can the influence of outliers be neglected?

Chapter 4

Uncertainty of Results

After fitting the given data in terms of observations y_i using the method of least-squares approximation, the experimenter knows the set of best[1] estimated parameters $\mathbf{a}$ of the utilised model function $f(\mathbf{x}|\mathbf{a})$. This chapter addresses the questions of whether (i) that best solution is adequate and (ii) the hypothesis is correct that the chosen model really represents the phenomena that have been measured in the experiment. In addition to this text, the reader can find help in the "Guide to the Expression of Uncertainty in Measurement" [GUM] with extensions and comments in [Cox06b] or in the German standard DIN 1319 [DIN96, DIN99], which is also available in English.

4.1 Goodness-of-fit, precision and accuracy

4.1.1 Consistence of statistical model and data

Let us first recapitulate the purpose of data fitting. In least-squares approximation, the idea is to minimise the following function

$$\chi^2 = \sum_{i=1}^{N} w_i \cdot [y_i - f(\mathbf{x}_i|\mathbf{a})]^2 \longrightarrow \text{Min}$$

for an experiment with N observations y_i, each measured under certain conditions represented by $\mathbf{x}_i$. The parameter vector $\mathbf{a}$ is the variable to be optimised, while the weights w_i should reflect the reliability of the corresponding observation (see Chapter 3).

Typically, data never exactly fit the chosen model function, regardless of whether the model is the correct one or not. The goodness-of-fit indicates how well a statistical model describes the data. This statistical model comprises both the chosen model function and the assumed deviation of the observations from that function.

[1] "best" means optimised with respect to the chosen minimisation criterion.

The goodness-of-fit is measured via[2]

$$g_{\text{fit}} = \frac{\chi^2}{N_{\text{free}}} = \frac{1}{N_{\text{free}}} \cdot \sum_{i=1}^{N} w_i \cdot [y_i - f(\mathbf{x}_i|\mathbf{a})]^2 \ , \tag{4.1}$$

where $N_{\text{free}} = N - M$ is the number of degrees of freedom for fitting N data points with M model parameters. A value of $g_{\text{fit}} \approx 1$ is an indicator that the statistical model is consistent with the data. This means that the weights w_i were correctly chosen equal to the reciprocal of the variance of the population from which their corresponding observation y_i was drawn ($w_i = 1/\sigma_i^2$) and that the model function is suitable. Nevertheless, $g_{\text{fit}} \approx 1$ does not rule out the existence of other, possibly more suitable models. If the model function is not appropriate for describing the observations, the deviations will be larger than estimated by the σ_is and thus $g_{\text{fit}} > 1$. On the other hand, a goodness-of-fit less than one does not necessarily signal a better fit; instead, it is likely that only the uncertainties σ_i are overestimated.

However, as already discussed in the previous chapter, there are many applications without available values for σ_i, which means the weights have to be determined in a different way. Ideally it is possible to estimate correct relative weights

$$w_i = \frac{k}{\sigma_i^2} \ . \tag{4.2}$$

The goodness-of-fit g_{fit} is then merely taken to be an estimate of the constant k.

4.1.2 Sample variance

The variance of the fit (sample variance) is equal to the estimate of the squared standard deviation σ_{y}^2 of the data and expressed by

$$\sigma_{\mathrm{y}}^2 = \frac{1}{N_{\text{free}}} \cdot \frac{\sum\limits_{i=1}^{N} w_i \cdot [y_i - f(\mathbf{x}_i|\mathbf{a})]^2}{\frac{1}{N}\sum\limits_{i=1}^{N} w_i} \ . \tag{4.3}$$

This formula is equivalent to equation (3.10) and it is also related to the goodness-of-fit (eq. 4.1) with additional normalisation of the weights. So it does not include complementary information about the quality of the fitting.

[2] There is no THE goodness-of-fit formula. Different textbooks assign unfortunately sometimes slightly different formulae. The symbol χ^2 points out the tight relation to the so-called χ^2-test (see elsewhere).

At this point, the use of the degrees of freedom $N_{\text{free}} = N - M$ for normalisation in (4.3) and other formulae for the *standard deviation* should be justified. When N is very large compared to the number M of parameters, the difference becomes negligible. Let us consider the straight-line example $y = a_1 \cdot x + a_2$ and let us assume that only $N = 2$ observations are made. It is obvious that it is *always* possible to draw a straight line through these two points. However, the probability that this is the correct line approaches zero and the uncertainty near infinite. Thus the denominator becomes $N_{\text{free}} = N - M = 2 - 2 = 0$ and the estimated deviation infinite as wanted. In general, this is related to the problem of overfitting. Using a model function with enough parameters, the fitting of any data (even noise) is possible. N_{free} prevents the experimenter to trust in such fits.

A standard deviation is only a measure of *precision*, one which results from random fluctuations of the observations. What the experimenter who collected the data also would like to know is the *accuracy* of the observations to lie in the true range [Bev92].

The difference between accuracy and precision of an experiment is shown in **Figure 4.1**. The accuracy expresses the closeness of observations to true values; it is a measure of correctness of the result (low systematic error). The precision is a measure of how well the observations fit the model function without regarding the true values; it is affected by random errors.

Unfortunately, it is very hard to detect errors of a systematic type and thus to make statements about the accuracy of the result. Systematic errors result from insufficiently calibrated equipment, from dependencies on not considered parameters influencing the measurement (e.g., the temperature etc.), or from bias on the part of the observer.

Example:
In an image-processing application, the observer can be a software algorithm for the detection of special objects, for example. The determination of the positions of these objects could be affected by the lighting conditions, introducing a systematic offset. Let us assume the simple case that bright objects have to be separated from a dark background via simple thresholding of the greyvalues and let us further assume that the centre of the object has subsequently to be determined from the remaining object area. If, for instance, the light source is shining from the right side, the object appears darker on the left side, which means the corresponding picture elements might be considered as background owing to the chosen threshold. The derived object centre would automatically be biased to the right.
□□□

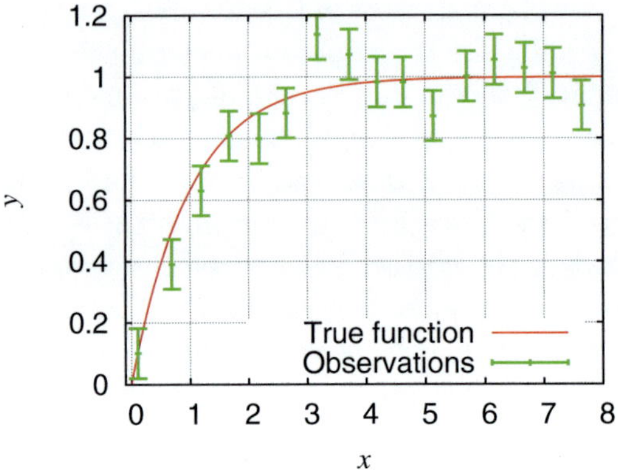

a)

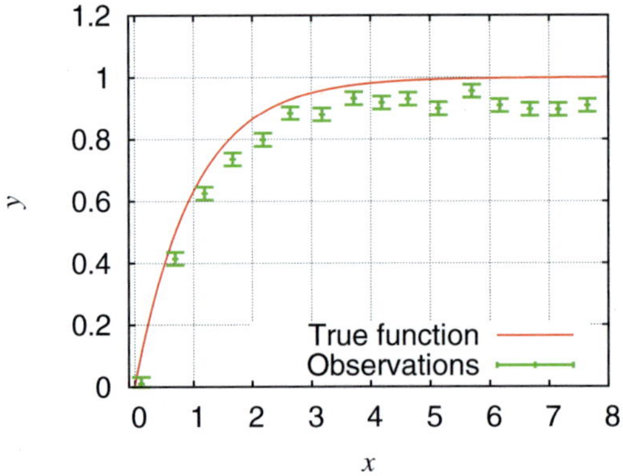

b)

Figure 4.1: Accuracy versus precision: a) accurate observations with low precision, b) precise but inaccurate data

Random errors are typically reduced by simply repeating the experiment to obtain more observations. If the random errors result from instrumental (or algorithmic) uncertainties, they could possibly be reduced by using more advanced instruments or algorithms. It should be noted that it does not make sense in either case to reduce the random error much below the systematic one.

4.2 Uncertainty of estimated parameters

In addition to determining the parameters of the model function, one should also make estimates of the uncertainties in those values. These estimates are usually expressed in terms of an *estimated standard deviation* (e.s.d) or *standard uncertainty*.

The standard uncertainty of parameters **a** can be derived directly from equation (2.6). The term

$$\mathbf{C}_a = \left(\mathbf{J}^{\mathrm{T}}\mathbf{W}\mathbf{J}\right)^{-1} \tag{4.4}$$

is equal to the variance-covariance matrix for all parameters [Rya97] describing how the single parameters are related to each other dependent on the chosen model function. The diagonal element c_{jj} contains the variance of parameter a_j. $\mathbf{C}_a$ is obviously independent of the actual observations, but influenced by the chosen conditions x_i and the weights w_i. If the weights merely represent relative uncertainties of the observations in terms of $w_i = k/\sigma_i^2$, $\mathbf{C}_a$ is commonly multiplied by the goodness-of-fit g_{fit} (eq. 4.1) [Pri92, Pre92]

$$\sigma_{a_j}^2 = g_{\text{fit}} \cdot c_{jj} \tag{4.5}$$

in order to compensate for the constant factor k (see Section 4.1). The uncertainty σ_{a_j} indicates the reliability of the model parameter a_j. It is often specified as a percentage of the estimated parameter value. If one or more of the parameters shows a relatively high uncertainty, this might point to an unsuitable model function.

While the diagonal elements of $\mathbf{C}_a$ correspond to the variances $\sigma_{a_j}^2$ of the model parameters a_j, the off-diagonal elements express the covariances σ_{a_j,a_k}^2, which could have also negative values.

The fact that the uncertainty of the estimated parameters is only dependent on the chosen conditions x_i and the weights w_i is somewhat surprisingly on first sight. That is why we should have a look at an example.

Example:
Figure 4.2a) shows a plot of six observations and the corresponding fit. The parent distributions of all observations have the same (and known) standard deviation $\sigma_i = \sigma = 0.15$. Based on the model function $y = a_1 \cdot x + a_2$, the estimated parameters are $a_1 = 0.9313$ and $a_2 = 0.5234$. The goodness-of-fit is $g_{\text{fit}} = 1.03$ and the uncertainties of the parameters are $\sigma_{a_1} = 0.1417, \sigma_{a_2} = 0.0364$ according to (4.5). Now let us investigate what will happen if the observations are taken at other conditions, Figure 4.2b). For a fair comparison, the same errors have been used. At visual inspection the observations seem to be moved along the same model function. The estimated parameters, however, are: $a_1 = 0.8376$, $a_2 = 0.5234$. The slope of line is the same, but the intersection with the y axis is different! It should be obvious that the more the observations move away from the y axis, the

a)

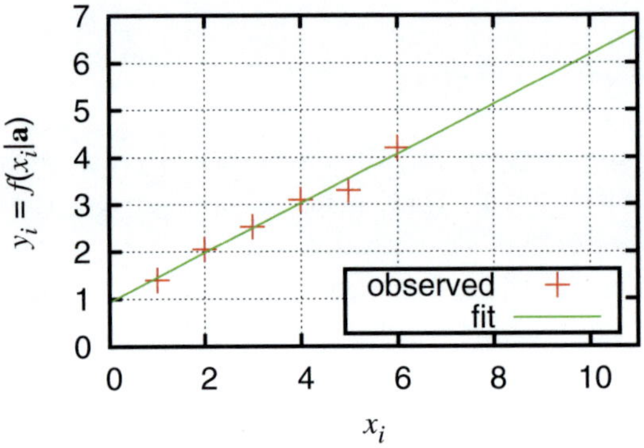

j	1	2
a_j	0.9313	0.5234
σ_{a_j}	0.1417	0.0364

b)

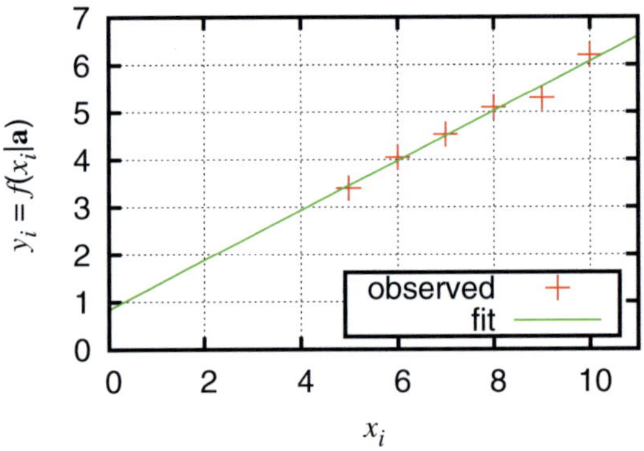

j	1	2
a_j	0.8376	0.5234
σ_{a_j}	0.2800	0.0364

c)

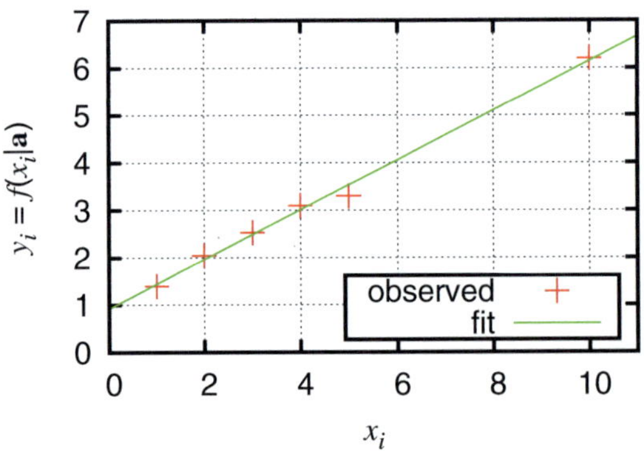

j	1	2
a_j	0.9185	0.5228
σ_{a_j}	0.0983	0.0004

Figure 4.2: Example plots and estimated values: a) small conditional values x_i within a small range; b) larger conditional values x_i within a small range; c) conditional values x_i within a larger range

greater is the influence of an incorrect estimation of a_2 on the error in a_1. The computed uncertainty is indeed larger: $\sigma_{a_1} = 0.2800$, while the uncertainty of a_2 remains the same.

If only a single observation is taken at a larger x_i, Figure 4.2c), both estimated parameters, the slope and the intersection, benefit from the larger range of conditions in terms of reliability.
□□□

4.3 Uncertainty of data approximation

The purpose of data fitting is not limited to the approximation of the data points. The data model can also be used to predict which values y would most likely be measured if other experimental conditions $\mathbf{x}$ were chosen. This is known in the literature as *model prediction.* Moreover, we are interested in the uncertainty of this model prediction.

Based on the uncertainty of parameters (eq. 4.5) and utilising the principle of error propagation (see Section 7.3, eq. 7.38), the uncertainty of the model prediction can be derived as

$$\sigma_f = \sqrt{\sum_j \sigma_{a_j}^2 \cdot \left(\frac{\partial f}{\partial a_j}\right)^2 + \sum_{j,k} \sigma_{a_j,a_k}^2 \cdot \left(\frac{\partial f}{\partial a_j}\right) \cdot \left(\frac{\partial f}{\partial a_k}\right)} \,.$$

For the example of straight-line fitting

$$f = f(\mathbf{x}|\mathbf{a}) = a_1 + a_2 \cdot x$$

we get as derivatives

$$\frac{\partial f}{\partial a_1} = 1 \qquad \frac{\partial f}{\partial a_2} = x$$

and accordingly

$$\sigma_f = \sqrt{\sigma_{a_1}^2 + \sigma_{a_2}^2 \cdot x^2 + 2 \cdot \sigma_{a_1,a_2}^2 \cdot x}$$

as the uncertainty of the modelled function. As you can see, this is a function of the condition x and increases in tendency with higher values of $|x|$. **Figure 4.3** visualizes this effect. The higher $|x|$ is, the more possible approximations deviate from the true model function.

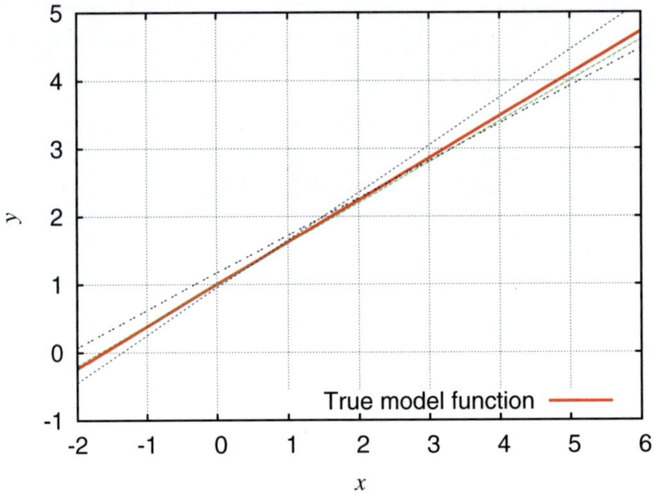

Figure 4.3: Model prediction based on a straight-line approximation. If the estimated model parameters are inaccurate, the prediction of the correct y value becomes more uncertain for a given condition x as $|x|$ increases (examples in green, blue, and black).

4.4 Inspection of plots

Numerical evaluation methods are essential, when the data fitting has to be an automatic process. In case of one-time experiments, it is additionally recommended to have a closer look at least at the plots of residuals ($y_i - f(\mathbf{x}_i|\mathbf{a})$) and preferably at the plot of the final model function against the observation. The latter yields an impression of whether the fitting procedure was successful in general while the former can reveal mismatches between the chosen model function and the true relation between conditions and observations.

Figure 4.4 shows the outcome of 31 measurements. The relationship between y and x seems to be of a linear nature. Thus,

$$y = f(\mathbf{x}|\mathbf{a}) = a_1 + a_2 \cdot x \tag{4.6}$$

is assumed to describe the system properly. The graph resulting from fitting a straight line is also depicted in Figure 4.4. At a first glance, the fit seems to be appropriate. An inspection of the deviates, however, reveals that they are not equally distributed over the range of conditions, but instead show a trend (**Figure 4.5**). This is a strong indication that the chosen model function is not

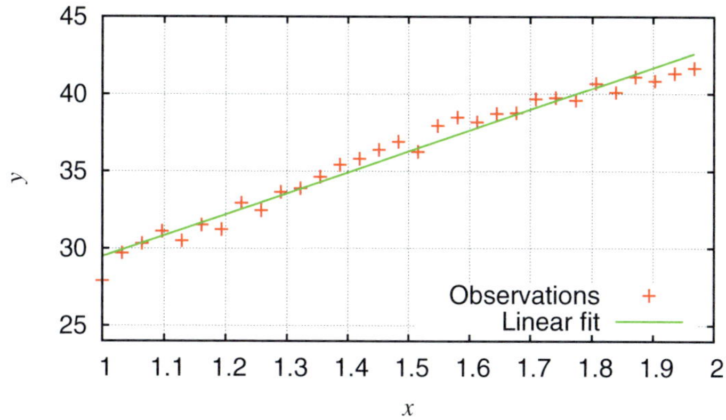

Figure 4.4: Result of data fitting with a straight-line model

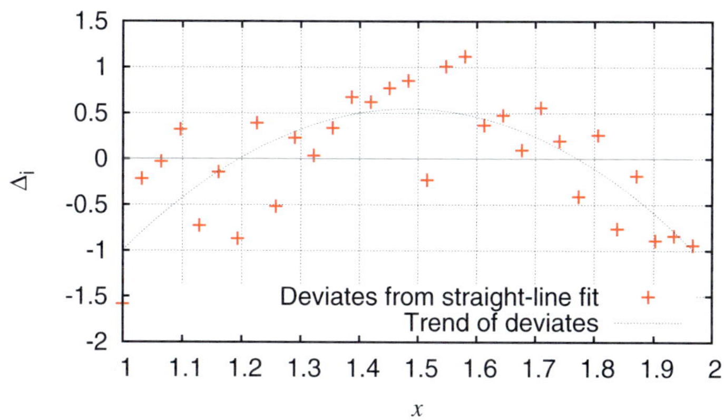

Figure 4.5: Plot of deviates $(y_i - f(\mathbf{x}_i|\mathbf{a}))$

sufficient to describe the process. Either polynomials of higher order or other functions have to be considered.

4.5 Example computations

The following brief subsections refer to the results of weighted data fitting reported in the previous chapter. Only three simple model functions are discussed here in detail, since the equations for all other models can be derived accordingly. The problem of model mismatch is then addressed.

4.5.1 Constant value

The minimisation problem for a constant value reduces to

$$\chi^2 = \sum_{i=1}^{N} w_i \cdot [y_i - f(\mathbf{x}_i|\mathbf{a})]^2 = \sum_{i=1}^{N} w_i \cdot [y_i - a_1]^2 \ .$$

Using the results of Subsection 2.6.1 we get the covariance matrix

$$\mathbf{C}_a = \left(\mathbf{J}^{\mathrm{T}}\mathbf{W}\mathbf{J}\right)^{-1} = \left(\sum_{i=1}^{N} w_i\right)^{-1} = c_{11} \ ,$$

which is simply a scalar in the current case. The goodness-of-fit is equal to

$$g_{\mathrm{fit}} = \frac{1}{N_{\mathrm{free}}} \cdot \sum_{i=1}^{N} w_i \cdot [y_i - a_1]^2 \ . \tag{4.7}$$

According to (4.3) and (4.5) the resulting variance of the parameter a_1 is therefore

$$\sigma_{a_1}^2 = g_{\mathrm{fit}} \cdot c_{11} = \frac{1}{N_{\mathrm{free}}} \cdot \frac{\sum_{i=1}^{N} w_i \cdot [y_i - a_1]^2}{\sum_{i=1}^{N} w_i} = \frac{\sigma_{\mathrm{y}}^2}{N} \ .$$

The uncertainty of the determined mean value a_1 is thus equal to $\sigma_{a_1} = \sigma_{\mathrm{y}}/\sqrt{N}$. Doubling the number of experiments reduces the uncertainty (the random error) by the factor $1/\sqrt{2}$.

The following table shows the improvement of the data fitting in the example discussed in Subsection 3.5.1 depending on using weights and outlier detection.

	χ^2	g_{fit}	a_1	σ_{a_1}	σ_{a_1}[%]	Outliers
Without weights	247.116	5.884	9.791	0.370	3.78	0
z-score	117.805	2.945	9.829	0.268	2.73	2
M-score	70.000	1.795	10.000	0.212	2.12	3
ClubOD	70.000	1.795	10.000	0.212	2.12	3
RANSAC	70.000	1.795	10.000	0.212	2.12	5
Estimated weights	28.867	0.687	9.961	0.153	1.53	0
Weights + z-score	23.857	0.645	9.942	0.149	1.50	5
Weights + M-score	25.867	0.663	9.965	0.150	1.51	3
Weights + ClubOD	25.867	0.663	9.965	0.150	1.51	3
Weights + RANSAC	25.867	0.663	9.965	0.150	1.51	3
+ weights reset	70.000	1.795	10.000	0.212	2.12	3

The upper half of the table contains results from ordinary least-squares approximation (OLS) with and without the detection of outliers. The detection is performed either using the z-score method (Subsection 3.4.1) or the ClubOD approach (Subsection 3.4.2). The lower half of the table lists results based on weighted least squares.

Which conclusions can be drawn from these values?

- χ^2 is the target of minimisation, so its decreasing value shows at least that the removal of outliers fulfils the optimisation criterion.
- The application of OLS (no weights) implies w_i = const. = 1.0, i.e. a standard uncertainty of $\sigma_i = \sqrt{1/w_i} = \sigma = 1$. This leads to following effects.
  - Introducing weights decreases χ^2 only, if their averaged value is smaller than 1.
  - The value $g_{\text{fit}} = 5.884$ indicates that the uncertainty of the observations is in fact higher than 1, presuming the model function was chosen correctly. After removing the contaminants, of course, the uncertainty decreases and also the goodness-of-fit does.
- The application of weights leads to goodness-of-fit values $g_{\text{fit}} < 1$, indicating an overestimation of the uncertainties of the observations, i.e., the weights are too small on average. This is a systematic effect caused by the selected weighting scheme.
- The standard uncertainty of the estimated parameter a_1, specified in percent of the parameter value, provides the most reliable information that the fitting process was successful. Of course, the uncertainty decreases if observations with high deviation from the model function are down-weighted or even removed.

4.5.2 Straight line

In the straight-line example, we have to determine the uncertainty of two parameters according to the model function

$$y = f(\mathbf{x}|\mathbf{a}) = a_1 + a_2 \cdot x \ .$$

The computation of the covariance matrix becomes a little bit more complex. Following equation (2.16) and applying (5.6) for matrix inversion we get

$$\mathbf{C}_a = \left(\mathbf{J}^{\mathrm{T}}\mathbf{W}\mathbf{J}\right)^{-1} = \begin{pmatrix} \sum\limits_{i=1}^{N} w_i & \sum\limits_{i=1}^{N} w_i \cdot x_i \\ \sum\limits_{i=1}^{N} w_i \cdot x_i & \sum\limits_{i=1}^{N} w_i \cdot x_i^2 \end{pmatrix}^{-1} = \begin{pmatrix} c_{11} & c_{12} \\ c_{21} & c_{22} \end{pmatrix}$$

$$= \frac{1}{\sum\limits_{i=1}^{N} w_i \cdot \sum\limits_{i=1}^{N} w_i \cdot x_i^2 - \sum\limits_{i=1}^{N} w_i \cdot x_i \cdot \sum\limits_{i=1}^{N} w_i \cdot x_i} \cdot \begin{pmatrix} \sum\limits_{i=1}^{N} w_i \cdot x_i^2 & -\sum\limits_{i=1}^{N} w_i \cdot x_i \\ -\sum\limits_{i=1}^{N} w_i \cdot x_i & \sum\limits_{i=1}^{N} w_i \end{pmatrix} .$$

From the matrix structure of $\mathbf{C}_a$ we can immediately conclude that there are in total four variances or covariances, respectively

$$\begin{pmatrix} \sigma_{a_1}^2 & \sigma_{a_1 a_2}^2 \\ \sigma_{a_2 a_1}^2 & \sigma_{a_2}^2 \end{pmatrix} = g_{\text{fit}} \cdot \begin{pmatrix} c_{11} & c_{12} \\ c_{21} & c_{22} \end{pmatrix} .$$

The influence of the weighting and outlier detection is shown in following tables, where $y^{(0)} \dots y^{(3)}$ refer to the examples discussed in Subsection 3.5.2.

$y^{(0)}$	χ^2	g_{fit}	a_1	a_2	σ_{a_1}[%]	σ_{a_2}[%]
True values			2.000	1.000		
Without weights	0.318	0.040	1.861	1.196	7.32	18.36
all methods	0.318	0.040	1.861	1.196	7.32	18.36
Estimated weights	7.539	0.942	2.000	1.111	4.62	13.19
Weights + z-score	5.479	0.913	2.016	1.101	4.61	13.38
Weights + M-score	6.500	0.928	2.020	1.084	4.64	13.62
Weights + ClubOD	4.387	0.877	2.016	1.112	4.52	13.02
Weights + RANSAC	6.509	0.930	1.996	1.128	4.60	12.98
+ reset (ClubOD)	0.059	0.012	2.004	1.126	4.59	13.13
+ reset (RANSAC)	0.201	0.029	2.031	0.964	7.03	22.67

$y^{(1)}$	χ^2	g_{fit}	a_1	a_2	σ_{a_1}[%]	σ_{a_2}[%]
Without weights	1.857	0.232	1.939	1.379	16.97	38.48
all methods	1.857	0.232	1.939	1.379	16.97	38.48
Estimated weights	6.240	0.780	1.718	1.554	7.02	13.49
all methods	4.218	0.703	1.712	1.546	6.73	12.99
+ weights reset	0.248	0.041	1.750	1.358	9.34	19.65

$y^{(2)}$	χ^2	g_{fit}	a_1	a_2	$\sigma_{a_1}[\%]$	$\sigma_{a_2}[\%]$
Without weights	1.932	0.242	2.436	−0.109	13.78	497.10
z-score	1.932	0.242	2.436	−0.109	13.78	497.10
M-score	0.452	0.065	2.135	0.711	8.65	46.12
ClubOD	1.932	0.242	2.436	−0.109	13.78	497.10
RANSAC	0.150	0.025	1.980	1.176	6.23	20.77
Estimated weights	4.263	0.533	2.021	1.058	2.50	10.46
all methods	0.221	0.055	2.015	1.084	0.81	3.39
+ weights reset	0.002	0.0004	2.015	1.084	0.81	3.39

$y^{(3)}$	χ^2	g_{fit}	a_1	a_2	$\sigma_{a_1}[\%]$	$\sigma_{a_2}[\%]$
Without weights	5.375	0.672	3.220	−0.952	17.39	94.82
all methods	5.375	0.672	3.220	−0.952	17.39	94.82
Estimated weights	6.269	0.784	2.220	0.460	9.90	80.21
all methods	3.209	0.642	2.161	0.561	9.30	60.40
+ weights reset	0.440	0.088	2.150	0.677	11.83	62.16

The conclusions regarding the values of χ^2, g_{fit} and $\sigma_{a_j}[\%]$ can be drawn accordingly to the statements in the previous subsection. The uncertainty of the observations is less than one, which is why χ^2 and g_{fit} become higher after the application of weights estimation in all four cases.

These examples of straight-line regression particularly illustrate the high relation of the standard uncertainty (in percent) of the estimated parameters and their correctness. Values of more than 50% definitely point to untrustworthy parameters or to inadequate model functions, respectively. If outliers are present (and not removed), the chosen model function is typically not appropriate to represent the observations and high relative standard uncertainties of parameters are observed. Values below 10% indicate parameters with reasonable values, however, the smallest uncertainty must not necessarily point to the parameter value closest to the true one as can be observed for a_1 and a_2 of $y^{(1)}$ as well as for a_2 of $y^{(3)}$, for example.

4.5.3 Cosine function

The approximation of the nonlinear model function

$$y = b_1 + b_2 \cdot \cos(x - b_3)$$

based on the observations used in example 2 (Subsection 3.5.6, page 94) leads to the following values:

Ex. 2	χ^2	g_{fit}	b_1	b_2	b_3	$\sigma_{b_1}[\%]$	$\sigma_{b_2}[\%]$	$\sigma_{b_3}[\%]$
True values			178.0	35.76	$-174.9°$			
Without weights	28638.	867.83	176.2	40.24	$-174.6°$	2.79	17.26	0.10
z-score	20637.	644.94	178.8	35.58	$-178.1°$	2.40	17.21	0.10
M-score	8206.7	273.56	174.2	31.43	$-174.9°$	1.65	13.03	0.07
ClubOD	8206.7	273.56	174.2	31.43	$-174.9°$	1.65	13.03	0.07
RANSAC	8206.7	273.56	174.2	31.43	$-174.9°$	1.65	13.03	0.07
Estimated weights	24.566	0.744	171.7	34.81	$-174.1°$	1.04	7.21	0.04
Weights + z-score	18.552	0.687	171.5	35.05	$-174.1°$	1.01	6.91	0.04
Weights + M-score	21.560	0.719	171.7	34.64	$-174.1°$	1.03	7.13	0.04
Weights + ClubOD	21.560	0.719	171.7	34.64	$-174.1°$	1.03	7.13	0.04
Weights + RANSAC	18.552	0.687	171.5	35.05	$-174.1°$	1.01	6.91	0.04
+ reset (ClubOD)	8206.7	273.56	174.2	31.43	$-174.9°$	1.65	13.03	0.07
+ reset (RANSAC)	3856.4	148.32	172.2	33.58	$-176.4°$	1.32	9.64	0.05

By definition, the uncertainties of the parameters in this fitting example are also smallest when weights are used (and outliers excluded).

Comparing the results with and without deletion of mavericks, the advantage of outlier detection becomes obvious in terms of the uncertainties of parameters as well.

4.5.4 Model mismatch

Example 1

Let us come back to the data set shown in Figures 4.4 – 4.5 on pages 121f. The fitting based on a straight-line model

$$y = f(\mathbf{x}|\mathbf{a}) = a_1 + a_2 \cdot x$$

and equal weights leads to the following values

χ^2	g_{fit}	$\sigma_{a_1}[\%]$	$\sigma_{a_2}[\%]$
13.011	0.449	3.96	3.07

The uncertainty of estimated parameters is reasonably low, leading to the assumption that the chosen model is suitable. The plot of residuals in Figure 4.5, however, gives indication of a model of higher order.

The goodness-of-fit is less than one, i.e., the uncertainty of the observations was over-estimated. One can interpret g_{fit} as factor k in equation (4.2), and since $w_i = w = 1$ was used, a squared uncertainty of $\sigma^2 = g_{\text{fit}}/w \approx 0.45$ can be derived. Actually, a variance of $\sigma^2 = 0.16$ was used for the generation of data via

$$y_i = f(\mathbf{x}_i|\mathbf{a}) = 10 + 20 \cdot x_i - 1 \cdot x_i^3 + \varepsilon_i \qquad \varepsilon_i \sim \mathcal{N}(0, 0.16) .$$

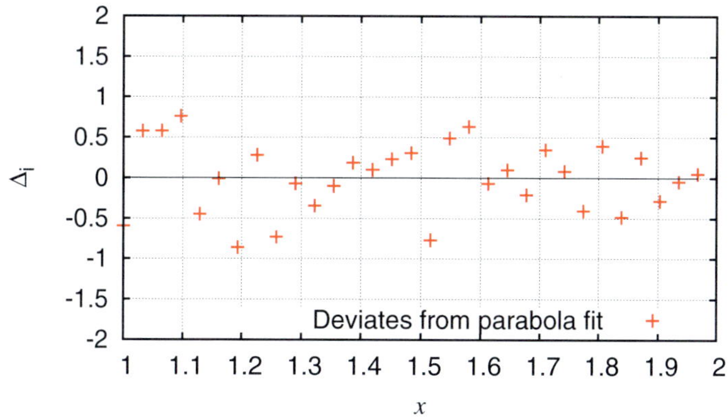

Figure 4.6: Plot of deviates ($y_i - f(\mathbf{x}_i|\mathbf{a})$) based on the parabola fit

The increased value of the estimated σ^2 (compared to the true one) is a result of the model mismatch.

Now let us check what happens when a parabola is used instead of a straight line. Applying the model function

$$y = f(\mathbf{x}|\mathbf{a}) = a_1 + a_2 \cdot x + a_3 \cdot x^2$$

the plot of residuals looks more reasonable (**Figure 4.6**). The results of the data-fitting evaluation are

χ^2	g_{fit}	$\sigma_{a_1}[\%]$	$\sigma_{a_2}[\%]$	$\sigma_{a_3}[\%]$
5.602	0.200	118.26	9.73	16.43

.

The goodness-of-fit has changed to $g_{\text{fit}} = 0.200$, leading to a squared estimated standard deviation of observations of $\sigma^2 = g_{\text{fit}}/w \approx 0.2$, which is already closer to the true value of 0.16. The uncertainties of parameters, however, have increased dramatically.

Using the true third-order model

$$y = f(\mathbf{x}|\mathbf{a}) = a_1 + a_2 \cdot x + a_3 \cdot x^2 + a_4 \cdot x^3$$

things get even worse. The values are

χ^2	g_{fit}	$\sigma_{a_1}[\%]$	$\sigma_{a_2}[\%]$	$\sigma_{a_3}[\%]$	$\sigma_{a_4}[\%]$
5.427	0.201	93.73	385.35	170.05	107.17

.

The estimated model parameters are very uncertain. Although this behaviour seems to be somewhat strange, everything is in accordance with the theory. The

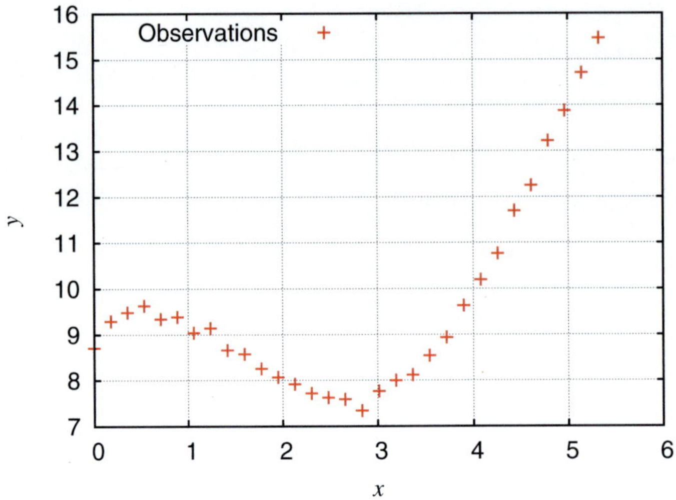

Figure 4.7: Harmonic oscillation with trend

reasons become clear when the estimated parameters are compared with the true ones

	a_1	a_2	a_3	a_4
True values	10.00	20.00	0.00	−1.00
Estimated values	14.17	7.24	11.25	−4.01

.

The estimated parameters are really far from the true values. The slope change of the curve to be fitted (Figure 4.4) is simply not pronounced enough in comparison with the random error in the data, making it impossible to estimate the parameters with a sufficient accuracy. In addition, the uncertainty of parameters generally increases as the number of parameters increases, while the number of observations is fixed.

Example 2

Let us examine another example of model mismatch. Given is a harmonic oscillation with a linear trend according to the model function

$$y_i = a_1 + a_2 \cdot x + a_3 \cdot \cos(x_i - a_4) + \varepsilon_i \qquad \varepsilon_i \sim \mathcal{N}(0, 0.01) \; . \tag{4.8}$$

The data points are slightly disturbed by random errors following a Gaussian distribution with ($\overline{\varepsilon_i} = 0, \sigma = 0.1$). **Figure 4.7** shows the corresponding plot. The model parameters are $a_1 = 6$, $a_2 = 1.5$, $a_3 = 5$, and $a_4 = 0.0873$ ($= 5°$).

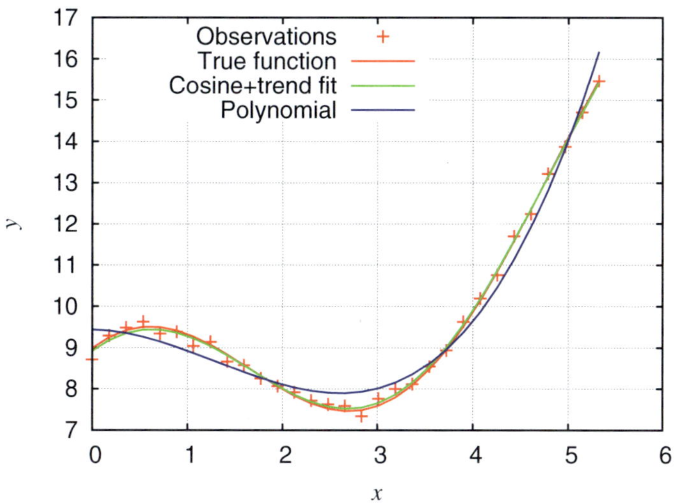

Figure 4.8: Fitting a harmonic oscillation with trend

Inspecting the plot could indicate a cubic polynomial, i.e.

$$y = a_1 + a_2 \cdot x + a_3 \cdot x^2 + a_4 \cdot x^3 \,.$$

As an alternative, a second-order trigonometric polynomial (see eq. (2.15) in Section 2.5) could be used as model function

$$y = f(x|\mathbf{a}) = a_1 + a_2 \cdot \cos(a_3 x - a_4) + a_5 \cdot \cos(2a_3 x - a_6) \,.$$

Fitting the data points with the true and the presumed polynomial model functions leads to the following numerical results

	a_1	a_2	a_3	a_4	a_5	a_6
True values	6.00	1.50	3.00	0.0873	−−	−−
Estimated cosine + trend	6.01	1.49	2.92	0.0843	−−	−−
Estimated polynomial	9.44	−0.026	−0.660	0.170	−−	−−
Trigonometric polynomial	−28.7	60.3	0.288	−5.23	23.2	−7.52

The fitted curve in **Figure 4.8**, corresponding to the polynomial model function, already reveals problems, which become even more visible when the plot of residuals is inspected (**Figure 4.9**). In addition, the uncertainties of the estimated parameters clearly point to a model mismatch, as can be seen in the following table

χ^2	g_{fit}	$\sigma_{a_1}[\%]$	$\sigma_{a_2}[\%]$	$\sigma_{a_3}[\%]$	$\sigma_{a_4}[\%]$
3.272	0.121	2.36	1436.30	24.61	11.81

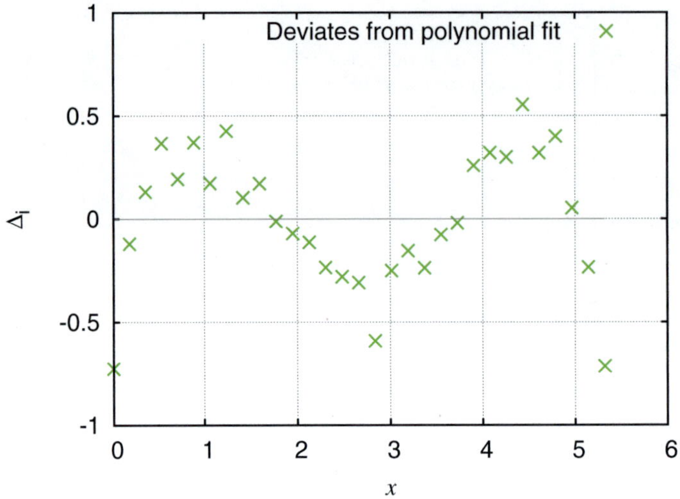

Figure 4.9: Plot of deviates of fitting a harmonic oscillation with trend using a third-order polynomial

Parameter a_2 is completely uncertain; in addition, the values of a_3 and a_4 seem not to be trustworthy.

The application of the trigonometric polynomial yields a better fit, mainly because it comprises more model parameters (**Figure 4.10**). The evaluation results are

χ^2	g_{fit}	$\sigma_{a_1}[\%]$	$\sigma_{a_2}[\%]$	$\sigma_{a_3}[\%]$	$\sigma_{a_4}[\%]$	$\sigma_{a_5}[\%]$	$\sigma_{a_6}[\%]$
0.288	0.012	269.3	183.5	38.54	10.21	145.69	10.14

The uncertainties of all parameters are relatively high. Although the uncertainties generally increase with decreasing value of M/N (number of parameters/ number of observations), it can still be concluded that the chosen trigonometric model function is not appropriate.

Fitting using the correct model results in

χ^2	g_{fit}	$\sigma_{a_1}[\%]$	$\sigma_{a_2}[\%]$	$\sigma_{a_3}[\%]$	$\sigma_{a_4}[\%]$
0.332	0.012	1.28	1.86	1.40	21.27

Here, only the phase shift (a_4) seems to be somewhat uncertain. This, however, is rather typical in fitting oscillations, especially because of the duality $\cos(\alpha) = \cos(2\pi - \alpha)$.

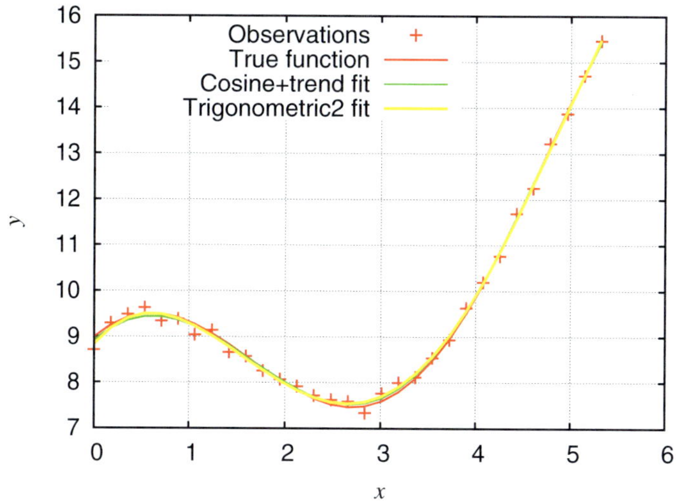

Figure 4.10: Fitting a harmonic oscillation with trend using a second-order trigonometric polynomial

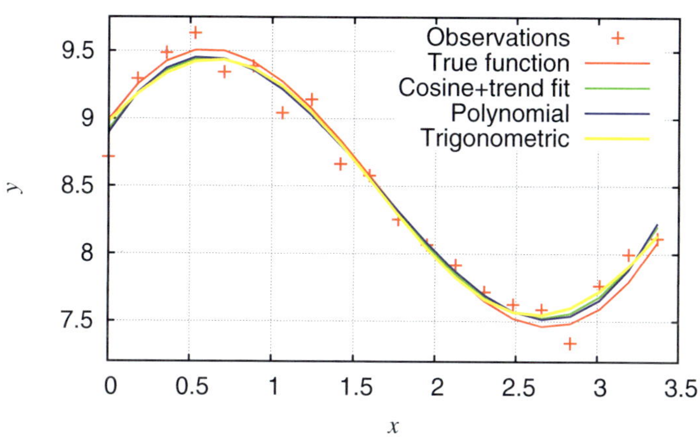

Figure 4.11: Fitting a harmonic oscillation with trend, limited range of conditions

Example 3

Unfortunately, the indication of model mismatch is not always such obvious as in Example 2. Let us take the same data points from the previous example, but using a more narrow range of conditions (**Figure 4.11**). The range is now cut

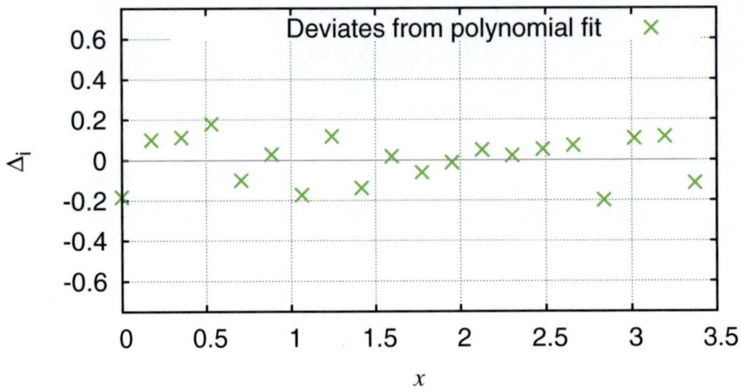

Figure 4.12: Plot of deviates of fitting a harmonic oscillation with trend using a third-order polynomial (limited range of conditions)

at about $x = 3.37$ decreasing the influence of the trend (parameter a_2 in eq. 4.8). The estimated parameters changes to

	a_1	a_2	a_3	a_4
True values	6.00	1.50	3.00	0.0873
Estimated cosine + trend	5.95	1.54	2.98	0.0747
Estimated polynomial	8.90	2.034	−2.073	0.419
Trigonometric 1st order	8.49	0.948	1.591	1.033

and their uncertainties to

	χ^2	g_{fit}	$\sigma_{a_1}[\%]$	$\sigma_{a_2}[\%]$	σ_{a_3}	$\sigma_{a_4}[\%]$
Cosine + trend	0.242	0.017	5.26	11.87	8.43	35.49
3rd order polynomial	0.255	0.016	1.06	12.27	8.42	8.12
Trigonometric 1st order	0.324	0.142	0.38	4.71	3.32	10.19

Although the estimated parameters of the correct model remain nearly the same, its advantage over the polynomial approaches has vanished. This is also reflected in the fitted curves (Figure 4.11). The plot of residual is not suspicious any more, either (**Figure 4.12**).

The results can be summarised in two statements

- It is possible to accurately fit data points stemming from an unknown model by a polynomial or trigonometric polynomial of adequate order, presuming the range of conditions is sufficiently narrow compared to that model order.

- The wider the range of condition, under which the observations are taken, the more reliably the test of model mismatch can be done.

4.6 Test questions

4.1 The goodness-of-fit indicates the consistency of the model function and data only if a certain prerequisite is fulfilled. What is the prerequisite?

4.2 Explain the difference between precision and accuracy.

4.3 Explain the difference between systematic and random errors?

4.4 How does the goodness-of-fit influence the standard uncertainty of the estimated model parameters?

4.5 On what does the uncertainty of model prediction depend?

4.6 What does the uncertainty of estimated model parameters tell us?

4.7 How can model mismatch be detected?

Part II

Mathematics, Optimisation Methods, and Add ons

"All through my life I've had this strange unaccountable feeling that something was going on in the world, something big, even sinister, and no one would tell me what it was."
"No," said the old man, "that's just perfectly normal paranoia. Everyone in the Universe has that." – Douglas Adams, The Hitchhiker's Guide to the Galaxy

Chapter 5

Matrix Algebra

This chapter briefly repeats basics and algorithms of linear algebra required for the understanding of the least-squares equations used in the previous chapters. For readers without any knowledge in linear algebra, we recommend to consult in addition textbooks about this topic as for instance "Linear algebra and its applications" by G. Strang [Str88].

5.1 Basics

A *matrix* is a rectangular array of numbers arranged in rows and columns. The elements of a matrix are referred using coordinates i and j. a_{ij} is the element in the ith row and the jth column of matrix $\mathbf{A}$. The number of rows is denoted by N, the number of columns by M

$$\mathbf{A} = \begin{pmatrix} a_{11} & a_{12} & \cdots & a_{1j} & \cdots & a_{1M} \\ a_{21} & a_{22} & \cdots & a_{2j} & \cdots & a_{2M} \\ \vdots & \vdots & \ddots & \cdots & \cdots & \vdots \\ a_{i1} & a_{i2} & \vdots & a_{ij} & \cdots & a_{iM} \\ \vdots & \vdots & \vdots & \vdots & \ddots & \vdots \\ a_{N1} & a_{N2} & \cdots & a_{Nj} & \cdots & a_{NM} \end{pmatrix}.$$

Matrices with the same number of rows and columns are *square*. If $a_{ij} = 0$ for all $i \neq j$, $\mathbf{A}$ is *diagonal*. Matrix $\mathbf{B}$ with $b_{ij} = a_{ji}$ for all i and j is called the *transpose* of $\mathbf{A}$ or $\mathbf{B} = \mathbf{A}^{\mathrm{T}}$. In the case of $a_{ij} = a_{ji}$, i.e. $\mathbf{A} = \mathbf{A}^{\mathrm{T}}$, $\mathbf{A}$ is called *symmetric*.

Examples:

$$\text{square matrix:} \quad \mathbf{A} = \begin{pmatrix} 1 & 3 & 4 \\ 2 & 2 & 1 \\ 5 & 7 & 3 \end{pmatrix}$$

$$\text{non-square matrix:} \quad \mathbf{A} = \begin{pmatrix} 6 & 3 & 4 \\ 2 & 8 & 1 \end{pmatrix}$$

$$\text{diagonal matrix:} \quad \mathbf{A} = \begin{pmatrix} 1 & 0 & 0 \\ 0 & 2 & 0 \\ 0 & 0 & 3 \end{pmatrix}$$

$$\text{matrix and its transpose:} \quad \mathbf{A} = \begin{pmatrix} 1 & 2 & 3 & 4 \\ 5 & 6 & 7 & 8 \end{pmatrix} \quad \mathbf{A}^{\mathrm{T}} = \begin{pmatrix} 1 & 5 \\ 2 & 6 \\ 3 & 7 \\ 4 & 8 \end{pmatrix}$$

$$\text{symmetric matrix:} \quad \mathbf{A} = \mathbf{A}^{\mathrm{T}} = \begin{pmatrix} 1 & 2 & 3 \\ 2 & 4 & 5 \\ 3 & 5 & 6 \end{pmatrix}$$

□□□

If the number of rows N is equal to one, the matrix is reduced to a *row vector* $\mathbf{a} = (a_1\ a_2\ \ldots\ a_M)$. When, in contrast, the number of columns M is equal to one, the matrix is called *column vector* $\mathbf{a} = (a_1\ a_2\ \ldots\ a_N)^{\mathrm{T}}$.

Examples:

$$\text{row vector:} \quad \mathbf{a} = \begin{pmatrix} 1 & 3 & 4 & 7 \end{pmatrix} \qquad \text{column vector} \quad \mathbf{b} = \mathbf{a}^{\mathrm{T}} = \begin{pmatrix} 1 \\ 3 \\ 4 \\ 7 \end{pmatrix}$$

□□□

A square matrix, whose diagonal elements ($i = j$) are equal to one and whose off-diagonal elements ($i \neq j$) are equal to zero, is an *identity matrix* (sometimes also called *unity matrix*), denoted by $\mathbf{I}$ with the property $\mathbf{A} \cdot \mathbf{I} = \mathbf{A}$.

Example: identity matrix

$$\mathbf{I} = \begin{pmatrix} 1 & 0 & \cdots & 0 & 0 \\ 0 & 1 & \cdots & 0 & 0 \\ \vdots & \vdots & \ddots & \vdots & \vdots \\ 0 & 0 & \cdots & 1 & 0 \\ 0 & 0 & \cdots & 0 & 1 \end{pmatrix}$$

□□□

A system of linear equations like

$$\begin{aligned}
a_{11}x_1 + a_{12}x_2 + \cdots + a_{1j}x_j + \cdots + a_{1M}x_M &= y_1 \\
a_{21}x_1 + a_{22}x_2 + \cdots + a_{2j}x_j + \cdots + a_{2M}x_M &= y_2 \\
&\vdots \\
a_{i1}x_1 + a_{i2}x_2 + \cdots + a_{ij}x_j + \cdots + a_{iM}x_M &= y_i \\
&\vdots \\
a_{N1}x_1 + a_{N2}x_2 + \cdots + a_{Nj}x_j + \cdots + a_{NM}x_M &= y_N
\end{aligned}$$

can be easily written as

$$\mathbf{A} \cdot \mathbf{x} = \mathbf{y} \,,$$

that is,

$$\begin{pmatrix} a_{11} & \cdots & a_{1M} \\ a_{21} & \cdots & a_{2M} \\ \vdots & \cdots & \vdots \\ a_{N1} & \cdots & a_{NM} \end{pmatrix} \cdot \begin{pmatrix} x_1 \\ \vdots \\ x_M \end{pmatrix} = \begin{pmatrix} y_1 \\ y_2 \\ \vdots \\ y_N \end{pmatrix} ,$$

where $\mathbf{x}$ and $\mathbf{y}$ are column vectors with M and N elements, respectively.

When $\mathbf{A}$ is a $N \times M$ matrix and $\mathbf{B}$ of size $M \times K$, then the $N \times K$ elements of the product $\mathbf{C}$ are computed according

$$\mathbf{C} = \mathbf{AB} \quad \longrightarrow \quad c_{ik} = \sum_{j=1}^{M} a_{ij} \cdot b_{jk} \,,$$

thus, c_{ik} is the inner product of the ith row of $\mathbf{A}$ and the kth column of $\mathbf{B}$. The number of columns in $\mathbf{A}$ must be equal to the number of rows of $\mathbf{B}$.

Example: matrix multiplication

$$\mathbf{A} = \begin{pmatrix} 1 & 1 & 2 \\ 2 & 2 & 1 \end{pmatrix} \qquad \mathbf{B} = \begin{pmatrix} 0 & 1 \\ 2 & 4 \\ 1 & 1 \end{pmatrix}$$

$$\mathbf{A} \cdot \mathbf{B} = \begin{pmatrix} 4 & 7 \\ 5 & 11 \end{pmatrix} \qquad \mathbf{B} \cdot \mathbf{A} = \begin{pmatrix} 2 & 2 & 1 \\ 10 & 10 & 8 \\ 3 & 3 & 3 \end{pmatrix}$$

□□□

The product of two matrices is in general dependent on their order, thus $\mathbf{AB} \neq \mathbf{BA}$.

Example: matrix multiplication

$$\mathbf{A} = \begin{pmatrix} 1 & 1 & 2 \\ 2 & 2 & 1 \\ 1 & 3 & 3 \end{pmatrix} \qquad \mathbf{B} = \begin{pmatrix} 0 & 0 & 1 \\ 2 & 4 & 0 \\ 1 & 1 & 2 \end{pmatrix}$$

$$\mathbf{A} \cdot \mathbf{B} = \begin{pmatrix} 4 & 6 & 5 \\ 5 & 9 & 4 \\ 9 & 15 & 7 \end{pmatrix} \qquad \mathbf{B} \cdot \mathbf{A} = \begin{pmatrix} 1 & 3 & 3 \\ 10 & 10 & 8 \\ 5 & 9 & 9 \end{pmatrix}$$

□□□

In addition, the multiplication is associative, i.e.,

$$(\mathbf{AB})\mathbf{C} = \mathbf{A}(\mathbf{BC}) \ . \tag{5.1}$$

Further, $(\mathbf{AB})^{\mathrm{T}} = \mathbf{B}^{\mathrm{T}}\mathbf{A}^{\mathrm{T}}$ holds.

Example:

$$\mathbf{A} = \begin{pmatrix} 1 & 3 \\ 2 & 2 \end{pmatrix} \qquad \mathbf{B} = \begin{pmatrix} 0 & 1 \\ 6 & 5 \end{pmatrix}$$

$$(\mathbf{A} \cdot \mathbf{B})^{\mathrm{T}} = \begin{pmatrix} 18 & 12 \\ 16 & 12 \end{pmatrix}$$

$$\mathbf{B}^{\mathrm{T}} \cdot \mathbf{A}^{\mathrm{T}} = \begin{pmatrix} 0 & 6 \\ 1 & 5 \end{pmatrix} \cdot \begin{pmatrix} 1 & 2 \\ 3 & 2 \end{pmatrix} = \begin{pmatrix} 18 & 12 \\ 16 & 12 \end{pmatrix}$$

□□□

If the determinant[1] $\det(\mathbf{A})$ of a square matrix $\mathbf{A}$ is equal to zero, then $\mathbf{A}$ is *singular* and no inverse of $\mathbf{A}$ exists, otherwise $\mathbf{A}$ is *nonsingular* and there is an *inverse* matrix $\mathbf{A}^{-1}$, such that $\mathbf{AA}^{-1} = \mathbf{A}^{-1}\mathbf{A} = \mathbf{I}$.

[1] see Section 5.2

5.2 Determinants

The *determinant* of a square $N \times N$ matrix $\mathbf{A}_N$

$$\det(\mathbf{A}_N) = \begin{vmatrix} a_{11} & \cdots & a_{1N} \\ a_{21} & \cdots & a_{2N} \\ \vdots & \ddots & \vdots \\ a_{N1} & \cdots & a_{NN} \end{vmatrix} \tag{5.2}$$

is a scalar value. The calculation of determinants of second and third order, i.e. of 2×2 and 3×3 matrices, is rather simple

$$\det(\mathbf{A}_2) = \begin{vmatrix} a_{11} & a_{12} \\ a_{21} & a_{22} \end{vmatrix} = a_{11} \cdot a_{22} - a_{12} \cdot a_{21} \tag{5.3}$$

and

$$\det(\mathbf{A}_3) = \begin{vmatrix} a_{11} & a_{12} & a_{13} \\ a_{21} & a_{22} & a_{23} \\ a_{31} & a_{32} & a_{33} \end{vmatrix} = \begin{matrix} a_{11}a_{22}a_{33} + a_{12}a_{23}a_{31} + a_{13}a_{21}a_{32} - \\ (a_{13}a_{22}a_{31} + a_{11}a_{23}a_{32} + a_{12}a_{21}a_{33}) \end{matrix} . \tag{5.4}$$

For determinants of higher order, *cofactors* A^{jk} of $\mathbf{A}$ have to be introduced. Cofactors are defined as $(-1)^{j+k}$ times the *minor* M^{jk} of $\mathbf{A}$. The minor M^{jk} is in turn the determinant of a subset of $\mathbf{A}$ given by cancelling all elements of the jth row and the kth column.

Example:
The cofactor A^{23} of a 3×3 matrix $\mathbf{A}$ is computed as follows.

At first the second row and the third column are deleted.

$$\mathbf{A} = \begin{pmatrix} a_{11} & a_{12} & a_{13} \\ a_{21} & a_{22} & a_{23} \\ a_{31} & a_{32} & a_{33} \end{pmatrix} \longrightarrow \begin{pmatrix} a_{11} & a_{12} & | \\ -- & -- & | \\ a_{31} & a_{32} & | \end{pmatrix}$$

The determinant of the remaining submatrix leads to the minor

$$M^{23} = \begin{vmatrix} a_{11} & a_{12} \\ a_{31} & a_{32} \end{vmatrix} = a_{11} \cdot a_{32} - a_{12} \cdot a_{31}$$

and, choosing the right sign, the cofacter can be derived as

$$A^{23} = (-1)^{2+3} \cdot M^{23} = -(a_{11} \cdot a_{32} - a_{12} \cdot a_{31}) .$$

□□□

Then, the determinant of arbitrary order is defined by a recursive process which is either an expansion along the jth column

$$\det(\mathbf{A}_N) = a_{1j} \cdot A^{1j} + a_{2j} \cdot A^{2j} + \cdots + a_{Nj} \cdot A^{Nj}$$

or an expansion along the ith row

$$\det(\mathbf{A}_N) = a_{i1} \cdot A^{i1} + a_{i2} \cdot A^{i2} + \cdots + a_{iN} \cdot A^{iN} .$$

Determinants equal to zero indicate the singularity of the corresponding matrix, as will become obvious in Subsection 5.3.1, equation (5.5). Nearly singular matrices have a very small determinant. A matrix is singular, for example, if one of its columns is equal to the multiple of another column.

Example:

$$\mathbf{A} = \begin{pmatrix} a_{11} & a_{12} & c \cdot a_{11} \\ a_{21} & a_{22} & c \cdot a_{21} \\ a_{31} & a_{32} & c \cdot a_{31} \end{pmatrix} \qquad \text{with } c = \text{const.}$$

The determinant is according to equation (5.4) equal to

$$\begin{aligned} \det(\mathbf{A}) = & \begin{array}{l} a_{11}a_{22}ca_{31} + a_{12}ca_{21}a_{31} + ca_{11}a_{21}a_{32} - \\ (ca_{11}a_{22}a_{31} + a_{12}a_{21}ca_{31} + a_{11}ca_{21}a_{32}) \end{array} \\ = & \begin{array}{l} c \cdot (a_{11}a_{22}a_{31} + a_{12}a_{21}a_{31} + a_{11}a_{21}a_{32}) - \\ c \cdot (a_{11}a_{22}a_{31} + a_{12}a_{21}a_{31} + a_{11}a_{21}a_{32}) \end{array} \\ = & \; 0 . \end{aligned}$$

□□□

The same holds true if one of the rows is a multiple of another row, as can be easily verified. However, this is not the sole reason of having a determinant equal to zero, as the following example shows

$$\mathbf{A} = \begin{pmatrix} 2 & 1 & 3 \\ 1 & 4 & 1 \\ 3 & 5 & 4 \end{pmatrix} .$$

In this case, the rows are not linearly independent of each other. The sum of the first two rows is equal to the third row.

5.3 Numerical solutions for matrix inversion

Solving least-squares problems using computer programs requires numerical solutions. The inversion of the square matrix $\mathbf{N} = \mathbf{J}^{\mathrm{T}}\mathbf{W}\mathbf{J}$ in equations (2.6) or (2.10) is the only problematical operation. There are different methods to invert matrices; four of them will be explained in the following.

5.3.1 Cofactor-matrix method

The cofactor-matrix approach is a basic method for inversion of square matrices, but should be used only for small matrices for different reasons.

Using the determinant $\det(\mathbf{A})$ and cofactors A^{jk} (see Section 5.2), the inverse matrix $\mathbf{A}^{-1}$ of $\mathbf{A}$ with the property $\mathbf{A}^{-1} \cdot \mathbf{A} = \mathbf{A} \cdot \mathbf{A}^{-1} = \mathbf{I}$ is defined by

$$\mathbf{A}^{-1} = \frac{1}{\det(\mathbf{A})} \cdot \begin{pmatrix} A^{11} & A^{21} & \dots & A^{M1} \\ A^{12} & A^{22} & \dots & A^{M2} \\ \vdots & \vdots & \dots & \vdots \\ A^{1M} & A^{2M} & \dots & A^{MM} \end{pmatrix} . \tag{5.5}$$

Thus the inverses of a 2×2 matrix $\mathbf{A}_2$ and a 3×3 matrix $\mathbf{A}_3$ are

$$(\mathbf{A}_2)^{-1} = \frac{1}{a_{11} \cdot a_{22} - a_{12} \cdot a_{21}} \cdot \begin{pmatrix} a_{22} & -a_{12} \\ -a_{21} & a_{11} \end{pmatrix} \tag{5.6}$$

and

$$(\mathbf{A}_3)^{-1} = \frac{1}{\det(\mathbf{A}_3)} \cdot \begin{pmatrix} (a_{22}a_{33}-a_{23}a_{32}) & -(a_{21}a_{33}-a_{23}a_{31}) & (a_{21}a_{32}-a_{22}a_{31}) \\ -(a_{12}a_{33}-a_{13}a_{32}) & (a_{11}a_{33}-a_{13}a_{31}) & -(a_{11}a_{32}-a_{12}a_{31}) \\ (a_{12}a_{23}-a_{13}a_{22}) & -(a_{11}a_{23}-a_{13}a_{21}) & (a_{11}a_{22}-a_{12}a_{21}) \end{pmatrix} . \tag{5.7}$$

The recursive development of determinants of higher order is disadvantageous in computational solutions and can be source of accumulation of round-off errors. That is why the cofactor method should not be used for matrices larger than 3×3.

The precomputed inverse of a 4×4 matrix $\mathbf{A}_4$ is

$$(\mathbf{A}_4)^{-1} = \frac{1}{\delta} \cdot \begin{pmatrix} \alpha_{11} & \alpha_{12} & \alpha_{13} & \alpha_{14} \\ \alpha_{21} & \alpha_{22} & \alpha_{23} & \alpha_{24} \\ \alpha_{31} & \alpha_{32} & \alpha_{33} & \alpha_{34} \\ \alpha_{41} & \alpha_{42} & \alpha_{43} & \alpha_{44} \end{pmatrix} \tag{5.8}$$

with

$$\begin{aligned}\delta =& a_{11}a_{22}a_{33}a_{44} - a_{11}a_{22}a_{34}a_{43} - a_{11}a_{32}a_{23}a_{44} + a_{11}a_{32}a_{24}a_{43} + \\ & a_{11}a_{42}a_{23}a_{34} - a_{11}a_{42}a_{24}a_{33} - a_{21}a_{12}a_{33}a_{44} + a_{21}a_{12}a_{34}a_{43} + \\ & a_{21}a_{32}a_{13}a_{44} - a_{21}a_{32}a_{14}a_{43} - a_{21}a_{42}a_{13}a_{34} + a_{21}a_{42}a_{14}a_{33} + \\ & a_{31}a_{12}a_{23}a_{44} - a_{31}a_{12}a_{24}a_{43} - a_{31}a_{22}a_{13}a_{44} + a_{31}a_{22}a_{14}a_{43} + \\ & a_{31}a_{42}a_{13}a_{24} - a_{31}a_{42}a_{14}a_{23} - a_{41}a_{12}a_{23}a_{34} + a_{41}a_{12}a_{24}a_{33} + \\ & a_{41}a_{22}a_{13}a_{34} - a_{41}a_{22}a_{14}a_{33} - a_{41}a_{32}a_{13}a_{24} + a_{41}a_{32}a_{14}a_{23}\end{aligned}$$

$$\begin{aligned}\alpha_{11} &= a_{22}\cdot(a_{33}a_{44} - a_{34}a_{43}) - a_{32}\cdot(a_{23}a_{44} - a_{24}a_{43}) + a_{42}\cdot(a_{23}a_{34} - a_{24}a_{33}) \\ \alpha_{12} &= -a_{12}\cdot(a_{33}a_{44} - a_{34}a_{43}) + a_{32}\cdot(a_{13}a_{44} - a_{14}a_{43}) - a_{42}\cdot(a_{13}a_{34} - a_{14}a_{33}) \\ \alpha_{13} &= a_{12}\cdot(a_{23}a_{44} - a_{24}a_{43}) - a_{22}\cdot(a_{13}a_{44} - a_{14}a_{43}) + a_{42}\cdot(a_{13}a_{24} - a_{14}a_{23}) \\ \alpha_{14} &= -a_{12}\cdot(a_{23}a_{34} - a_{24}a_{33}) + a_{22}\cdot(a_{13}a_{34} - a_{14}a_{33}) - a_{32}\cdot(a_{13}a_{24} - a_{14}a_{23})\end{aligned}$$

$$\begin{aligned}\alpha_{21} &= -a_{21}\cdot(a_{33}a_{44} - a_{34}a_{43}) + a_{31}\cdot(a_{23}a_{44} - a_{24}a_{43}) - a_{41}\cdot(a_{23}a_{34} - a_{24}a_{33}) \\ \alpha_{22} &= a_{11}\cdot(a_{33}a_{44} - a_{34}a_{43}) - a_{31}\cdot(a_{13}a_{44} - a_{14}a_{43}) + a_{41}\cdot(a_{13}a_{34} - a_{14}a_{33}) \\ \alpha_{23} &= -a_{11}\cdot(a_{23}a_{44} - a_{24}a_{43}) + a_{21}\cdot(a_{13}a_{44} - a_{14}a_{43}) - a_{41}\cdot(a_{13}a_{24} - a_{14}a_{23}) \\ \alpha_{24} &= a_{11}\cdot(a_{23}a_{34} - a_{24}a_{33}) - a_{21}\cdot(a_{13}a_{34} - a_{14}a_{33}) + a_{31}\cdot(a_{13}a_{24} - a_{14}a_{23})\end{aligned}$$

$$\begin{aligned}\alpha_{31} &= a_{21}\cdot(a_{32}a_{44} - a_{34}a_{42}) - a_{31}\cdot(a_{22}a_{44} - a_{24}a_{42}) + a_{41}\cdot(a_{22}a_{34} - a_{24}a_{32}) \\ \alpha_{32} &= -a_{11}\cdot(a_{32}a_{44} - a_{34}a_{42}) + a_{31}\cdot(a_{12}a_{44} - a_{14}a_{42}) - a_{41}\cdot(a_{12}a_{34} - a_{14}a_{32}) \\ \alpha_{33} &= a_{11}\cdot(a_{22}a_{44} - a_{24}a_{42}) - a_{21}\cdot(a_{12}a_{44} - a_{14}a_{42}) + a_{41}\cdot(a_{12}a_{24} - a_{14}a_{22}) \\ \alpha_{34} &= -a_{11}\cdot(a_{22}a_{34} - a_{24}a_{32}) + a_{21}\cdot(a_{12}a_{34} - a_{14}a_{32}) - a_{31}\cdot(a_{12}a_{24} - a_{14}a_{22})\end{aligned}$$

$$\begin{aligned}\alpha_{41} &= -a_{21}\cdot(a_{32}a_{43} - a_{33}a_{42}) + a_{31}\cdot(a_{22}a_{43} - a_{23}a_{42}) - a_{41}\cdot(a_{22}a_{33} - a_{23}a_{32}) \\ \alpha_{42} &= a_{11}\cdot(a_{32}a_{43} - a_{33}a_{42}) - a_{31}\cdot(a_{12}a_{43} - a_{13}a_{42}) + a_{41}\cdot(a_{12}a_{33} - a_{13}a_{32}) \\ \alpha_{43} &= -a_{11}\cdot(a_{22}a_{43} - a_{23}a_{42}) + a_{21}\cdot(a_{12}a_{43} - a_{13}a_{42}) - a_{41}\cdot(a_{12}a_{23} - a_{13}a_{22}) \\ \alpha_{44} &= a_{11}\cdot(a_{22}a_{33} - a_{23}a_{32}) - a_{21}\cdot(a_{12}a_{33} - a_{13}a_{32}) + a_{31}\cdot(a_{12}a_{23} - a_{13}a_{22})\end{aligned}$$

5.3.2 Inversion via Gauss-Jordan elimination

The Gauss-Jordan elimination[2] is a version of the Gauss elimination[3] that produces not a upper triangular matrix but a diagonal one. It can be straightforwardly used for the inversion of square matrices.

[2] named after Carl-Friedrich Gauss (1777 – 1855) and Wilhelm Jordan (1842 – 1899)

[3] see, for instance, reference [Str88]

The basic idea is to augment the matrix with an identity matrix of proper size.

$$\mathbf{A} = \begin{pmatrix} a_{11} & a_{12} & a_{13} \\ a_{21} & a_{22} & a_{23} \\ a_{31} & a_{32} & a_{33} \end{pmatrix} \quad \Longrightarrow \quad [\mathbf{AI}] = \left(\begin{array}{ccc|ccc} a_{11} & a_{12} & a_{13} & 1 & 0 & 0 \\ a_{21} & a_{22} & a_{23} & 0 & 1 & 0 \\ a_{31} & a_{32} & a_{33} & 0 & 0 & 1 \end{array}\right)$$

This new matrix has to be modified in order to find a form

$$[\mathbf{IB}] \quad \text{with} \quad \mathbf{A}^{-1} = \mathbf{B}$$

The processing steps are as follows

1. Select the first column from the left, which has at least one non-zero element.
2. If the top element of the selected column equal to zero, then exchange the first row with another one that does not contain a zero element in this column. The desired elements on the diagonal are the *pivots* and the exchange of rows is called *pivoting*.
3. Divide the first row by the top element of the selected column.
4. Multiples of the first row must be subtracted from all remaining rows in order to produce zeros at the first position of all rows.
5. Reduce the matrix by deleting the first row and the first column.
6. Repeat steps 1 to 5 until the entire matrix is triangular.
7. Do the same procedure in order to produce a diagonal (identity) matrix.

Example:

$$\mathbf{A} = \begin{pmatrix} 0 & 1 & 2 \\ 2 & 2 & 1 \\ 2 & 1 & 0 \end{pmatrix} \quad \Longrightarrow \quad [\mathbf{AI}] = \left(\begin{array}{ccc|ccc} 0 & 1 & 2 & 1 & 0 & 0 \\ 2 & 2 & 1 & 0 & 1 & 0 \\ 2 & 1 & 0 & 0 & 0 & 1 \end{array}\right)$$

The first column has non-zero elements, but the top element is zero. Therefore, the first two rows must be exchanged (pivoting)

$$\left(\begin{array}{ccc|ccc} 2 & 2 & 1 & 0 & 1 & 0 \\ 0 & 1 & 2 & 1 & 0 & 0 \\ 2 & 1 & 0 & 0 & 0 & 1 \end{array}\right).$$

Then, we divide the first row by its most-left element

$$\left(\begin{array}{ccc|ccc} 1 & 1 & 1/2 & 0 & 1/2 & 0 \\ 0 & 1 & 2 & 1 & 0 & 0 \\ 2 & 1 & 0 & 0 & 0 & 1 \end{array}\right).$$

The second row already looks okay, but we have to subtract the first row two times from the third row

$$\left(\begin{array}{ccc|ccc} 1 & 1 & 1/2 & 0 & 1/2 & 0 \\ 0 & 1 & 2 & 1 & 0 & 0 \\ 0 & -1 & -1 & 0 & -1 & 1 \end{array}\right).$$

Subtract second row multiplied with -1 from third row

$$\left(\begin{array}{ccc|ccc} 1 & 1 & 1/2 & 0 & 1/2 & 0 \\ 0 & 1 & 2 & 1 & 0 & 0 \\ 0 & 0 & 1 & 1 & -1 & 1 \end{array}\right).$$

Subtract third row two times from second row. Subtract third row, times 1/2, from first row

$$\left(\begin{array}{ccc|ccc} 1 & 1 & 0 & -1/2 & 1 & -1/2 \\ 0 & 1 & 0 & -1 & 2 & -2 \\ 0 & 0 & 1 & 1 & -1 & 1 \end{array}\right).$$

Subtract second from first row

$$\left(\begin{array}{ccc|ccc} 1 & 0 & 0 & 1/2 & -1 & 3/2 \\ 0 & 1 & 0 & -1 & 2 & -2 \\ 0 & 0 & 1 & 1 & -1 & 1 \end{array}\right) = [\mathbf{IA}^{-1}].$$

The result is

$$\mathbf{A}^{-1} = \frac{1}{2} \cdot \begin{pmatrix} 1 & -2 & 3 \\ -2 & 4 & -4 \\ 2 & -2 & 2 \end{pmatrix} = \frac{1}{\det(\mathbf{A})} \cdot \begin{pmatrix} -1 & 2 & -3 \\ 2 & -4 & 4 \\ -2 & 2 & -2 \end{pmatrix}.$$

□□□

5.3.3 Inversion via LU decomposition

Basic Idea

The idea of the LU approach is to split a square matrix **A** into two triangular matrices **L** and **U** such that

$$\mathbf{A} = \mathbf{L} \cdot \mathbf{U} \tag{5.9}$$

in which $\mathbf{L}$ is *lower triangular* and $\mathbf{U}$ is *upper triangular*

$$\mathbf{L} = \begin{pmatrix} \lambda_{11} & 0 & \cdots & 0 \\ \lambda_{21} & \lambda_{22} & \cdots & 0 \\ \vdots & \vdots & \ddots & \vdots \\ \lambda_{M1} & \lambda_{M2} & \cdots & \lambda_{MM} \end{pmatrix} \qquad \mathbf{U} = \begin{pmatrix} \upsilon_{11} & \upsilon_{12} & \cdots & \upsilon_{1M} \\ 0 & \upsilon_{22} & \cdots & \upsilon_{2M} \\ \vdots & \vdots & \ddots & \vdots \\ 0 & 0 & \cdots & \upsilon_{MM} \end{pmatrix} \tag{5.10}$$

with typically $\lambda_{jj} = 1$.

What is the advantage of this triangular representation? Let us consider again the case of a linear equation system

$$\mathbf{A} \cdot \mathbf{x} = \mathbf{y} \,, \tag{5.11}$$

where the result for $\mathbf{x}$ has to be found via matrix inversion

$$\mathbf{x} = \mathbf{A}^{-1} \cdot \mathbf{y} \,.$$

Using the LU decomposition we write instead

$$\mathbf{A} \cdot \mathbf{x} = (\mathbf{L} \cdot \mathbf{U}) \cdot \mathbf{x} = \mathbf{L} \cdot (\mathbf{U} \cdot \mathbf{x}) = \mathbf{L} \cdot \mathbf{u} = \mathbf{y} \tag{5.12}$$

in order to find a solution for $\mathbf{u}$ first. In a second step, $\mathbf{U} \cdot \mathbf{x} = \mathbf{u}$ has to be solved. Both cases can benefit from the properties of the triangular matrices as follows. Based on

$$\mathbf{L} \cdot \mathbf{u} = \begin{pmatrix} \lambda_{11} & 0 & \cdots & 0 \\ \lambda_{21} & \lambda_{22} & \cdots & 0 \\ \vdots & \vdots & \ddots & \vdots \\ \lambda_{M1} & \lambda_{M2} & \cdots & \lambda_{MM} \end{pmatrix} \cdot \begin{pmatrix} u_1 \\ u_2 \\ \vdots \\ u_M \end{pmatrix} = \begin{pmatrix} y_1 \\ y_2 \\ \vdots \\ y_M \end{pmatrix} , \tag{5.13}$$

the first element of $\mathbf{u}$ is simply

$$u_1 = \frac{y_1}{\lambda_{11}} \,,$$

the second element is computed via

$$\lambda_{21} \cdot u_1 + \lambda_{22} \cdot u_2 = y_2 \quad \rightsquigarrow \quad u_2 = \frac{y_2 - \lambda_{21} \cdot u_1}{\lambda_{22}} \,,$$

the third one by

$$\lambda_{31} \cdot u_1 + \lambda_{32} \cdot u_2 + \lambda_{33} \cdot u_3 = y_3 \quad \rightsquigarrow \quad u_3 = \frac{y_3 - \lambda_{21} \cdot u_1 - \lambda_{32} \cdot u_2}{\lambda_{33}} \,,$$

and so on and so forth. The general formula is

$$u_i = \frac{1}{\lambda_{ii}} \cdot \left[y_i - \sum_{j=1}^{i-1} \lambda_{ij} \cdot u_j \right] \qquad i = 2, 3, \ldots, M . \tag{5.14}$$

As soon as **u** is known, equation

$$\mathbf{U} \cdot \mathbf{x} = \begin{pmatrix} v_{11} & v_{12} & \cdots & v_{1M} \\ 0 & v_{22} & \cdots & v_{2M} \\ \vdots & \vdots & \ddots & \vdots \\ 0 & 0 & \cdots & v_{MM} \end{pmatrix} \cdot \begin{pmatrix} x_1 \\ x_2 \\ \vdots \\ x_M \end{pmatrix} = \begin{pmatrix} u_1 \\ u_2 \\ \vdots \\ u_M \end{pmatrix} \tag{5.15}$$

is also solvable using the same procedure with the single difference that the last element x_M is calculated first via

$$v_{MM} \cdot x_M = u_M \quad \rightsquigarrow \quad x_M = \frac{u_M}{v_{MM}} ,$$

the next element via

$$\begin{aligned} v_{M-1,M-1} \cdot x_{M-1} + v_{M-1,M} \cdot x_M &= u_{M-1} \\ \rightsquigarrow \quad x_{M-1} &= \frac{u_{M-1} - v_{M-1,M} \cdot x_M}{v_{M-1,M-1}} , \end{aligned}$$

or in general

$$x_i = \frac{1}{v_{ii}} \cdot \left[u_i - \sum_{j=i+1}^{M} v_{ij} \cdot x_j \right] \qquad i = 2, 3, \ldots, M .$$

In principle, the inverse of **A** is not required to solve (5.11), since **x** results from equations (5.13) and (5.15). However, given the LU decomposition of **A**, its inverse can be computed column by column. Using a vector **y** with all elements equal to zero apart from $y_i = 1$, the resulting vector **x** contains the ith column of $\mathbf{A}^{-1}$. This is explained below using a concrete example.

The decomposition of **A** into **L** and **U** is performed via the Gaussian elimination method (see [Str88], for instance) or Crout's algorithm (see [Pre92]).

Example for LU decomposition

The LU decomposition of $\mathbf{A}$ is computed in an iterative fashion. Let us define $\mathbf{A}^{(0)} = \mathbf{A}$ with

$$\mathbf{A} = \begin{pmatrix} 1 & 5 & 3 & 2 \\ 4 & 2 & 3 & 2 \\ 2 & 3 & 4 & 1 \\ 3 & 0 & 1 & 2 \end{pmatrix} .$$

The matrix elements below the main diagonal in the n-th column ($n = 1, 2, \ldots, N-1$) of $\mathbf{A}^{(n-1)}$ are eliminated by adding the n-th row multiplied by

$$b_{i,n} = -\frac{a_{i,n}^{(n-1)}}{a_{n,n}^{(n-1)}} \qquad a_{n,n}^{(n-1)} \neq 0$$

to the i-th row of this matrix for $i = n+1, 2, \ldots, N$. Please note that $(n-1)$ is not an exponent but indicates the matrix $\mathbf{A}^{(n-1)}$ at the iteration number $n-1$. In case that the element $a_{n,n}^{(n-1)}$ is zero, the n-th row must be interchanged with another row below before continuing. This corresponds to the procedure described in Subsection 5.3.2, steps 1.– 6.

For $n = 1$, the factors are

$$b_{2,1} = -\frac{a_{2,1}^{(0)}}{a_{1,1}^{(0)}} = -\frac{4}{1} \qquad b_{3,1} = -\frac{a_{3,1}^{(0)}}{a_{1,1}^{(0)}} = -\frac{2}{1} \qquad b_{4,1} = -\frac{a_{4,1}^{(0)}}{a_{1,1}^{(0)}} = -\frac{3}{1}$$

and the operations result in matrix

$$\mathbf{A}^{(1)} = \begin{pmatrix} 1 & 5 & 3 & 2 \\ 0 & -18 & -9 & -6 \\ 0 & -7 & -2 & -3 \\ 0 & -15 & -8 & -4 \end{pmatrix} .$$

This operation is equal to

$$\mathbf{A}^{(n)} = \mathbf{B}_n \cdot \mathbf{A}^{(n-1)}$$

with

$$\mathbf{B}_n = \begin{pmatrix} 1 & \cdots & 0 & \cdots & 0 \\ & \ddots & 0 & & 0 \\ & 0 & 1 & & \vdots \\ & 0 & b_{n+1,n} & \ddots & 0 \\ & 0 & \vdots & 0 \quad \ddots & \vdots \\ & 0 & b_{N,n} & 0 & 1 \end{pmatrix} .$$

Using the example matrix, we have

$$\mathbf{A}^{(1)} = \mathbf{B}_1 \cdot \mathbf{A}^{(0)} = \begin{pmatrix} 1 & 0 & 0 & 0 \\ -4 & 1 & 0 & 0 \\ -2 & 0 & 1 & 0 \\ -3 & 0 & 0 & 1 \end{pmatrix} \cdot \begin{pmatrix} 1 & 5 & 3 & 2 \\ 4 & 2 & 3 & 2 \\ 2 & 3 & 4 & 1 \\ 3 & 0 & 1 & 2 \end{pmatrix}$$

$$= \begin{pmatrix} 1 & 5 & 3 & 2 \\ 0 & -18 & -9 & -6 \\ 0 & -7 & -2 & -3 \\ 0 & -15 & -8 & -4 \end{pmatrix}.$$

The next iteration $n = 2$ yields

$$b_{3,2} = -\frac{a_{3,2}^{(1)}}{a_{2,2}^{(1)}} = -\frac{-7}{-18} = -\frac{7}{18} \qquad b_{4,2} = -\frac{a_{4,2}^{(1)}}{a_{2,2}^{(1)}} = -\frac{-15}{-18} = -\frac{15}{18}$$

and

$$\mathbf{A}^{(2)} = \mathbf{B}_2 \cdot \mathbf{A}^{(1)} = \begin{pmatrix} 1 & 0 & 0 & 0 \\ 0 & 1 & 0 & 0 \\ 0 & -7/18 & 1 & 0 \\ 0 & -15/18 & 0 & 1 \end{pmatrix} \cdot \begin{pmatrix} 1 & 5 & 3 & 2 \\ 0 & -18 & -9 & -6 \\ 0 & -7 & -2 & -3 \\ 0 & -15 & -8 & -4 \end{pmatrix}$$

$$= \begin{pmatrix} 1 & 5 & 3 & 2 \\ 0 & -18 & -9 & -6 \\ 0 & 0 & 3/2 & -2/3 \\ 0 & 0 & -1/2 & 1 \end{pmatrix}.$$

The third and last iteration reads as

$$b_{4,3} = -\frac{a_{4,3}^{(2)}}{a_{3,3}^{(2)}} = -\frac{-1/2}{3/2} = \frac{1}{3}$$

and

$$\mathbf{A}^{(3)} = \mathbf{B}_3 \cdot \mathbf{A}^{(2)} = \begin{pmatrix} 1 & 0 & 0 & 0 \\ 0 & 1 & 0 & 0 \\ 0 & 0 & 1 & 0 \\ 0 & 0 & 1/3 & 1 \end{pmatrix} \cdot \begin{pmatrix} 1 & 5 & 3 & 2 \\ 0 & -18 & -9 & -6 \\ 0 & 0 & 3/2 & -2/3 \\ 0 & 0 & -1/2 & 1 \end{pmatrix}$$

$$= \underline{\underline{\begin{pmatrix} 1 & 5 & 3 & 2 \\ 0 & -18 & -9 & -6 \\ 0 & 0 & 3/2 & -2/3 \\ 0 & 0 & 0 & 7/9 \end{pmatrix}}}.$$

The final upper triangular matrix $\mathbf{A}^{(N-1)}$ is equal to $\mathbf{U}$ from equation (5.9).

The lower triangular matrix $\mathbf{L}$ can be determined based on the given matrix $\mathbf{A}$ and the corresponding upper triangular matrix $\mathbf{U}$ via

$$\mathbf{A} = \mathbf{L} \cdot \mathbf{U}$$

$$\begin{pmatrix} 1 & 5 & 3 & 2 \\ 4 & 2 & 3 & 2 \\ 2 & 3 & 4 & 1 \\ 3 & 0 & 1 & 2 \end{pmatrix} = \begin{pmatrix} \lambda_{11} & 0 & 0 & 0 \\ \lambda_{21} & \lambda_{22} & 0 & 0 \\ \lambda_{31} & \lambda_{32} & \lambda_{33} & 0 \\ \lambda_{41} & \lambda_{42} & \lambda_{43} & \lambda_{44} \end{pmatrix} \cdot \begin{pmatrix} 1 & 5 & 3 & 2 \\ 0 & -18 & -9 & -6 \\ 0 & 0 & 3/2 & -2/3 \\ 0 & 0 & 0 & 7/9 \end{pmatrix}$$

$$\begin{aligned} a_{i1} =& \lambda_{i1} \cdot v_{11} \quad \rightsquigarrow \quad \lambda_{i1} = \frac{a_{i1}}{v_{11}} \\ a_{i2} =& \lambda_{i1} \cdot v_{12} + \lambda_{i2} \cdot v_{22} \quad \rightsquigarrow \quad \lambda_{i2} = \frac{a_{i2} - \lambda_{i1} \cdot v_{12}}{v_{22}} \\ a_{i3} =& \lambda_{i1} \cdot v_{13} + \lambda_{i2} \cdot v_{23} + \lambda_{i3} \cdot v_{33} \\ & \rightsquigarrow \quad \lambda_{i3} = \frac{a_{i3} - (\lambda_{i1} \cdot v_{13} + \lambda_{i2} \cdot v_{23})}{v_{33}} \end{aligned}$$

or in general

$$\lambda_{ij} = \frac{1}{v_{jj}} \cdot \left[a_{ij} - \sum_{k=1}^{j-1} \lambda_{ik} \cdot v_{k,j} \right] \qquad i \geq j \tag{5.16}$$

leading to

$$\mathbf{L} = \begin{pmatrix} \lambda_{11} & 0 & 0 & 0 \\ \lambda_{21} & \lambda_{22} & 0 & 0 \\ \lambda_{31} & \lambda_{32} & \lambda_{33} & 0 \\ \lambda_{41} & \lambda_{42} & \lambda_{43} & \lambda_{44} \end{pmatrix} = \underline{\underline{\begin{pmatrix} 1 & 0 & 0 & 0 \\ 4 & 1 & 0 & 0 \\ 2 & 7/18 & 1 & 0 \\ 3 & 5/6 & -1/3 & 1 \end{pmatrix}}}.$$

The LU decomposition of $\mathbf{A}$ is finally

$$\begin{aligned} \mathbf{A} &= \begin{pmatrix} 1 & 5 & 3 & 2 \\ 4 & 2 & 3 & 2 \\ 2 & 3 & 4 & 1 \\ 3 & 0 & 1 & 2 \end{pmatrix} \\ &= \begin{pmatrix} 1 & 0 & 0 & 0 \\ 4 & 1 & 0 & 0 \\ 2 & 7/18 & 1 & 0 \\ 3 & 5/6 & -1/3 & 1 \end{pmatrix} \cdot \begin{pmatrix} 1 & 5 & 3 & 2 \\ 0 & -18 & -9 & -6 \\ 0 & 0 & 3/2 & -2/3 \\ 0 & 0 & 0 & 7/9 \end{pmatrix}. \end{aligned}$$

Example for matrix inversion via LU decomposition

Based on the LU decomposition example from previous Subsection

$$\mathbf{A} = \mathbf{L} \cdot \mathbf{U}$$

$$\begin{pmatrix} 1 & 5 & 3 & 2 \\ 4 & 2 & 3 & 2 \\ 2 & 3 & 4 & 1 \\ 3 & 0 & 1 & 2 \end{pmatrix} = \begin{pmatrix} 1 & 0 & 0 & 0 \\ 4 & 1 & 0 & 0 \\ 2 & 7/18 & 1 & 0 \\ 3 & 5/6 & -1/3 & 1 \end{pmatrix} \cdot \begin{pmatrix} 1 & 5 & 3 & 2 \\ 0 & -18 & -9 & -6 \\ 0 & 0 & 3/2 & -2/3 \\ 0 & 0 & 0 & 7/9 \end{pmatrix},$$

the inverse matrix $\mathbf{A}^{-1}$ has to be determined via equations (5.12), (5.13) and (5.15).

The determination i-th column of $\mathbf{A}^{-1}$ is based on a vector $\mathbf{y}$ with all elements equal to zero apart from $y_i = 1$. Here, only the first column of $\mathbf{A}^{-1}$ shall be computed.

Based on

$$\mathbf{L} \cdot \mathbf{u} = \mathbf{y} \tag{5.17}$$

$$\begin{pmatrix} 1 & 0 & 0 & 0 \\ 4 & 1 & 0 & 0 \\ 2 & 7/18 & 1 & 0 \\ 3 & 5/6 & -1/3 & 1 \end{pmatrix} \cdot \begin{pmatrix} u_1 \\ u_2 \\ u_3 \\ u_4 \end{pmatrix} = \begin{pmatrix} 1 \\ 0 \\ 0 \\ 0 \end{pmatrix}, \tag{5.18}$$

the vector $\mathbf{u}$ has firstly to be determined via eq. (5.14) leading to

$$\mathbf{u} = \begin{pmatrix} 1 & -4 & -4/9 & 5/27 \end{pmatrix}^{\mathrm{T}} .$$

Now $\mathbf{x}$, i.e. the first column of $\mathbf{A}^{-1}$, can be computed

$$\mathbf{U} \cdot \mathbf{x} = \mathbf{u}$$

$$\begin{pmatrix} 1 & 5 & 3 & 2 \\ 0 & -18 & -9 & -6 \\ 0 & 0 & 3/2 & -2/3 \\ 0 & 0 & 0 & 7/9 \end{pmatrix} \cdot \begin{pmatrix} x_1 \\ x_2 \\ x_3 \\ x_4 \end{pmatrix} = \begin{pmatrix} 1 \\ -4 \\ -4/9 \\ 5/27 \end{pmatrix},$$

leading to

$$\mathbf{x} = (a_{i1}) = \begin{pmatrix} -2/21 & 5/21 & -4/21 & 5/21 \end{pmatrix}^{\mathrm{T}} .$$

All other columns of $\mathbf{A}^{-1}$ are derived in the same manner, however, with shifted positions of the value of 1 in $\mathbf{y}$ of equation (5.18)

$$\mathbf{A}^{-1} = \frac{1}{21} \cdot \begin{pmatrix} -2 & 23 & -12 & -15 \\ 5 & 16 & -12 & -15 \\ -4 & -17 & 18 & 12 \\ 5 & -26 & 9 & 27 \end{pmatrix} .$$

Determinant computation via LU decomposition

The determinant of matrix $\mathbf{A}$ can be easily computed using the LU decomposition of $\mathbf{A}$. In general, if $\mathbf{A} = \mathbf{B} \cdot \mathbf{C}$ then $\det(\mathbf{A}) = \det(\mathbf{B}) \cdot \det(\mathbf{C})$. As the determinant of a triangular matrix is simply the product of its elements on the main diagonal,

$$\det(\mathbf{A}) = \det(\mathbf{L}) \cdot \det(\mathbf{U}) = 1 \cdot \det(\mathbf{U}) = \prod_{i=1}^{M} \upsilon_{ii}$$

holds with υ_{ii} according to equation (5.10).

5.3.4 Inversion by singular value decomposition (SVD)

The method of singular value decomposition enables the direct inversion even of non-square matrices. Each rectangular $N \times M$ matrix can be decomposed into a $N \times M$ matrix $\mathbf{U}$, a diagonal matrix $\mathbf{S}$ containing the so-called singular values, and a $M \times M$ matrix $\mathbf{V}$

$$\Bigg(\mathbf{A} \Bigg) = \Bigg(\mathbf{U} \Bigg) \cdot \Big(\mathbf{S} \Big) \cdot \Big(\mathbf{V}^{\mathrm{T}} \Big) \tag{5.19}$$

with

$$\mathbf{S} = \begin{pmatrix} s_1 & & & \\ & s_2 & & \\ & & \ddots & \\ & & & s_M \end{pmatrix} . \tag{5.20}$$

A value of $s_j = 0$ indicates a singular matrix $\mathbf{A}$, while one or more very small values of s_j signal indicate a nearly singular matrix.

The matrix **U** is orthogonal in its columns, i.e.,

$$\sum_{i=1}^{N} u_{i,j} \cdot u_{i,k} = \begin{cases} 0 & \text{for} \quad j \neq k \\ C & \text{for} \quad j = k \end{cases} \qquad j,k = 1,\ldots,M \;. \tag{5.21}$$

such that $\mathbf{U}^{\mathrm{T}} \cdot \mathbf{U} = \mathbf{I}$ holds. **V** is a square matrix with orthogonality in both, columns and rows; its inverse is identical to its transpose $\mathbf{V}^{\mathrm{T}} \cdot \mathbf{V} = \mathbf{I} = \mathbf{V} \cdot \mathbf{V}^{\mathrm{T}}$.

Using these properties, the solution of the equation system is straightforward

$$\begin{aligned} \mathbf{A} \cdot \mathbf{x} &= \mathbf{y} \\ \mathbf{U} \cdot \mathbf{S} \cdot \mathbf{V}^{\mathrm{T}} \cdot \mathbf{x} &= \mathbf{y} \\ \mathbf{S} \cdot \mathbf{V}^{\mathrm{T}} \cdot \mathbf{x} &= \mathbf{U}^{\mathrm{T}} \cdot \mathbf{y} \\ \mathbf{V}^{\mathrm{T}} \cdot \mathbf{x} &= \mathbf{S}^{-1} \cdot \mathbf{U}^{\mathrm{T}} \cdot \mathbf{y} \\ \mathbf{x} &= \mathbf{V} \cdot \mathbf{S}^{-1} \cdot \mathbf{U}^{\mathrm{T}} \cdot \mathbf{y} \;. \end{aligned}$$

The inverse of a diagonal matrix is another diagonal with its reciprocal values as elements

$$\mathbf{S}^{-1} = \begin{pmatrix} 1/s_1 & & & \\ & 1/s_2 & & \\ & & \ddots & \\ & & & 1/s_M \end{pmatrix} .$$

Finally we get

$$\mathbf{A}^{-1} = \left(\mathbf{U} \cdot \mathbf{S} \cdot \mathbf{V}^{\mathrm{T}}\right)^{-1} = \mathbf{V} \cdot \mathbf{S}^{-1} \cdot \mathbf{U}^{\mathrm{T}} \;, \tag{5.22}$$

that is,

$$\left(\quad \mathbf{A}^{-1} \quad \right) = \left(\;\mathbf{V}\; \right) \cdot \left(\;\mathbf{S}^{-1}\; \right) \cdot \left(\quad \mathbf{U}^{\mathrm{T}} \quad \right) \tag{5.23}$$

as the inverse of an arbitrarily shaped matrix **A**.

The singular value decomposition is described for instance in [Bjo96, Law95].

5.4 Test questions

5.1 To which row of a matrix does the element a_{43} belong?

5.2 What are the properties of a diagonal matrix?

5.3 The elements of matrix **B** have following property: $b_{ij} = b_{ji}$. To which special group of matrices does **B** belong?

5.4 What is a column vector?

5.5 How is the matrix **I** called?

5.6 Which of the following statements is true: $\mathbf{C} \cdot \mathbf{D} = \mathbf{D} \cdot \mathbf{C}$ or $\mathbf{C} \cdot \mathbf{B} \neq \mathbf{B} \cdot \mathbf{C}$?

5.7 If **D** and **E** have to be multiplied, which requirement has to be fulfilled?

5.8 How can the singularity of a matrix be tested?

5.9 What is the result of $\mathbf{DD}^{-1}$?

5.10 How can the determinant of a 3×3 matrix easily be computed?

5.11 Explain the matrix inversion using the cofactor-matrix method.

5.12 What is the principle idea of matrix inversion via LU decomposition?

5.13 Compute the inverse of

$$\mathbf{A} = \begin{pmatrix} 2 & 1 & 1 & 1 \\ 1 & 4 & 1 & 3 \\ 2 & 2 & 1 & 3 \\ 3 & 12 & 4 & 2 \end{pmatrix}$$

via

- Gauss-Jordan elimination
- LU decomposition.

5.14 How is the singular value decomposition defined?

Chapter 6

The Idea behind Least Squares

6.1 Normal distribution

The *normal distribution* or *Gaussian distribution* of continuous values plays a fundamental role for the principle of least-squares approximation, as will be shown in the next section. Many random numbers are distributed normally in real life, i.e., they follow a probability density function given by

$$f(x) = \frac{1}{\sigma\sqrt{2\pi}} \cdot \exp\left\{-\frac{1}{2} \cdot \left[\frac{x-\mu}{\sigma}\right]^2\right\} \qquad -\infty < x < \infty\,. \tag{6.1}$$

with

$$\int_{-\infty}^{\infty} f(x)\,dx = 1\,. \tag{6.2}$$

Figure 6.1 depicts the graph of the normal distribution. Because of its characteristic shape, it is also called a *Gaussian bell curve.* The distribution is determined by two parameters, namely the mean (or expectation) value μ and the standard deviation σ of the variable x. The notation is typically $x \sim \mathcal{N}(\mu;\sigma^2)$. The mean value defines the position of the curve's peak; the parameter σ influences the width of the distribution, as well as the maximum of the distribution, as can be seen in **Figure 6.2**. The area below the graph of the Gaussian bell is equal to one, due to the normalisation factor $1/(\sigma\sqrt{2\pi})$ in eq. (6.1). 68.26 percent of the area lies within the range of $\mu-\sigma < x < \mu+\sigma$, i.e., approximately 68 out of 100 random numbers drawn from a Gaussian distribution deviate not more than the standard deviation. 95.44% of all random values are located within the limits of $\pm 2\sigma$. Only 0.27% of all normal distributed values are outside the range of $\mu - 3\sigma < x < \mu + 3\sigma$. That means values of x deviating more than 3σ from the mean μ are very unlikely, but of course not impossible.

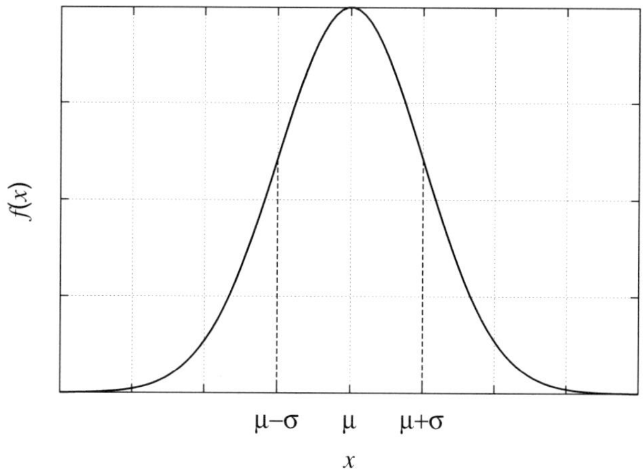

Figure 6.1: Probability density function of the normally distributed variable x

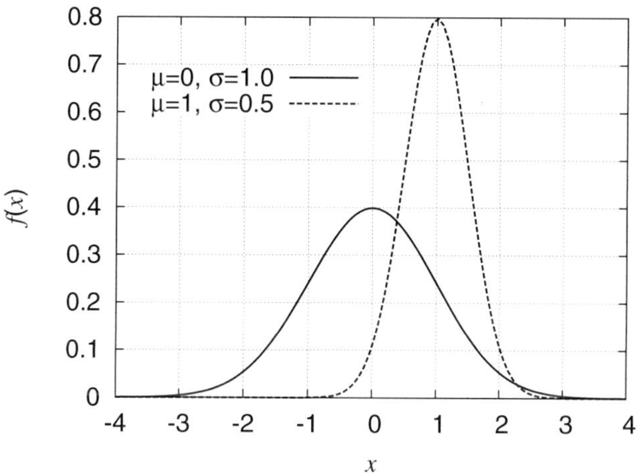

Figure 6.2: Normal probability density function in dependence on its parameters μ and σ

6.2 Maximum likelihood principle

The method of least squares can be considered as a special case of the more general *method of maximum likelihood* ([Bev92, Gel68, Tay97]).

Given an experiment with the functional connection

$$y = f(\mathbf{x}|\mathbf{a}) \ ,$$

for a certain vector of conditions $\mathbf{x}_i$ and a model-parameter vector $\mathbf{a}$, the experimental outcome might be y_i. It is assumed that the individual observation y_i is drawn from a Gaussian distribution (see previous section) with mean equal to $f(\mathbf{x}_i|\mathbf{a})$ and a standard deviation σ_i, i.e., the errors $(f(\mathbf{x}_i|\mathbf{a}) - y_i)$ are expected to be normally distributed. Instead of asking about the probability that a particular vector of parameters $\mathbf{a}$ is correct, we have to consider the probability that the observation y_i could have occurred for the given set of parameters. Thus, according to eq.(6.1), the probability P_i for making the single measurement y_i is

$$P_i = P_{y_i|\mathbf{a}} = \frac{1}{\sigma_i\sqrt{2\pi}} \cdot \exp\left\{-\frac{1}{2} \cdot \left[\frac{y_i - f(\mathbf{x}_i|\mathbf{a})}{\sigma_i}\right]^2\right\} \ . \tag{6.3}$$

Performing the experiment N times and assuming that the single observations are not influenced by others, i.e., the data can be assumed to be uncorrelated, the measurement probability dependent on chosen model parameters $\mathbf{a}$ is the product of the probabilities of each single observation

$$\begin{aligned} P|_{\mathbf{a}} &= \prod_{i=1}^{N} P_i = \prod_{i=1}^{N} \left(\frac{1}{\sigma_i\sqrt{2\pi}}\right) \cdot \exp\left\{-\frac{1}{2} \cdot \sum_{i=1}^{N} \left[\frac{y_i - f(\mathbf{x}_i|\mathbf{a})}{\sigma_i}\right]^2\right\} \\ &= \prod_{i=1}^{N} \left(\frac{1}{\sigma_i\sqrt{2\pi}}\right) \cdot \exp\left\{-\chi^2\left(\mathbf{a}\right)\right\} \ . \end{aligned}$$

The product of the exponentials is expressed by the sum of the arguments. Since $P|_{\mathbf{a}}$ has to be maximised, the best choice of the parameter vector $\mathbf{a}$ is given when $\chi^2(\mathbf{a})$ is minimised

$$\chi^2\left(\mathbf{a}\right) = \frac{1}{2} \cdot \sum_{i=1}^{N} \left[\frac{y_i - f(\mathbf{x}_i|\mathbf{a})}{\sigma_i}\right]^2 \ . \tag{6.4}$$

In contrast to equation (2.2) on page 26, an additional factor of 1/2 is used for convenience in the following. It has no influence on the fitting procedure. The factors $1/\sigma_i^2$ act as weights in order to adjust the influence of single observations according to their importance.

This derivation is based on the initial assumption that the measurement errors follow a Gaussian (normal) distribution. "Taken too literally, this requirement would practically eliminate least squares as a method of computation; for in nearly every set of measurements there is presumably a certain amount of systematic or constant error" [Bir32] or the random error follows divergent (non-Gaussian) distributions. Nevertheless, the least-squares approach is a handy tool that yields sufficient model parameters in the majority of cases, especially if proper weighting and outlier detection is used as described in Chapter 3.

In applications with more than one observation per measurement, a vector $\mathbf{y}$ must be used instead of a scalar y and the probability in equation (6.3) has to be calculated based on a multidimensional Gaussian function accordingly.

6.3 Fitting of linear model functions

The principle formula for solving linear data fitting problems in data fitting was given in Section 2.4 (eqs. 2.10 and 2.11) with

$$\mathbf{a} = \left(\mathbf{J}^{\mathrm{T}} \cdot \mathbf{W} \cdot \mathbf{J}\right)^{-1} \cdot \mathbf{J}^{\mathrm{T}} \cdot \mathbf{W} \cdot \mathbf{y}\,, \tag{6.5}$$

i.e., the optimal parameter vector $\mathbf{a}$ depends on the Jacobian matrix $\mathbf{J}$ (see pages 29ff), the matrix of weights $\mathbf{W}$, and the vector of observations $\mathbf{y}$.

It had become apparent at closer inspection that the problem, ignoring the weights, can be expressed simply by the multiplication of the Jacobian matrix with the vector of parameters resulting in the set of observations $\mathbf{y}$ (eq. (2.12))

$$\mathbf{y} = f(\mathbf{x}|\mathbf{a}) = \mathbf{J} \cdot \mathbf{a}\,.$$

The solution for $\mathbf{a}$ is obviously

$$\mathbf{a} = \mathbf{J}^{-1} \cdot \mathbf{y}\,.$$

The following subsections deal with the inversion of the Jacobian matrix $\mathbf{J}$.

6.3.1 Standard approach

Since the Jacobian matrix $\mathbf{J}$ is not a square matrix, $\mathbf{J}^{-1}$ is mostly written as $\mathbf{J}^{+}$ and called *pseudoinverse*, *generalised inverse*, or Moore-Penrose inverse [Wei98].

Where equation (6.5) does comes from? In order to make the matrices square, both sides of the equation are multiplied by the transpose of $\mathbf{J}$

$$\begin{aligned}\mathbf{J} \cdot \mathbf{a} &= \mathbf{y} \\ \mathbf{J}^{\mathrm{T}} \cdot \mathbf{J} \cdot \mathbf{a} &= \mathbf{J}^{\mathrm{T}} \cdot \mathbf{y}\,.\end{aligned} \tag{6.6}$$

On the left side of the equation, the term $\mathbf{J}^{\mathrm{T}} \cdot \mathbf{J}$ has to be removed in order to get a result for $\mathbf{a}$. This is performed through multiplication with the corresponding inverse expression yielding the formula for solving linear data fitting by ordinary least squares

$$\mathbf{a} = (\mathbf{J}^{\mathrm{T}} \cdot \mathbf{J})^{-1} \cdot \mathbf{J}^{\mathrm{T}} \cdot \mathbf{y} \,. \tag{6.7}$$

Since $\mathbf{J}^{\mathrm{T}} \cdot \mathbf{J}$ is a square matrix, its inversion is straightforward (see Section 5.3). The pseudo-inverse of $\mathbf{J}$ is obviously equal to $(\mathbf{J}^{\mathrm{T}} \cdot \mathbf{J})^{-1} \cdot \mathbf{J}^{\mathrm{T}}$.

Now, the weights w_i, i.e. the standard uncertainties of the single observations, have to be incorporated. According to the maximum-likelihood principle (eq. 6.4), both, $f(\mathbf{x}_i|\mathbf{a})$ and y_i, are divided by the standard deviation σ_i . Since $w_i = 1/\sigma_i^2$, (eq. 2.5 on page 48), equation (6.6) changes to

$$\mathbf{J}' \cdot \mathbf{a} = \mathbf{y}' \tag{6.8}$$

with

$$\mathbf{J}' = \mathbf{\Sigma} \cdot \mathbf{J} \qquad \mathbf{y}' = \mathbf{\Sigma} \cdot \mathbf{y}$$

and

$$\mathbf{\Sigma} = \begin{pmatrix} \sqrt{w_1} & 0 & 0 & \cdots & 0 \\ 0 & \sqrt{w_2} & 0 & \cdots & 0 \\ 0 & 0 & \ddots & \cdots & \vdots \\ \vdots & \vdots & \vdots & \ddots & 0 \\ 0 & 0 & \cdots & 0 & \sqrt{w_N} \end{pmatrix} = \begin{pmatrix} 1/\sigma_1 & 0 & 0 & \cdots & 0 \\ 0 & 1/\sigma_2 & 0 & \cdots & 0 \\ 0 & 0 & \ddots & \cdots & \vdots \\ \vdots & \vdots & \vdots & \ddots & 0 \\ 0 & 0 & \cdots & 0 & 1/\sigma_N \end{pmatrix} .$$

$\mathbf{J}'$ is also called a *design matrix*. Substituting $\mathbf{J}$ with $\mathbf{J}'$ and $\mathbf{y}$ with $\mathbf{y}'$ in equation (6.7), the parameter vector is computed as

$$\begin{aligned} \mathbf{a} &= (\mathbf{J}'^{\mathrm{T}} \cdot \mathbf{J}')^{-1} \cdot (\mathbf{J}')^{\mathrm{T}} \cdot \mathbf{y}' \\ &= \left((\mathbf{\Sigma} \cdot \mathbf{J})^{\mathrm{T}} \cdot \mathbf{\Sigma} \cdot \mathbf{J}\right)^{-1} \cdot (\mathbf{\Sigma} \cdot \mathbf{J})^{\mathrm{T}} \cdot \mathbf{\Sigma} \cdot \mathbf{y} \\ &= \left(\mathbf{J}^{\mathrm{T}} \cdot \mathbf{\Sigma}^{\mathrm{T}} \cdot \mathbf{\Sigma} \cdot \mathbf{J}\right)^{-1} \cdot \mathbf{J}^{\mathrm{T}} \cdot \mathbf{\Sigma}^{\mathrm{T}} \cdot \mathbf{\Sigma} \cdot \mathbf{y} \\ &= \left(\mathbf{J}^{\mathrm{T}} \cdot \mathbf{W} \cdot \mathbf{J}\right)^{-1} \cdot \mathbf{J}^{\mathrm{T}} \cdot \mathbf{W} \cdot \mathbf{y} \,, \end{aligned} \tag{6.9}$$

which is identical to formula (6.5).

6.3.2 Solution using singular value decomposition (SVD)

Using the technique of *singular value decomposition* (see Subsection 5.3.4) non-squared matrices can also be inverted directly. Therefore, we can write, based on eq. (6.8)

$$\begin{aligned}\mathbf{a} &= \left(\mathbf{J}'\right)^{-1} \cdot \mathbf{y}' \\ &= (\mathbf{\Sigma} \cdot \mathbf{J})^{-1} \cdot \mathbf{\Sigma} \cdot \mathbf{y} \,.\end{aligned}$$

The term $\mathbf{\Sigma} \cdot \mathbf{J}$ is decomposed into $\mathbf{U} \cdot \mathbf{S} \cdot \mathbf{V}^{\mathrm{T}}$ (compare eq. (5.19))

$$\mathbf{a} = \left(\mathbf{U} \cdot \mathbf{S} \cdot \mathbf{V}^{\mathrm{T}}\right)^{-1} \cdot \mathbf{\Sigma} \cdot \mathbf{y} \,. \tag{6.10}$$

Since $\left(\mathbf{U} \cdot \mathbf{S} \cdot \mathbf{V}^{\mathrm{T}}\right)^{-1} = \mathbf{V} \cdot \mathbf{S}^{-1} \cdot \mathbf{U}^{\mathrm{T}}$ (see equation 5.22 on page 154), the weighted linear fitting problem can be solved now via

$$\mathbf{a} = \mathbf{V} \cdot \begin{pmatrix} 1/s_1 & & & \\ & 1/s_2 & & \\ & & \ddots & \\ & & & 1/s_M \end{pmatrix} \cdot \mathbf{U}^{\mathrm{T}} \cdot \mathbf{\Sigma} \cdot \begin{pmatrix} y_1 \\ y_2 \\ \vdots \\ y_N \end{pmatrix} \,. \tag{6.11}$$

As the reciprocal singular values are required, the values of s_j have to be checked in advance. If s_j is very close to zero, then matrix $\mathbf{\Sigma} \cdot \mathbf{J}$ is nearly singular and its inversion becomes obviously critical. The value $1/s_j$ should be set to zero.

[Pre92] recommends SVD as the method of choice for solving most linear least-squares tasks, since it can deal with problematic, e.g. singular or nearly singular matrices.

Compared to the solution via equation (6.9), where the normal matrix $\mathbf{J}^{\mathrm{T}} \cdot \mathbf{W} \cdot \mathbf{J}$ must be inverted, equation (6.10) merely requires the decomposition of $\mathbf{\Sigma} \cdot \mathbf{J}$. This increases the numerical stability. Least-squares approximation based on polynomials of high order, for example, can lead to ill-conditioned matrices, which cannot be processed accurately with (6.9). Subsection B.7 provides some examples.

Uncertainty of parameters

One minor disadvantage in using SVD is the missing covariance matrix $\mathbf{C} = (\mathbf{J}^{\mathrm{T}} \cdot \mathbf{W} \cdot \mathbf{J})^{-1}$ of the model parameters (eq. 4.4 on page 117). Thus, the uncertainties of the estimated parameters have to be determined differently from what was shown in Section 4.2. Instead, the properties of **V** are exploited. The covariance matrix is computed based on the matrices **V** and **S**

$$\begin{aligned}\mathbf{C} &= \mathbf{V}\mathbf{S}^{-1} \cdot \left(\mathbf{V}\mathbf{S}^{-1}\right)^{\mathrm{T}} = \mathbf{V}\mathbf{S}^{-1} \cdot \left(\mathbf{S}^{-1}\right)^{\mathrm{T}} \mathbf{V}^{\mathrm{T}} \\ &= \mathbf{V}\mathbf{S}^{-2}\mathbf{V}^{\mathrm{T}}\end{aligned} \tag{6.12}$$

and the variances of the estimated parameters a_j can be derived via

$$\sigma_{a_j}^2 = g_{\text{fit}} \cdot c_{jj}$$

as in the standard approach.

6.3.3 Scaling of conditions

The numerical stability of the matrix inversion in equations (6.7) and (6.10) depends, among other things, on the dynamic range of the matrix elements. If the matrix contains very small and very large elements, the CPU precision might not be sufficient to allow a correct inversion. Polynomial functions like

$$y_i = f(\mathbf{x}_i|\mathbf{a}) = a_1 + \sum_{j=2}^{M} a_j \cdot x_i^{j-1} + \varepsilon_i \qquad i = 1, 2, \ldots, N \tag{6.13}$$

are typical examples in which the elements of the Jacobian matrix

$$\mathbf{J} = \begin{pmatrix} 1 & x_1 & x_1^2 & \cdots & x_1^{M-1} \\ 1 & x_2 & x_2^2 & \cdots & x_2^{M-1} \\ \vdots & \vdots & \vdots & \cdots & \vdots \\ 1 & x_N & x_N^2 & \cdots & x_N^{M-1} \end{pmatrix},$$

show a high dynamic range, dependent on the order $M-1$ of the polynomial. The range becomes even larger when computing the normal matrix $\mathbf{J}^{\mathrm{T}} \cdot \mathbf{J}$. If, for example, $x_1 = 10$, then the entries of the first row of **J** are $(1, 10, 100, 1000, 10000, \ldots)$.

In order to avoid this exponential increase of values, it would be desired to keep the conditions in the range $[-1, 1]$. Let us assume that $\tilde{x} = \max_i(|x_i|)$ is the maximum absolute condition value. Then we can re-formulate $x_i = x_i' \cdot \tilde{x}$ and equation (6.13) becomes

$$\begin{aligned}
y_i = f(\mathbf{x}_i|\mathbf{a}) &= a_1 + \sum_{j=2}^{M} a_j \cdot (x'_i \cdot \tilde{x})^{j-1} + \varepsilon_i \\
= f(\mathbf{x}_i|\mathbf{a}) &= a_1 + \sum_{j=2}^{M} a_j \cdot \tilde{x}^{j-1} \cdot (x'_i)^{j-1} + \varepsilon_i \\
= f(\mathbf{x}_i|\mathbf{a}) &= b_1 + \sum_{j=2}^{M} b_j \cdot (x'_i)^{j-1} + \varepsilon_i \quad \text{with} \quad a_j = \frac{b_j}{\tilde{x}^{j-1}},\ x'_i = \frac{x_i}{\tilde{x}} \ .
\end{aligned}$$

The values x'_i are within the range of $[-1, 1]$ and the matrix inversion should be more stable. The true model parameters a_j must be derived from the intermediate values b_j.

Other model functions can be treated in the same manner, as long as each x is multiplicatively combined with one model parameter. Subsection B.7.1 demonstrates the effectiveness of this normalisation technique.

6.4 Fitting of nonlinear model functions

6.4.1 Error-surface approximation

As already described in Chapter 2, the fitting of models that are nonlinear for at least one model parameter has to be performed iteratively. The set of M parameters in $\mathbf{a}$ spans an M-dimensional space. Each point in this space is characterised by a corresponding value of $\chi^2(\mathbf{a})$ according to equation (6.4). The entity of all points in that space is called the *hyper-surface* (see also Section 2.1). Starting from an initial position $\mathbf{a}$, the fitting algorithm is expected to walk towards the global minimum of $\chi^2(\mathbf{a})$.

In order to determine the right direction, the principle of expansion by Taylor series [Gel68] is used. Any function $f(x)$ that is defined in the vicinity of $f(x_0)$ and has infinite continuous derivatives $f^{(\nu)}(x)$ can be expanded in the following fashion

$$f(x) = \sum_{\nu=0}^{\infty} \frac{f^{(\nu)}(x_0)}{\nu!} \cdot (x - x_0)^{\nu}$$

or, when expressing x as sum of the point x_0 and an offset h by $x = x_0 + h$,

$$f(x_0 + h) = \sum_{\nu=0}^{\infty} \frac{f^{(\nu)}(x_0)}{\nu!} \cdot h^{\nu} \ .$$

For small offsets h, the first three terms including the first and second derivative of $f(x)$ at x_0 provide a fairly good approximation

$$f(x_0 + h) \simeq f(x_0) + \frac{f'(x_0)}{1!} \cdot h + \frac{f''(x_0)}{2!} \cdot h^2 \,. \tag{6.14}$$

Now let us analyse how this approximation can be utilised to determine the steps on the hyper-surface, i.e. the improvement of the parameter vector $\mathbf{a}$.

6.4.2 Gauss-Newton method

The Gauss-Newton method is an approach for finding the minimum in an one- or multi-dimensional signal. With respect to data fitting via least squares, the average quadratic error between the observations and the model function has to be minimised, i.e., the minimum of the hyper-surface $\chi^2(\mathbf{a})$ has to be found. This is achieved by approximating the target function by a Taylor series of second order.

The parameter vector $\mathbf{a} = (a_1 \; a_2 \; \dots \; a_j \; \dots \; a_M)^{\mathrm{T}}$ determines the dimensionality of the minimisation problem. The variables from equation (6.14) must be substituted as follows. The independent variable we are looking for is $\mathbf{a}$, thus

$$x_0 \longrightarrow \mathbf{a}$$

and the corresponding step size is

$$h \longrightarrow \mathbf{\Delta a} \,.$$

The function that has to be approximated via Taylor series is

$$f(x_0) \longrightarrow \chi^2(\mathbf{a})$$

and its derivatives are

$$f'(x_0) \longrightarrow \frac{\mathrm{d}\chi^2(\mathbf{a})}{\mathrm{d}\mathbf{a}} = \sum_{j=1}^{M} \frac{\partial \chi^2(\mathbf{a})}{\partial a_j}$$

and

$$f''(x_0) \longrightarrow \frac{\mathrm{d}^2\chi^2(\mathbf{a})}{\mathrm{d}\mathbf{a}^2} = \sum_{j=1}^{M} \sum_{k=1}^{M} \frac{\partial^2 \chi^2(\mathbf{a})}{\partial a_j \partial a_k} \,.$$

Corresponding to equation (6.14), this leads to an approximation of the error surface at the new position $\mathbf{a} + \mathbf{\Delta a}$

$$\begin{aligned}\chi^2(\mathbf{a} + \mathbf{\Delta a}) &\simeq \chi^2(\mathbf{a}) + \sum_{j=1}^{M} \frac{\partial \chi^2(\mathbf{a})}{\partial a_j} \cdot \Delta a_j + \frac{1}{2} \cdot \sum_{j=1}^{M} \sum_{k=1}^{M} \frac{\partial^2 \chi^2(\mathbf{a})}{\partial a_j \partial a_k} \cdot \Delta a_j \cdot \Delta a_k \\ &\simeq F(\mathbf{a} + \mathbf{\Delta a}) \ .\end{aligned}$$

Using matrices, this reads as

$$F(\mathbf{a} + \mathbf{\Delta a}) = \chi^2(\mathbf{a}) + \mathbf{g}^{\mathrm{T}} \cdot \mathbf{\Delta a} + \frac{1}{2} \cdot (\mathbf{\Delta a})^{\mathrm{T}} \cdot \mathbf{H} \cdot \mathbf{\Delta a} \tag{6.15}$$

with a gradient vector

$$\mathbf{g} = \begin{pmatrix} \frac{\partial \chi^2(\mathbf{a})}{\partial a_1} & \frac{\partial \chi^2(\mathbf{a})}{\partial a_2} & \cdots & \frac{\partial \chi^2(\mathbf{a})}{\partial a_M} \end{pmatrix}^{\mathrm{T}} . \tag{6.16}$$

The elements g_j of $\mathbf{g}$ are, using the equation (6.4) for χ^2 and $w_i = 1/\sigma_i^2$, equal to

$$g_j = \frac{\partial \chi^2(\mathbf{a})}{\partial a_j} = -\sum_{i=1}^{N} w_i \cdot [y_i - f(\mathbf{x}_i|\mathbf{a})] \cdot \frac{\partial f(\mathbf{x}_i|\mathbf{a})}{\partial a_j} . \tag{6.17}$$

$\mathbf{H}$ in eq.(6.15) is the so-called *Hessian* matrix[1]

$$\mathbf{H} = \begin{pmatrix} \frac{\partial^2 \chi^2(\mathbf{a})}{\partial a_1 \partial a_1} & \frac{\partial^2 \chi^2(\mathbf{a})}{\partial a_1 \partial a_2} & \cdots & \frac{\partial^2 \chi^2(\mathbf{a})}{\partial a_1 \partial a_M} \\ \frac{\partial^2 \chi^2(\mathbf{a})}{\partial a_2 \partial a_1} & \frac{\partial^2 \chi^2(\mathbf{a})}{\partial a_2 \partial a_2} & \cdots & \frac{\partial^2 \chi^2(\mathbf{a})}{\partial a_2 \partial a_M} \\ \vdots & \vdots & \ddots & \vdots \\ \frac{\partial^2 \chi^2(\mathbf{a})}{\partial a_M \partial a_1} & \frac{\partial^2 \chi^2(\mathbf{a})}{\partial a_M \partial a_2} & \cdots & \frac{\partial^2 \chi^2(\mathbf{a})}{\partial a_M \partial a_M} \end{pmatrix} . \tag{6.18}$$

[1] Named after the German mathematician Ludwig Otto Hesse (1811-1874).

The elements of $\mathbf{H}$ are computed using the product rule for derivations $(u \cdot v)' = u' \cdot v + u \cdot v'$

$$
\begin{aligned}
H_{jk} = \frac{\partial^2 \chi^2(\mathbf{a})}{\partial a_j \partial a_k} = \frac{\partial g_j}{\partial a_k} &= \frac{\partial \left(-\sum_{i=1}^{N} w_i \cdot [y_i - f(\mathbf{x}_i|\mathbf{a})] \cdot \frac{\partial f(\mathbf{x}_i|\mathbf{a})}{\partial a_j} \right)}{\partial a_k} \\
&= -\sum_{i=1}^{N} \frac{\partial \left(w_i \cdot [y_i - f(\mathbf{x}_i|\mathbf{a})] \right)}{\partial a_k} \cdot \frac{\partial f(\mathbf{x}_i|\mathbf{a})}{\partial a_j} - \sum_{i=1}^{N} w_i \cdot [y_i - f(\mathbf{x}_i|\mathbf{a})] \cdot \frac{\partial^2 f(\mathbf{x}_i|\mathbf{a})}{\partial a_j \partial a_k} \\
&= \sum_{i=1}^{N} w_i \cdot \frac{\partial f(\mathbf{x}_i|\mathbf{a})}{\partial a_k} \cdot \frac{\partial f(\mathbf{x}_i|\mathbf{a})}{\partial a_j} - \sum_{i=1}^{N} w_i \cdot [y_i - f(\mathbf{x}_i|\mathbf{a})] \cdot \frac{\partial^2 f(\mathbf{x}_i|\mathbf{a})}{\partial a_j \partial a_k} \,. \qquad (6.19)
\end{aligned}
$$

In order to minimise $\chi^2(\mathbf{a} + \mathbf{\Delta a}) \simeq F(\mathbf{a} + \mathbf{\Delta a})$, we have to look for a point where the gradient of $F(\mathbf{a} + \mathbf{\Delta a})$ (eq. (6.15)) is zero

$$
\frac{\mathrm{d}\, F(\mathbf{a} + \mathbf{\Delta a})}{\mathrm{d}\, \mathbf{\Delta a}} = F'(\mathbf{a} + \mathbf{\Delta a}) = \mathbf{g} + \mathbf{H} \cdot (\mathbf{\Delta a}) = \mathbf{0}
$$

with the result of

$$
\mathbf{\Delta a} = -\mathbf{H}^{-1} \cdot \mathbf{g} \,. \qquad (6.20)
$$

Voilá! The step $\mathbf{\Delta a}$ can be computed. However, compared to (2.6), equation (6.20) seems to be very different. Let us take a closer look at the elements of $\mathbf{g}$ and $\mathbf{H}$ (eqs. 6.17 and 6.19). It turns out that

$$
\mathbf{g} = -\mathbf{J}^{\mathrm{T}} \cdot \mathbf{W} \cdot \mathbf{r} \qquad \text{and} \qquad \mathbf{H} = \mathbf{J}^{\mathrm{T}} \cdot \mathbf{W} \cdot \mathbf{J} + \mathbf{Q} \qquad (6.21)
$$

with $\mathbf{J}$, $\mathbf{r}$, and $\mathbf{W}$, already defined in equations (2.3), (2.4), and (2.5), respectively

$$
\mathbf{J} = \begin{pmatrix}
\frac{\partial f(\mathbf{x}_1|\mathbf{a})}{\partial a_1} & \frac{\partial f(\mathbf{x}_1|\mathbf{a})}{\partial a_2} & \cdots & \frac{\partial f(\mathbf{x}_1|\mathbf{a})}{\partial a_M} \\
\frac{\partial f(\mathbf{x}_2|\mathbf{a})}{\partial a_1} & \frac{\partial f(\mathbf{x}_2|\mathbf{a})}{\partial a_2} & \cdots & \frac{\partial f(\mathbf{x}_2|\mathbf{a})}{\partial a_M} \\
\vdots & \vdots & \ddots & \vdots \\
\frac{\partial f(\mathbf{x}_N|\mathbf{a})}{\partial a_1} & \frac{\partial f(\mathbf{x}_N|\mathbf{a})}{\partial a_2} & \cdots & \frac{\partial f(\mathbf{x}_N|\mathbf{a})}{\partial a_M}
\end{pmatrix} ,
$$

$$
\mathbf{r} = \begin{pmatrix}
y_1 - f(\mathbf{x}_1|\mathbf{a}) \\
y_2 - f(\mathbf{x}_2|\mathbf{a}) \\
y_3 - f(\mathbf{x}_3|\mathbf{a}) \\
\vdots \\
y_N - f(\mathbf{x}_N|\mathbf{a})
\end{pmatrix} ,
$$

and

$$\mathbf{W} = \begin{pmatrix} w_1 & 0 & 0 & \cdots & 0 \\ 0 & w_2 & 0 & \cdots & 0 \\ 0 & 0 & \ddots & \cdots & \vdots \\ \vdots & \vdots & \vdots & \ddots & 0 \\ 0 & 0 & \cdots & 0 & w_N \end{pmatrix} .$$

The term $\mathbf{Q}$ contains the second-order derivatives in equation (6.19), which are ignored in many texts – on the one hand because it is zero for linear problems, and on the other hand because the multiplying term $[y_i - f(\mathbf{x}_i|\mathbf{a})]$ in equation (6.19) is merely a random measurement error of each point. This error can be either positive or negative as soon as $\mathbf{a}$ is close to its optimum. If the multiplying term is uncorrelated with the second derivative of $f(\mathbf{x}_i|\mathbf{a})$, these terms tend to cancel each other out when summed over i [Pre92] and we get

$$\mathbf{\Delta a} = \left(\mathbf{J}^{\mathrm{T}} \cdot \mathbf{W} \cdot \mathbf{J}\right)^{-1} \cdot \mathbf{J}^{\mathrm{T}} \cdot \mathbf{W} \cdot \mathbf{r} . \tag{6.22}$$

This procedure of minimisation is called the *Gauss-Newton method.* Fast convergence to the closest minimum is its main advantage. The success of finding the minimum, however, depends on the curvature of the target function. Without special treatment, the algorithm gets stuck in saddle points and also can walk in the wrong direction (see Subsection 6.4.5).

6.4.3 Gradient-descent method

Although the Gauss-Newton method is quite efficient, it is possible that it does not convert to the desired minimum. Especially when the initial estimate of parameter vector $\mathbf{a} = (a_1\ a_2\ \ldots\ a_j\ \ldots\ a_M)^{\mathrm{T}}$ is not close enough to the target point, the approximation using limited Taylor series can affect the procedure adversely, in terms of too large update steps $\mathbf{\Delta a}$ or wrong directions. In these cases, it is preferable to walk strictly downhill using a *gradient-descent method.*

Utilising the expansion by a Taylor series with use of the first derivatives only (see equations (6.14) and (6.15)), it can be stated

$$\chi^2(\mathbf{a} + \mathbf{\Delta a}) \simeq \chi^2(\mathbf{a}) + \mathbf{g}^{\mathrm{T}} \cdot \mathbf{\Delta a} \tag{6.23}$$

with $\mathbf{g}$ defined by (6.16).

In order to decrease $\chi^2(\mathbf{a})$, $\mathbf{g}^{\mathrm{T}} \cdot \mathbf{\Delta a}$ must be obviously less than zero, say

$$\mathbf{g}^{\mathrm{T}} \cdot \mathbf{\Delta a} = -\gamma \qquad \gamma > 0$$

leading to

$$\begin{aligned}\mathbf{g}^{\mathrm{T}} \cdot \boldsymbol{\Delta}\mathbf{a} \cdot \mathbf{g} &= -\gamma \cdot \mathbf{g} \\ \boldsymbol{\Delta}\mathbf{a} &= -\gamma \cdot \frac{\mathbf{g}}{||\mathbf{g}||^2} \\ \boldsymbol{\Delta}\mathbf{a} &= -\beta \cdot \frac{\mathbf{g}}{||\mathbf{g}||} \;. \end{aligned} \tag{6.24}$$

The factor β determines the step width and $\mathbf{g}$ only controls the direction of move, due to the normalisation with the norm of the gradient vector.

This method is rather stable when the current position in the parameter space is far from the final solution. However, it is difficult to estimate β and its value has to be kept relatively small. Close to the target minimum, the gradient-descent method becomes unfavourable. Without the normalisation by $||\mathbf{g}||$ in (6.24), the convergence process would slow down due to the very small gradients in this region; in addition, saddle points would be a barrier for the gradient descent. When applying the normalisation as in eq. (6.24), the value of β must be lowered as soon as the position of the minimum has passed over and the algorithm starts to oscillate. That is why the Gauss-Newton approach should be used in the vicinity of the minimum instead (see the next two subsections).

6.4.4 Levenberg-Marquardt method

The *Levenberg-Marquardt method* combines the advantages of the gradient-descent and Gauss-Newton methods, introducing a damping factor μ [Lev44, Mar63]. This factor modifies equation (6.22) to

$$\boldsymbol{\Delta}\mathbf{a} = \left(\mathbf{J}^{\mathrm{T}} \cdot \mathbf{W} \cdot \mathbf{J} + \mu \cdot \mathbf{I}\right)^{-1} \cdot \mathbf{J}^{\mathrm{T}} \cdot \mathbf{W} \cdot \mathbf{r} \;. \tag{6.25}$$

Positive values of μ guarantee that $\boldsymbol{\Delta}\mathbf{a}$ is a descent direction. For large μ, the introduced term $\mu \cdot \mathbf{I}$ becomes dominant and the solution is

$$\boldsymbol{\Delta}\mathbf{a} = \frac{1}{\mu} \cdot \mathbf{J}^{\mathrm{T}} \cdot \mathbf{W} \cdot \mathbf{r} = -\frac{1}{\mu} \cdot \mathbf{g} \;,$$

which corresponds to the gradient-descent method (6.24).Very small values of μ force steps similar to the Gauss-Newton method, supporting final convergence at the global minimum [Mad04].

The adaptation of μ is as follows. Let $\mathbf{N} = \mathbf{J}^{\mathrm{T}} \cdot \mathbf{W} \cdot \mathbf{J}$. Starting with a very small value, say μ equal to 0.001 times the maximum of all diagonal elements of $\mathbf{N}$ ($\mu = 0.001 \cdot \max_j(N_{jj})$), the decrease (or increase) of $\chi^2(\mathbf{a})$ is observed.

When $\chi^2(\mathbf{a} + \mathbf{\Delta a}) < \chi^2(\mathbf{a})$, the algorithm is on the right path. The position in the parameter space has to be updated via $\mathbf{a} \leftarrow \mathbf{a} + \mathbf{\Delta a}$ and μ can be decreased. Otherwise, the gradient descent seems to be more suitable; the update step is skipped and μ increased by a substantial factor. The choice of μ could be also linked directly to the conformance between the predicted value of χ^2 (based on the Taylor expansion) and the real χ^2 after the update step ([Mad04]).

6.4.5 Example of finding the minimum

Let us assume the one-dimensional error-surface is described by

$$\chi^2(a) = a^4 - 3 \cdot a^3 + 10 . \tag{6.26}$$

Using the Taylor expansion (see also Subsection 6.4.1) we can approximate this function around a certain (initial) point a_0 with arbitrary accuracy. The zero-order approximation would simply be a constant

$$\chi^2(a_0 + \Delta a) \approx \chi^2(a_0) . \tag{6.27}$$

Approximations of the first and second order correspond to a straight line and a parabola, respectively

$$\chi^2(a_0 + \Delta a) \approx \chi^2(a_0) + \frac{\mathrm{d}\chi^2(a_0)}{\mathrm{d}a} \cdot \Delta a \tag{6.28}$$

and

$$\chi^2(a_0 + \Delta a) \approx \chi^2(a_0) + \frac{\mathrm{d}\chi^2(a_0)}{\mathrm{d}a} \cdot \Delta a + \frac{1}{2} \cdot \frac{\mathrm{d}^2\chi^2(a_0)}{\mathrm{d}a^2} \cdot (\Delta a)^2 . \tag{6.29}$$

Figure 6.3 shows the error-surface example $\chi^2(a)$ (solid line) and its approximated versions $\chi^2(a_0 + \Delta a)$ for $a_0 = 3.2$. As can be seen, the approximated versions touch the original function at $(a_0; \chi^2(a_0))$ and slightly diverge with increasing distance from this position.

Two things can be derived from this plot. First, the parabola approximation is, in accordance with the theory of Taylor series, more accurate than the straight-line approximation. Secondly, the parabola shows a turning point, which can assist in searching for the minimum of $\chi^2(a)$. This is exactly what the Gauss-Newton method is doing. It takes the second-order approximation, determines the turning point, and assumes that the turning point is closer to the desired minimum position than the current operating point a_0. Then, the new position is taken as operating point and so on until the operating point and turning point of the parabola are identical.

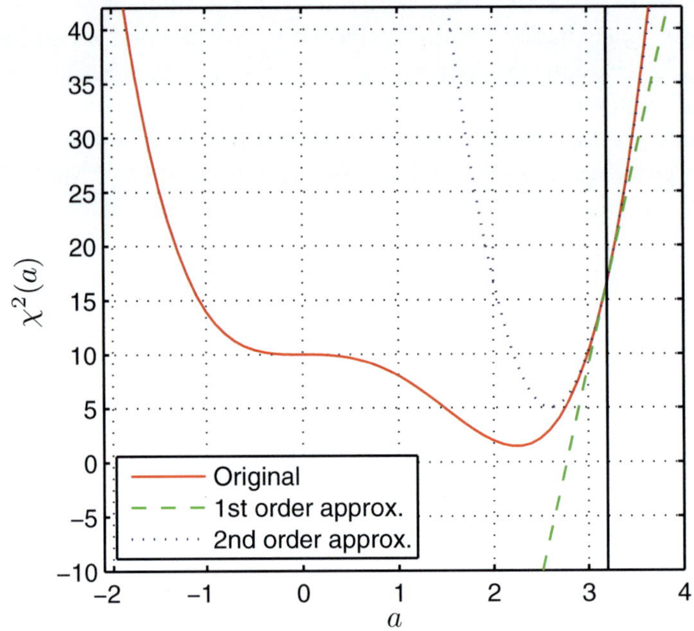

Figure 6.3: Approximation of a curve using Taylor expansion. The vertical line indicates the operating point $a_0 = 3.2$.

In application to the example function (6.26), we have

$$\chi^2(a_0 + \Delta a) \approx \chi^2(a_0) + \frac{\mathrm{d}\chi^2(a_0)}{\mathrm{d}a} \cdot \Delta a + \frac{1}{2} \cdot \frac{\mathrm{d}^2\chi^2(a_0)}{\mathrm{d}a^2} \cdot (\Delta a)^2$$

$$\chi^2(a_0 + \Delta a) \approx (a_0^4 - 3 \cdot a_0^3 + 10) + (4 \cdot a_0^3 - 9 \cdot a_0^2) \cdot \Delta a + \frac{1}{2} \cdot (12 \cdot a_0^2 - 18 \cdot a_0) \cdot (\Delta a)^2 \; . \tag{6.30}$$

The course of the Gauss-Newton method results in a series of operating points. **Figure 6.4** shows the progress of convergence. Six iterations are sufficient in this example.

The first-order approximation corresponds to the gradient-descent method. If the gradient of the first-order approximation is positive, the value of a_0 must be decreased; if the gradient is negative, a_0 must be increased in order to walk towards the minimum. The minimum is reached when the gradient is zero.

As already pointed out in Subsection 2.3, the choice of the initial operation point a_0 is a crucial task in nonlinear data fitting. **Figure 6.5** illustrates what happens

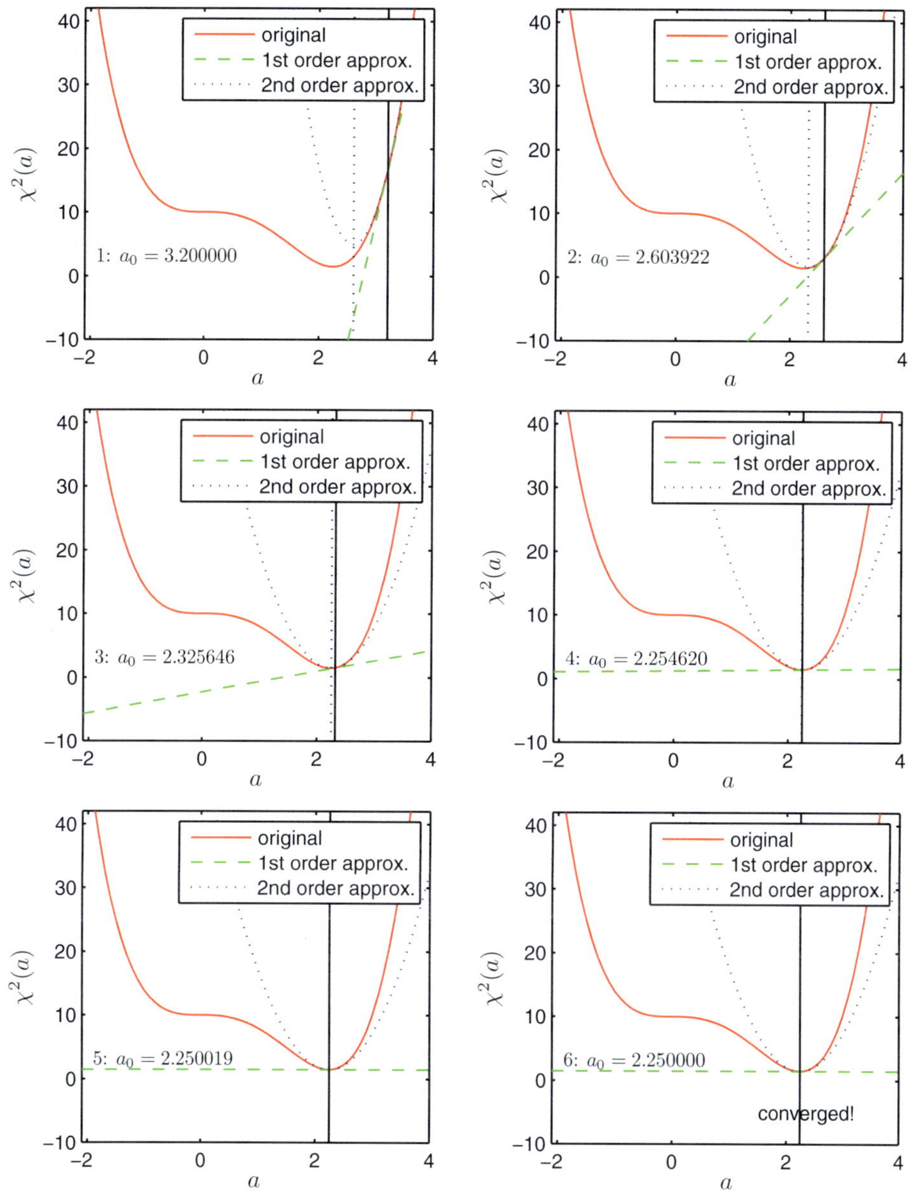

Figure 6.4: Searching the global minimum using the Gauss-Newton method. The initial starting point is $a_0 = 3.2$.

if the operation point (the model parameter) is initialised with $a_0 = -1.5$ and the Gauss-Newton method is applied. Now, 25 iterations are required, however, the algorithm fails to reach the global minimum and converges instead at the saddle point. It also can be concluded from the iteration numbers that the convergence process slows down in the vicinity of the saddle point. The incremental steps Δa become very small. This problem could only be solved using a momentum that pushes the algorithm a little bit beyond the critical position.

Let us assume that the initialisation is slightly more accurate and that the starting point is $a_0 = 1.2$ (**Figure 6.6**). What happens? Since the curvature of the original function is different, the turning point of the parabola becomes a maximum instead of a minimum. The Gauss-Newton method accordingly walks in the wrong direction, increasing the value of $\chi^2(a)$.

This is exactly the reason why the Levenberg-Marquardt method should be used instead the plain Gauss-Newton approach. The parabola enables more exact steps in most cases. However, it can also misguide the minimisation process, whereas the gradient (first-order approximation) always points to the minimum (**Figure 6.6**, left plot). As soon as the value of $\chi^2(a)$ increases, the minimisation procedure must therefore omit this step and increase the influence of the gradient-descent method, forcing the process to walk strictly downhill.

Since the gradient-descent method always guides the minimisation process in the right direction, why not use it in general? There are at least two reasons. First, it is difficult to determine a proper step size Δa. It has to be kept relatively small in the beginning as it is unknown how far away the minimum is. Secondly, when the vicinity of the minimum is reached, the method will jump over this minimum, regardless how small the step size is, and the algorithm runs into oscillation.

In order to overcome these problems, the step size could be increased as long as the downhill walk does not change its direction. This speeds up the convergence, especially if the initial step size was too small. When a change in the direction is detected (i.e., the oscillation starts), the step size has to be decreased as long as the oscillation persists. The procedure of lowering the modifications from cycle to cycle is also called *simulated annealing*. Typically, the step size also should be independent of the absolute value of the actual gradient. This helps to escape long valleys or saddle points.

For the tests with the error surface defined by (6.26), the following rule is used

$$\Delta a = \begin{cases} 0 & \text{if} \quad |g| < 10^{-15} \\ -\beta \cdot \frac{g}{||g||} & \text{otherwise} \end{cases} \qquad \text{with} \quad g = \frac{\mathrm{d}\chi^2(a_0)}{\mathrm{d}a} \,. \tag{6.31}$$

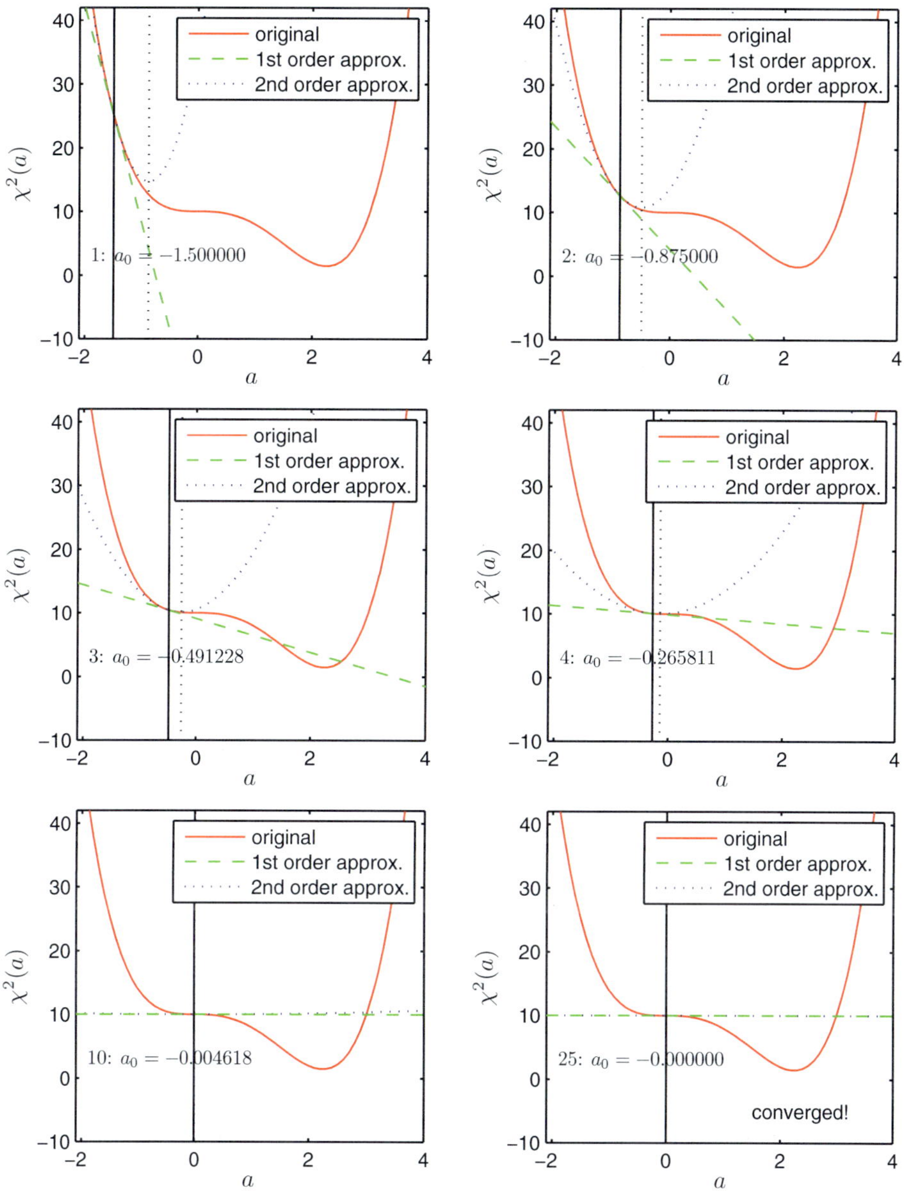

Figure 6.5: Searching the global minimum using the Gauss-Newton method. The initial starting point is $a_0 = -1.5$.

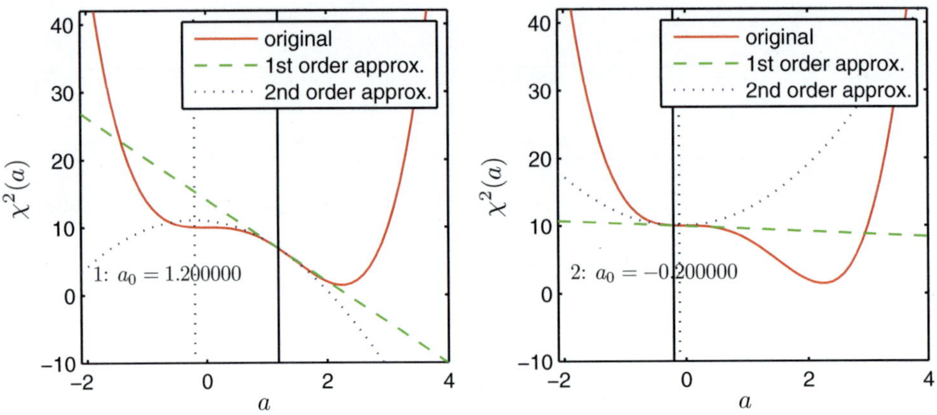

Figure 6.6: Searching the global minimum using the Gauss-Newton method. The initial starting point is $a_0 = 1.2$.

The normalisation with the absolute value of the gradient avoids convergence at saddle points. The value of β is initially set to 0.2 in this example and is increased by 10% after each successful iteration. As soon as the operating point has passed the minimum and the walking direction is reversed, β is divided by 5 and only 25% of the computed step size is applied, in order to damp the oscillation. **Figure 6.7** shows the result with an initial operation point $a_0 = 3.2$. The entire process of convergence takes 45 iterations, while the Gauss-Newton method requires only six. One can conclude from the plots that major efforts (iterations $10 \ldots 45$) are required to finally approach the target.

We should also test the course of convergence when starting at $a_0 = -1.5$ (**Figure 6.8**). Since the gradient is normalised by its absolute value (eq. 6.31), the step size is only defined by β and increases consistently, speeding up the process of walking towards the minimum. For the same reason, the saddle point is easily passed over (iterations 6 to 8) and the operation point moves steadily until the minimum is also passed (iteration 13). Then, again, oscillation starts with decreasing values of β, i.e., decreasing step sizes. The majority of iterations is spent reaching the equilibrium. This underlines that the gradient-descent approach is helpful to a certain extent, however, convergence is achieved much faster when using the Gauss-Newton method.

Now let us investigate the behaviour of the Levenberg-Marquardt method. When starting at $a_0 = -1.5$ (**Figure 6.9**), the algorithm walks towards the saddle point and the parabola becomes more flat. The parameter μ is very small at this point and so the influence of the gradient descent method. Since also the

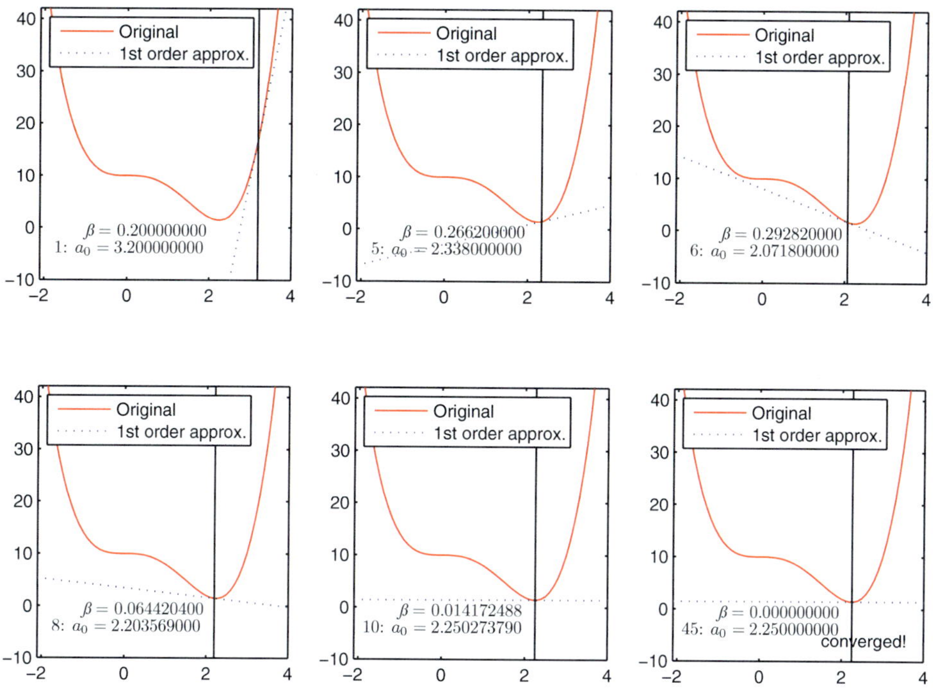

Figure 6.7: Searching the global minimum using the gradient-descent method. The initial starting point is $a_0 = 3.2$. The value of β corresponds in equation (6.31).

gradient g converges to zero the Levenberg-Marquardt method get stuck in the same way as the Gauss-Newton method. In two- or multidimensional problems the procedure probably will move along another parameter a_j circumnavigating this saddle point. In our one-dimensional problem, we need a trick to pass this critical point. One option is to operate with a modified gradient value g', as, for instance,

$$g' = \begin{cases} g + 0.001 & \text{if} \quad g > 0 \\ g - 0.001 & \text{otherwise} \end{cases} \tag{6.32}$$

The result is shown in **Figure 6.10**. It has to be mentioned that the artificial modification of the gradient g might put a penalty on the convergence process at other positions of $\mathbf{a}$.

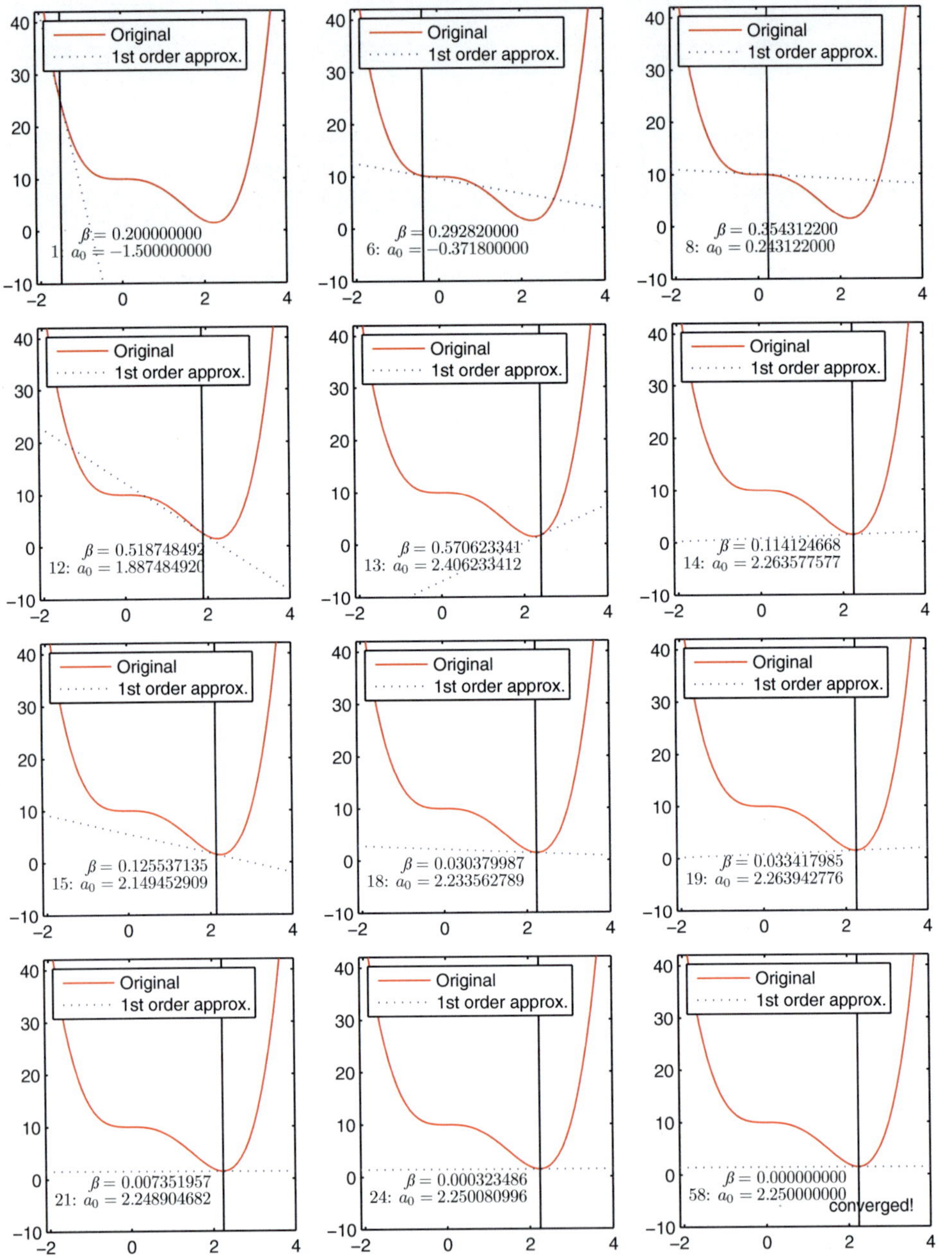

Figure 6.8: Searching the global minimum using the gradient-descent method. The initial starting point is $a_0 = -1.5$. The value of β corresponds to equation (6.31).

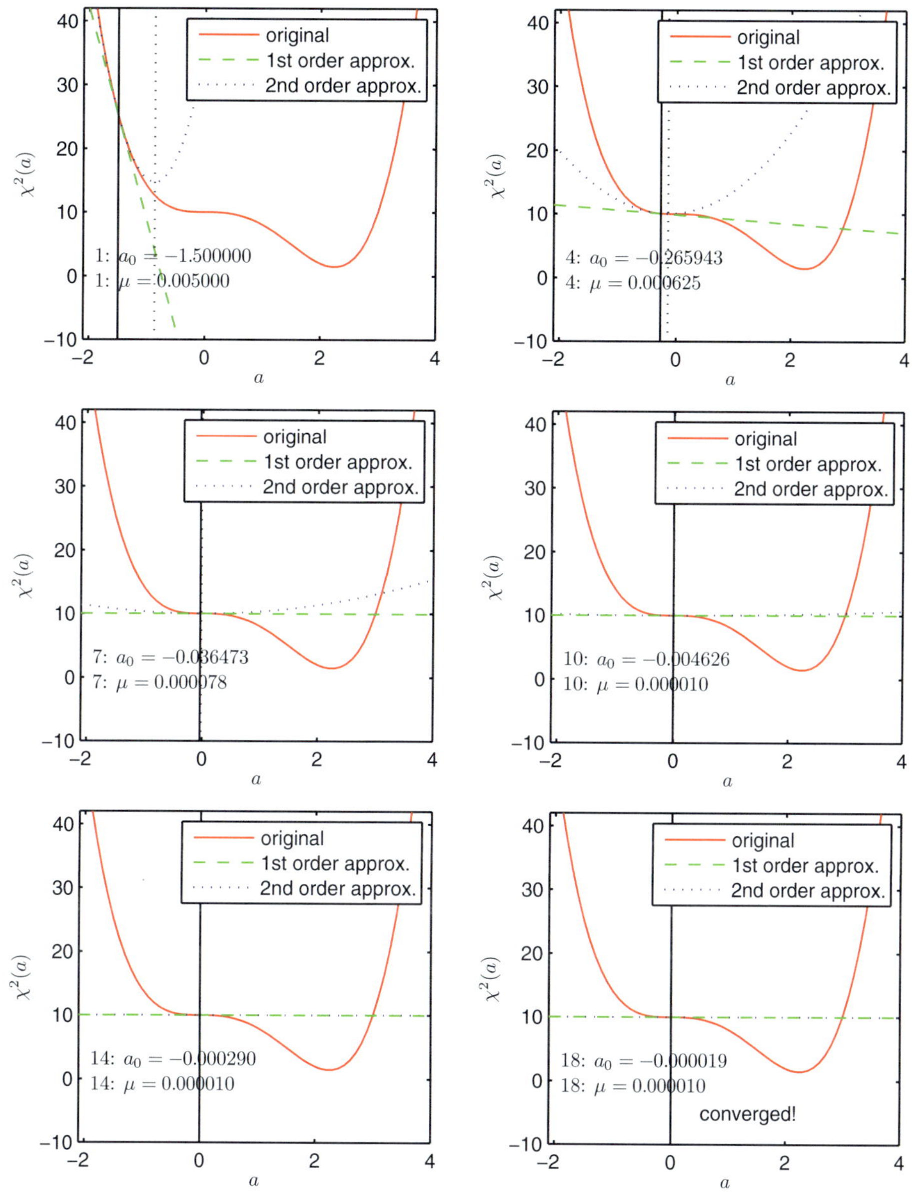

Figure 6.9: Searching the global minimum using the Levenberg-Marquardt method. The initial starting point is $a_0 = -1.5$.

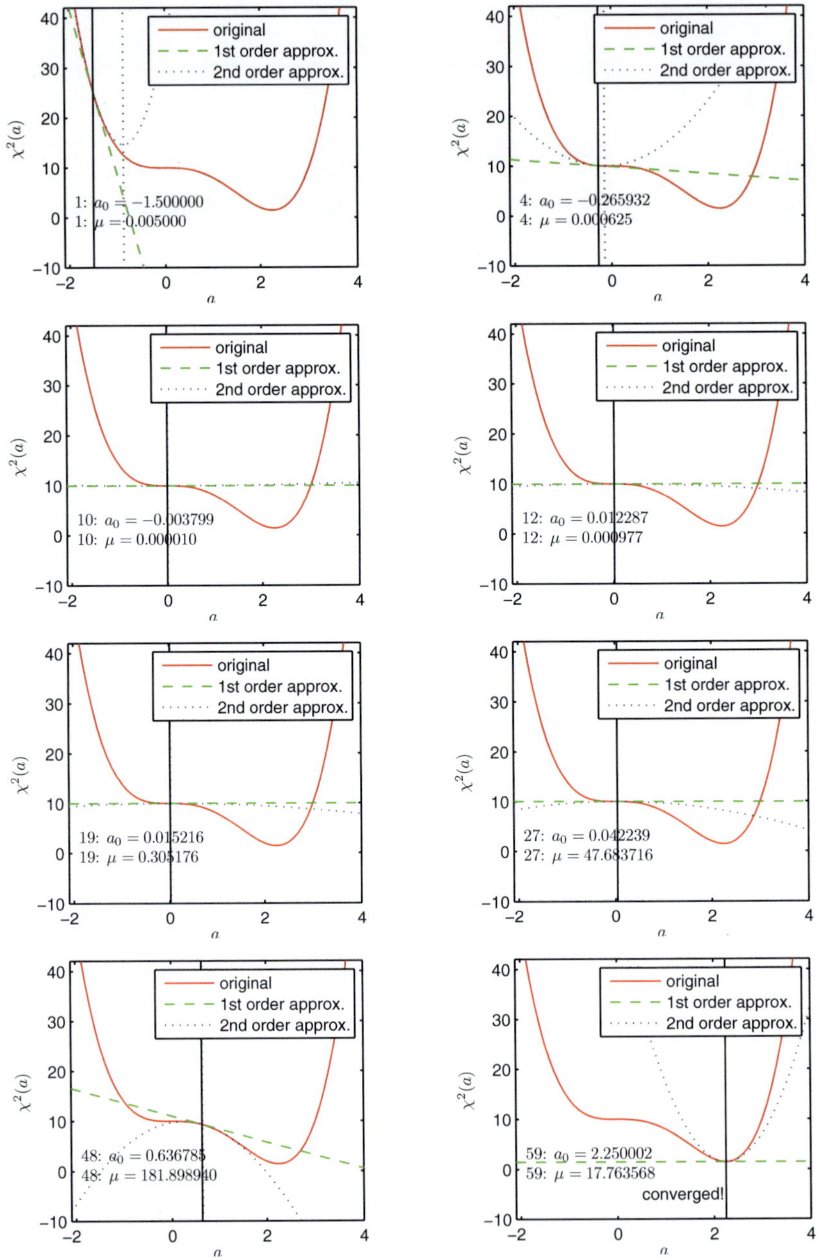

Figure 6.10: Searching the global minimum using the Levenberg-Marquardt method and a modified gradient according to (6.32). The initial starting point is $a_0 = -1.5$.

6.5 Test questions

6.1 What are the parameters of the Gaussian distribution and how do they influence its characteristics?

6.2 How are the least-squares method and the maximum-likelihood principle related to each other?

6.3 How can the SVD be used to fit linear models?

6.4 Give the definition of the expansion of an arbitrary function by Taylor series.

6.5 How many derivations are used in the Gaussian-Newton method for minimisation problems?

6.6 What is a Hessian matrix?

6.7 How can the Hessian matrix be approximated?

6.8 Explain the gradient-descent method.

6.9 What is the principle of the Levenberg-Marquardt method?

6.10 Discuss the advantages and disadvantages of the Gauss-Newton method based on an arbitrary example.

6.11 Discuss the advantages and disadvantages of the gradient-descent method based on an arbitrary example.

Chapter 7

Supplemental Tools and Methods

7.1 Alternative parameter estimation

Besides the least-squares methods explained in this book so far, there is a variety of alternatives for optimisation (see for instance [Pre92]), which might be favourable for particular problems. Three examples are briefly discussed in this section to give the reader an overview.

7.1.1 Recursive adaptation of parameters

Let us consider the case that the model parameters of a given linear system are not constant but vary slightly over n (e.g. over time or any other condition). Obviously, there is no optimal set of parameters, but it has to be estimated on the fly. This problem occurs, for instance, in applications that predict signals. The typical task of prediction is to forecast a signal value $x[n]$ (or more abstract: a symbol) from previous ones. The estimated value of $x[n]$ is denoted with $\hat{x}[n]$ (compare eq.(1.23))

$$\begin{aligned}\hat{x}[n] &= f(\mathbf{x}_n|\mathbf{a}) = a_1 \cdot x[n-1] + a_2 \cdot x[n-2] + a_3 \cdot x[n-3] + \dots \\ &= \sum_k a_k \cdot x[n-k] \qquad k > 0\,.\end{aligned}$$

Signal values that have already been processed serve as vector of conditions $\mathbf{x}_n$, which is itself dependent on n. Here, $x[n]$ has to be interpreted as a certain value of a time-discrete signal $\{x[n]\}$. The prediction error is $e[n] = x[n] - \hat{x}[n]$.

The general set-up according to equation (2.2) on page 26 aims at a global optimisation. If one reduces the process to the last and single observation, the summation in (2.2) must be ignored. The minimisation now concerns the square error $(e[n])^2$

$$E[n] = (e[n])^2 = [y_i - f(\mathbf{x}_i|\mathbf{a})]^2 \longrightarrow \text{Min}$$
$$= \left(x[n] - \sum_k a_k \cdot x[n-k]\right)^2 \longrightarrow \text{Min}\,. \tag{7.1}$$

The partial derivatives of $E[n]$ correspond to the ascent of $E[n]$ depending on the parameters a_k

$$\frac{\partial E[n]}{\partial a_j} = \frac{\partial}{\partial a_j}\left\{\left(x[n] - \sum_k a_k \cdot x[n-k]\right)^2\right\}\,. \tag{7.2}$$

Using the chain rule (outer derivative times inner derivative) yields[1]

$$\frac{\partial E[n]}{\partial a_j} = 2 \cdot \left(x[n] - \sum_k a_k \cdot x[n-k]\right) \cdot (-x[n-j])$$
$$= -2 \cdot e[n] \cdot x[n-j]\,. \tag{7.3}$$

If the ascent $\partial E[n]/\partial a_j$ is negative with respect to a_j, then a_j must be increased by a certain amount in order to decrease the square error (downhill towards the minimum)

$$a_j[n+1] = a_j[n] + \mu \cdot e[n] \cdot x[n-j]\,. \tag{7.4}$$

As a result, a given set of initial parameters can be updated step-by-step depending on the prediction error $e[n]$, the condition $x[n-j]$ and the current parameter value $a_j[n]$. The value of μ determines the speed of adaptation. In some contexts, it is also called the *learning rate*. Its optimal value depends on both the speed of changes to signal properties and the values of the signal themselves. In the case of a too large value of μ, oscillation could happen. Rules of thumb recommend $0 < \mu < 1/\sigma_{\mathrm{x}}^2$ or $0 < \mu < 1/(x_{\max})^2$, however, the choice should be carefully made.

This kind of adaptation is related to the theory of adaptive digital filters, more precisely, least-mean-square (LMS) filters. The values $x[n]$ should ideally have a mean of zero. For the update process in (7.4) several variants have been proposed. In [Str14], for instance, the signal values (i.e., the conditions $x[n-j]$)

[1] Within the parenthesis of (7.2), there is a single term containing a_j. So, the inner derivative is merely $x[n-j]$.

are normalised to the mean of absolute values $\overline{|x[n]|}$, which itself is recursively computed based on already processed values. Equation (7.4) becomes:

$$a_j[n+1] = a_j[n] + \mu \cdot e[n] \cdot \frac{x[n-j]}{\overline{|x[n]|} + \epsilon} \quad .$$

The value $\epsilon > 0$ simply avoids a possible division by zero.

7.1.2 Gradient descent by trial and error

This method of best parameter determination utilises similar ideas like the gradient-descent method discussed in Subsection 6.4.3. Knowing the error hypersurface $\chi^2(\mathbf{a})$ and a set of initial parameters, it is possible to slightly vary the elements a_j of $\mathbf{a}$ one after each other in order to determine the direction of descent without implicitly knowing the gradient itself

$$a_j[n+1] = \begin{cases} a_j[n]{+}\Delta a_j & \text{if} \quad \chi^2(\mathbf{a}[n]{+}\Delta) \le \min\left(\chi^2(\mathbf{a}[n]), \chi^2(\mathbf{a}[n]{-}\Delta)\right) \\ a_j[n]{-}\Delta a_j & \text{if} \quad \chi^2(\mathbf{a}[n]-\Delta) \le \min\left(\chi^2(\mathbf{a}[n]), \chi^2(\mathbf{a}[n]{+}\Delta)\right) \\ a_j[n] & \text{otherwise} \end{cases} \quad .$$

In words: for each parameter a_j, the algorithm tests the steps in both directions and takes the one in a downhill direction, or stays at its position if the position corresponds to a saddle point or a minimum. Unfortunately, the optimal step sizes Δa_j are not known in advance and have to be kept relatively small. The convergence to the desired parameter values can take many iterations in unfortunate cases. Besides this, at each iteration both, $\chi^2[\mathbf{a}(n)+\Delta]$ and $\chi^2[\mathbf{a}(n)-\Delta]$, have to be evaluated. On the other hand, nonlinear systems are also supported and arbitrary hyper-surfaces $\chi^2(\mathbf{a})$ are tractable, enabling even the principle of total-least-squares fitting, when the shortest distance between the measured point $(y_i; \mathbf{x}_i)$ and the model function $(f(\mathbf{x}_i^{\text{TLS}}|\mathbf{a}), \mathbf{x}_i^{\text{TLS}})$ is evaluated instead of the simple difference $y_i - f(\mathbf{x}_i|\mathbf{a})$ (see Figure 7.9 in Section 7.7).

The choice of Δa_j can be combined with the method of simulated annealing, i.e. initially using high absolute step values in order to escape local minima and then slowly decreasing the step sizes for each iteration. An alternation between positive and negative steps is an indication of oscillation and the corresponding step sizes must be shortened.

A popular variant of this technique is the so-called Metropolis algorithm [Met53]. If applied to the optimisation problems focussed on here, it can be described as follows. The algorithm starts at an arbitrary point $\mathbf{a}$ on the error hyper-surface. There are two variables, which will change over the course of the iterative process:

the maximum step size $\Delta a_{\max}$ for moving the single parameters a_j and a certain probability p_{acc}.

For each a_j, a step Δa_j in the range $-\Delta a_{\max} \leq \Delta a_j \leq \Delta a_{\max}$ is randomly selected. If $\chi^2(\mathbf{a}+\Delta a_j)$ is smaller than the error $\chi^2(\mathbf{a})$ at the current position, the step is performed, i.e., $\mathbf{a} \leftarrow \mathbf{a}+\Delta a_j$. This is quite similar to the approach above. The innovation comes into play when $\chi^2(\mathbf{a}+\Delta a_j) > \chi^2(\mathbf{a})$ holds. In these cases, this step is accepted up to a certain acceptance probability p_{acc}. This allows the algorithm to walk uphill and supports the escape of local minima at the beginning of the iterative process. As the process continues, the probability p_{acc} is gradually reduced forcing the downhill walk. The maximum step size $\Delta a_{\max}$ is also lowered (simulated annealing) allowing for equilibrium at the minimum.

7.1.3 Evolutionary approach

The optimisation of model parameters can also be regarded as an evolutionary process. All parameters a_j are considered as genes, which are subject to mutation, selection, and cross-over. Mutations can be introduced via randomisation. An entire population of K randomised parameter vectors $\mathbf{a}_k$ $(k = 1, 2, \ldots, K)$ must be generated. A certain percentage of those vectors yielding the best fits, i.e. smallest values of $\chi^2(\mathbf{a}_k)$, is selected for further investigations. By chance, single elements a_j of different $\mathbf{a}_k$ are exchanged (cross-over). These steps are iterated until convergence.

Obviously, this approach is rather computationally expensive. Besides this, success is sensitive to the chosen options, that is, strength of randomisation in mutation and cross-over, total size of the population, portion of surviving individuals (parameter vectors $\mathbf{a}_k$), etc.

Of course, the evolutionary approach (also known as 'genetic algorithm') may also be combined with the method of simulated annealing.

7.1.4 Training of neural networks using backpropagation

Preliminary remarks

Artificial neural networks belong to the class of learning systems. Their parameters are optimised in a training step and remain fixed afterwards. Generally, an input vector $\mathbf{x}$ containing a finite number of elements x_n $(n = 1, 2, \ldots, N)$ is mapped to an output vector $\mathbf{y}$ with K elements.

Figure 7.1 shows a feed-forward net with three layers (two hidden layers, and one output layer). The attribute 'hidden' indicates that the processed values

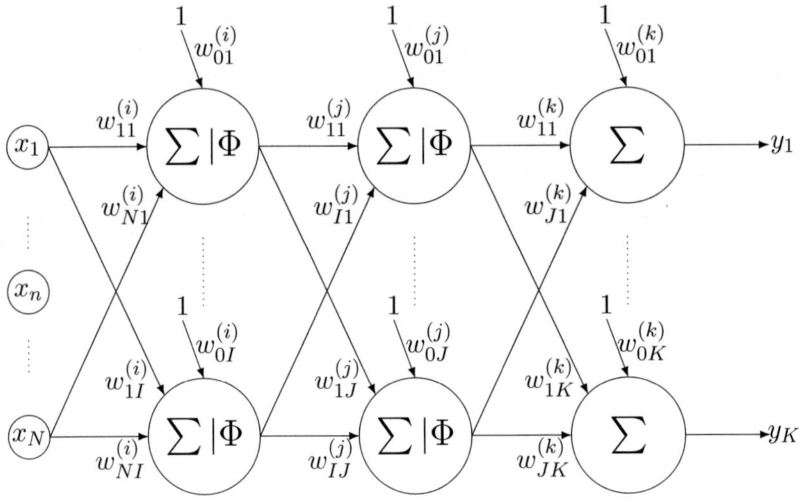

Figure 7.1: Artificial neural network with N input units, two hidden layers (i) and (j), and an output layer (k)

from these layers are not visible at the output of the network. The parameters of a neural net are called weights w_{ij}. Each neuron of layer j processes weighted output values of the neurons of the preceding layer i. In Figure 7.1, the notations $^{(i)}$, $^{(j)}$, and $^{(k)}$ indicate to which layer the weights belong. When using variable indices $i = 1, 2, \ldots, I$, $j = 1, 2, \ldots, J$, $k = 1, 2, \ldots, K$, however, one can omit this indication: $w_{ni}^{(i)} = w_{ni}$, $w_{ij}^{(j)} = w_{ij}$, $w_{jk}^{(k)} = w_{jk}$, and so on.

According to the biological model, at first the weighted values are accumulated

$$a_j = \sum_i x_i \cdot w_{ij} \; . \tag{7.5}$$

While x_n is regarded as the nth input value of the entire network, x_i is here the output value from the ith neuron in layer (i) and contemporaneously the ith input value with respect to layer (j). In literature about neural networks, net_j is often used instead of a_j and o_i (for output) is used instead of x_i. The calculation of a_j implicitly presumes that one vector element is fixed $x_0 = 1$. This allows the operation point of the neuron to be adjusted based on the weight w_{0j}.

The final output value of the neuron depends on an activation function $\Phi(a_j)$. Considering the learning process, it should be possible to continually differentiate this activation function. In the following, the hyperbolic tangent is used as activation function (**Fig. 7.2**)

$$x_j = \Phi(a_j) = \tanh(a_j) \; . \tag{7.6}$$

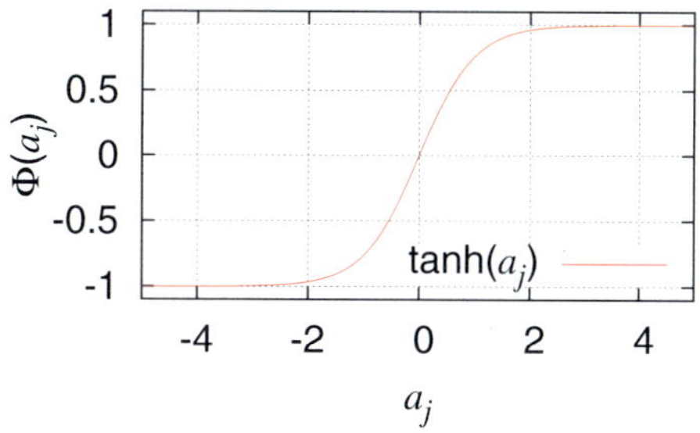

Figure 7.2: Hyperbolic tangent as activation function

The hyperbolic tangent belongs to the class of sigmoid (S-shaped) functions. The plot in Figure 7.2 shows that the gradient of the function is almost zero at arguments $|a_j| > 4$. Hence, the output of the neuron is limited to the range $-1 < x_j < 1$.

The output layer k is mostly realised without this activation function and the output value is equal to the sum of the weighted input values

$$x_k = a_k = \sum_j x_j \cdot w_{jk} \,. \tag{7.7}$$

The desired output value (the target) at neuron k is called t_k. In the training or learning mode phase, the first layer is presented different input vectors $\mathbf{x}_k$ and it is expected that the output layer responds with vector $\mathbf{t}_k$. At the beginning of the training, in particular, when the weights still have more or less arbitrary values, the error magnitude

$$|\mathbf{e}_k| = |\mathbf{t}_k - \mathbf{x}_k| \tag{7.8}$$

is relatively high.

In the following, it will be derived step-by-step, how the weights of the neural network can be recursively optimised using the method of least squares.

Set-up for output neurons without a special activation function

In this section we regard a single neuron of layer k and its connections to the preceding layer j. The output value x_j of a neuron of layer j is weighted with w_{jk}. The receiving neuron accumulates all weighted values resulting in a_k. Since the activation function is $\Phi(a_k) = a_k$ (identity), the output value x_k of output neuron k can be written as a function of weights w_{jk} in the form of

$$x_k = a_k = \sum_j x_j \cdot w_{jk} \ .$$

The energy of the error at the output of this neuron k

$$\begin{aligned} E_k &= (e_k)^2 = (t_k - x_k)^2 = \left(t_k - \sum_j x_j \cdot w_{jk}\right)^2 \\ &= t_k^2 - 2 \cdot t_k \cdot \sum_j w_{jk} \cdot x_j + \left(\sum_j x_j \cdot w_{jk}\right)^2 \ . \end{aligned} \tag{7.9}$$

has to be minimised. The minimisation principle is the same as for the adaptation of least-mean-square filters (LMS filters), see Section 7.1.1.

The partial derivatives of E_k indicate the ascent of E_k as a function of the coefficients w_{jk}. Looking at a concrete weight w_{nk} from the set of all w_{jk}, it holds

$$\begin{aligned} \frac{\partial E_k}{\partial w_{nk}} &= \frac{\partial}{\partial w_{nk}} \left\{ t_k^2 - 2 \cdot t_k \cdot \sum_j w_{jk} \cdot x_j + \left(\sum_j x_j \cdot w_{jk}\right)^2 \right\} \\ &= -2 \cdot t_k \cdot x_n + \left[2 \cdot \left(\sum_j x_j \cdot w_{jk}\right) \cdot \frac{\partial}{\partial w_{nk}} \left(\sum_j x_j \cdot w_{jk}\right) \right] \\ &= -2 \cdot t_k \cdot x_n + 2 \cdot \left(\sum_j x_j \cdot w_{jk}\right) \cdot x_n \\ &= -2 \cdot x_n \cdot \left[t_k - \left(\sum_j x_j \cdot w_{jk}\right) \right] \\ \frac{\partial E_k}{\partial w_{nk}} &= -2 \cdot x_n \cdot e_k \ . \end{aligned}$$

If the slope is negative with respect to w_{nk}, then w_{nk} has to be increased by a certain amount, so that the energy is decreased (walking downhill towards the minimum)

$$w_{jk} \leftarrow w_{jk} + \mu \cdot x_j \cdot e_k \ . \tag{7.10}$$

The learning rate $0 < \mu \ll 1$ influences the speed of learning and should damp possible oscillations. This adaptation corresponds to the gradient-descent method, see also Sections 6.4.3 and 7.1.1.

Set-up for output neurons with tanh() activation function

Now we investigate how the adaptation for weight w_{jk} changes when the corresponding neuron from layer k has an activation function

$$x_k = \Phi(a_k) = \tanh(a_k) \qquad \text{with} \qquad a_k = \sum_j x_j \cdot w_{jk} \,. \tag{7.11}$$

The first derivative of $\Phi(a_k) = \tanh(a_k)$ with respect to a_k is

$$\frac{\mathrm{d}\,\Phi(a_k)}{\mathrm{d}\,a_k} = [1 - \tanh^2(a_k)] \,. \tag{7.12}$$

As before, we are looking for the optimal weight w_{jk} or a rule, how this weight can successively be adapted. The output value of the kth neuron is calculated by

$$x_k = \tanh\left(\sum_j x_j \cdot w_{jk}\right) = \tanh(a_k)$$

and the error at the neuron output is

$$e_k = t_k - x_k \,.$$

The squared error results from

$$E_k \;=\; (e_k)^2 = \left[t_k - \tanh\left(\sum_j x_j \cdot w_{jk}\right)\right]^2 . \tag{7.13}$$

The minimum of E_k with respect to w_{nk} has to be found. The indexing with n only clarifies the reference to a concrete weight from the set of all w_{jk}. It can be seen that the differentiation of E_k with respect to w_{nk} requires several steps. Using the chain rule (outer derivative times inner derivative) we can write

$$\frac{\partial E_k}{\partial w_{nk}} = \frac{\partial E_k}{\partial a_k} \cdot \frac{\partial a_k}{\partial w_{nk}} \,.$$

The outer derivative also requires the chain rule. In combination with equation (7.12), we get

$$\begin{aligned} \frac{\partial E_k}{\partial a_k} &= \frac{\partial}{\partial a_k}\left[t_k - \tanh(a_k)\right]^2 \\ &= 2 \cdot [t_k - \tanh(a_k)] \cdot (-1) \cdot \left[1 - \tanh^2(a_k)\right] \\ &= -2 \cdot [1 - \tanh^2(a_k)] \cdot e_k \,. \end{aligned} \tag{7.14}$$

The inner derivative uses (7.11)

$$\begin{aligned} \frac{\partial a_k}{\partial w_{nk}} &= \frac{\partial}{\partial w_{nk}} \left\{ \sum_j x_j \cdot w_{jk} \right\} \\ &= x_n \end{aligned} \tag{7.15}$$

Combining (7.14) and (7.15), the differentiation of the error energy with respect to w_{nk} finally yields

$$\frac{\partial E_k}{\partial w_{nk}} = -2 \cdot [1 - \tanh^2(a_k)] \cdot e_k \cdot x_n \, .$$

If the ascent is negative with respect to w_{nk}, an increasing weight w_{nk} minimises the error energy

$$\begin{aligned} w_{jk} &\leftarrow w_{jk} + \mu \cdot [1 - \tanh^2(a_k)] \cdot e_k \cdot x_j \\ w_{jk} &\leftarrow w_{jk} + \mu \cdot \delta_k \cdot x_j \, . \end{aligned} \tag{7.16}$$

In comparison to (7.10) it can be seen that the change of the weight is additionally combined with the derivative of the activation function.

Set-up for hidden neurons with tanh() activation function

While for the neurons of the output layer k the target values t_k are known, there are no target values available for a hidden layer j and the error e_j has to be determined indirectly using the output layer. The entire function of the neural network must be incorporated, when optimising the weights w_{ij}.

For the output value of neuron j, it holds

$$x_j = \tanh(a_j) = \tanh\left(\sum_i x_i \cdot w_{ij} \right)$$

and for a neuron of layer k we have

$$\begin{aligned} x_k = \tanh(a_k) &= \tanh\left(\sum_j x_j \cdot w_{jk} \right) \\ &= \tanh\left\{ \sum_j \left[\tanh\left(\sum_i x_i \cdot w_{ij} \right) \right] \cdot w_{jk} \right\} \, . \end{aligned}$$

In order to adjust the weights w_{ij}, the derivative of

$$E_k = (e_k)^2 = (t_k - x_k)^2 \tag{7.17}$$

with respect to w_{ij} is needed. As already shown in the preceding sections, it is reasonable to split the whole problem into sub-problems using the chain rule. The weight w_{mn} connects the mth neuron of layer i with the nth neuron of layer j. All variables i and j have accordingly to be substituted by m and n for this derivation. The differentiation of the error function is now given by

$$\frac{\partial E_k}{\partial w_{mn}} = \frac{\partial E_k}{\partial a_k} \cdot \frac{\partial a_k}{\partial w_{mn}} = \frac{\partial E_k}{\partial a_k} \cdot \frac{\partial a_k}{\partial x_n} \cdot \frac{\partial x_n}{\partial w_{mn}} = \frac{\partial E_k}{\partial a_k} \cdot \frac{\partial a_k}{\partial x_n} \cdot \frac{\partial x_n}{\partial a_n} \cdot \frac{\partial a_n}{\partial w_{mn}} \,.$$

The first derivative on the far-left is, according to (7.14), equal to

$$\frac{\partial E_k}{\partial a_k} = -2 \cdot [1 - \tanh^2(a_k)] \cdot e_k \,. \tag{7.18}$$

In addition we have

$$\frac{\partial a_k}{\partial x_n} = \frac{\partial}{\partial x_n} \sum_j x_j \cdot w_{jk} = w_{nk} \,, \tag{7.19}$$

$$\frac{\partial x_n}{\partial a_n} = \frac{\partial}{\partial a_n} \tanh(a_n) = [1 - \tanh^2(a_n)] \tag{7.20}$$

and the last derivative is, analogous to (7.15),

$$\frac{\partial a_n}{\partial w_{mn}} = \frac{\partial}{\partial w_{mn}} \left\{ \sum_i x_i \cdot w_{in} \right\} = x_m \,. \tag{7.21}$$

Finally, we get

$$\frac{\partial E_k}{\partial w_{mn}} = -2 \cdot [1 - \tanh^2(a_k)] \cdot e_k \cdot w_{nk} \cdot [1 - \tanh^2(a_n)] \cdot x_m \,.$$

Considering the error e_k at the output of neuron k, the weight has to be updated with

$$w_{ij} \leftarrow w_{ij} + \mu \cdot [1 - \tanh^2(a_k)] \cdot e_k \cdot w_{jk} \cdot [1 - \tanh^2(a_j)] \cdot x_i \,. \tag{7.22}$$

At closer inspection, we see that the term $[1 - \tanh^2(a_k)] \cdot e_k$ corresponds to the value δ_k from equation (7.16)

$$w_{ij} \leftarrow w_{ij} + \mu \cdot \delta_k \cdot w_{jk} \cdot [1 - \tanh^2(a_j)] \cdot x_i \,. \tag{7.23}$$

If there are different neurons in layer k, which are connected with neuron j, then the contributions are additively superimposed and $\delta_k \cdot w_{jk}$ has to be substituted by the sum over all k

$$w_{ij} \leftarrow w_{ij} + \mu \cdot \left(\sum_k \delta_k \cdot w_{jk} \right) \cdot [1 - \tanh^2(a_j)] \cdot x_i \,. \tag{7.24}$$

Using δ_k, the knowledge of e_k is not needed. Hence, the changes of weights can be recursively computed starting from the output layer towards the first layer. This is called backpropagation.

Final remarks

The advantage of the backpropagation method lies in the relatively simple adjustment of the weights independent of the complexity of the entire neural network. The gradient-descent method is the underlying optimisation technique. Consequently, the adjustment of the weights is rather time-consuming.

The convergence result depends heavily on the initial weights, when using backpropagation. The adaptation of the weights is often supplemented by a momentum avoiding fast changes of the adaptation direction and allowing uphill walks. The rule

$$w_{ij}[n+1] \leftarrow w_{ij}[n] + \Delta[n]$$

changes to

$$w_{ij}[n+1] \leftarrow w_{ij}[n] + \Delta[n] + \nu \cdot \Delta[n-1] \quad \text{mit} \quad 0 < \nu < 1 \,. \tag{7.25}$$

7.2 Chauvenet's criterion for outlier detection

Given a model function $f(\mathbf{x}|\mathbf{a})$ and observations y_i, Chauvenet's criterion [Cha71] for outlier detection examines the deviates

$$\Delta_i = y_i - f(\mathbf{x}_i|\hat{\mathbf{a}}) = y_i - \hat{y}_i \,,$$

wherein y_i denotes the measurements observed under the conditions $\mathbf{x}$ and $\hat{y}_i$ denotes their estimates based on the fitting of a model function $f(\mathbf{x}|\hat{\mathbf{a}})$ with the estimated parameter vector $\hat{\mathbf{a}} = (a_1 \; a_2 \; \ldots \; a_M)$.

It is assumed that the values of the deviates are normally distributed with a mean value of zero and a deviation of σ_y

$$f(\Delta) = \frac{1}{\sqrt{2\pi} \cdot \sigma_\mathrm{y}} \cdot \exp\left[-0.5 \cdot \left(\frac{\Delta}{\sigma_\mathrm{y}} \right)^2 \right] \tag{7.26}$$

with

$$\int_{-\infty}^{+\infty} f(\Delta)\, d\Delta = 1\,. \tag{7.27}$$

The standard uncertainty σ_{y} under consideration of N observations y_i and using M model parameters is

$$\sigma = \sigma_{\mathrm{y}} = \sqrt{\frac{1}{N-M} \cdot \sum_{i=1}^{N} [y_i - f(\mathbf{x}_i|\hat{\mathbf{a}})]^2}\,. \tag{7.28}$$

As already pointed out in Section 6.1, the majority of all observations drawn from a normal distribution are not more distant to the mean than a certain multiple of its standard deviation. Chauvenet's criterion utilises this fact and investigates how likely a value with a certain distance to the mean might occur.

The *a priori* probability P_{in} that all deviates are inside the range of $-\kappa\cdot\sigma \leq \Delta \leq \kappa\cdot\sigma$ is given by

$$\begin{aligned} P_{\text{in}} &= \int_{-\kappa\cdot\sigma}^{+\kappa\cdot\sigma} \frac{1}{\sqrt{2\pi}\cdot\sigma} \cdot \exp\left[-0.5\cdot\left(\frac{\Delta}{\sigma}\right)^2\right] d\Delta \\ &= 2\cdot \int_{0}^{\kappa\cdot\sigma} \frac{1}{\sqrt{2\pi}\cdot\sigma} \cdot \exp\left[-0.5\cdot\left(\frac{\Delta}{\sigma}\right)^2\right] d\Delta \\ &= 2\cdot\left[\frac{1}{\sqrt{2\pi}\cdot\sigma}\cdot\sqrt{\frac{\pi}{2}}\cdot\sigma\cdot\operatorname{erf}\left(\frac{\Delta}{\sqrt{2}\cdot\sigma}\right)\right]_0^{\kappa\cdot\sigma} = \operatorname{erf}\left(\frac{\kappa}{\sqrt{2}}\right). \end{aligned} \tag{7.29}$$

$P_{\text{out}} = 1 - P_{\text{in}}$ is related to all possible values of Δ outside the range defined by κ

$$P_{\text{out}} = 1 - \operatorname{erf}\left(\frac{\kappa}{\sqrt{2}}\right)$$

(see **Figure 7.3** a and b).

The question is how likely a deviate outside this range can occur, when a certain number N of measurements are made. Scaled to the number of observations N, $\nu_0 = N\cdot P_{\text{out}}$ expresses the average number of deviates outside $-\kappa\cdot\sigma \leq \Delta \leq \kappa\cdot\sigma$.

Since we are interested in the threshold $\kappa\cdot\sigma$ separating the good observations from the contaminants, κ must be expressed dependent on N and ν_0. It is

$$\begin{aligned} \kappa &= \sqrt{2}\cdot[\operatorname{inverf}(1-P_{\text{out}})] \\ &= \sqrt{2}\cdot\left[\operatorname{inverf}\left(1-\frac{\nu_0}{N}\right)\right]\,. \end{aligned} \tag{7.30}$$

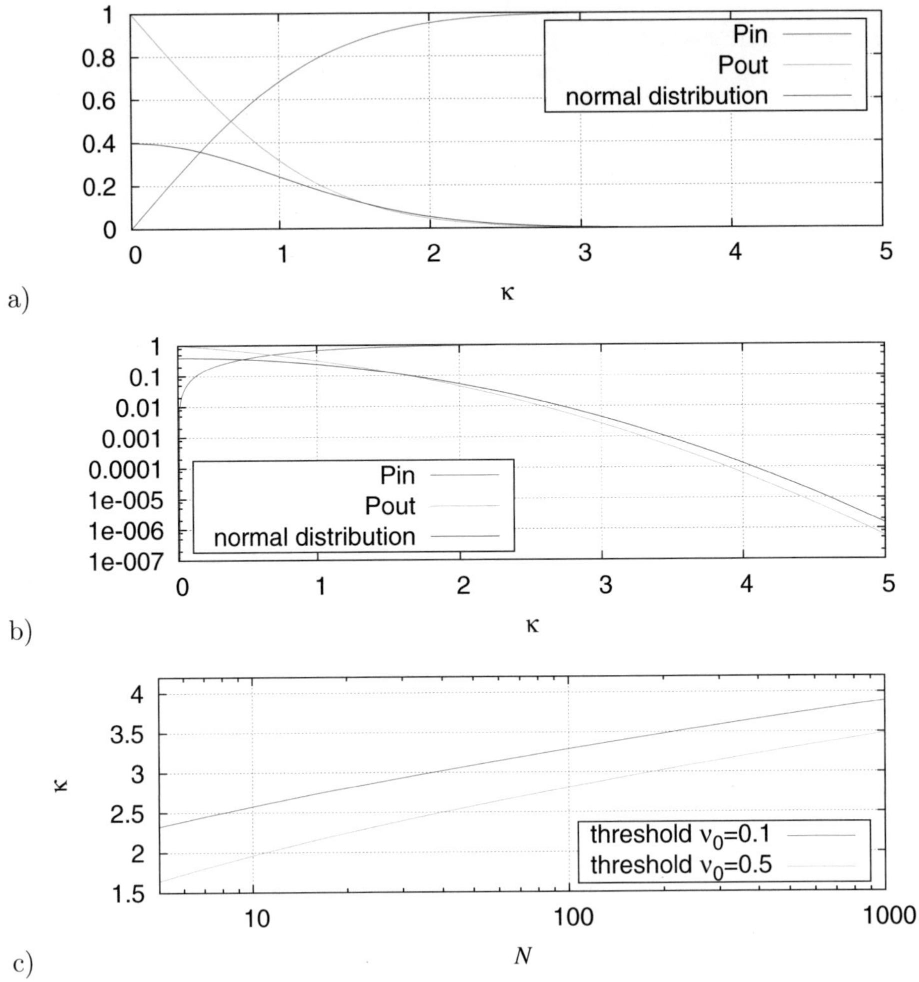

Figure 7.3: Chauvenet's criterion; a) a priori probability P_{in} that all deviates are inside the range of $-\kappa \cdot \sigma \leq \Delta \leq \kappa \cdot \sigma$; b) same as a) with logarithmic scale on y-axis; c) parameter κ for outlier detection as function of the number of observations

Using ν_0, the threshold can be adjusted to the wanted sensitivity.

Chauvenet's criterion is one of the first attempts to identify contaminants and was used from the beginning also by other scientist (e.g. [Wri84]). It yields a quite plain and logical measure for outlier detection. Chauvenet had proposed a portion of $\nu_0 = 0.5$. This, however, would imply 0.5 outliers per data set on average. In practice, the outliers would not be evenly distributed over all possible data sets. Less than fifty percent of all data sets would be declared containing outliers, but some would contain more than a single one. The value $\nu_0 = 0.5$ is obviously much too high. In principle, it is up to the implementer to choose another value ν_0. In order to tighten the limits for outliers, a lower value ν_0 should be used (**Figure 7.3** c). In many applications, κ is simply set to a value of about 3.

In the context of regression analysis, a variety of approaches to outlier detection exists, mostly tailored to the distribution of the observations or assumptions of the estimated number of possible contaminants [Bar94, Car88, Ham86]. Some scientists believe that observations should never be rejected without reliable evidence that the measurement in question is incorrect. Barnett and Lewis [Bar94] even call Chauvenet's criterion as "primitive", "unreasonable", "misguided" and "having evident short comings". The major weakness lies in the estimation of σ_{y} (7.28). Enough data must be available to have considerable confidence in σ_{y}. Besides this, the assumption of normal distributed deviates is not always fulfilled. However, dealing with computational measurements for instance, it is nevertheless a suitable method for outlier detection. For a more detailed discussion see, for instance, [Tay97].

7.3 Propagation of errors

In addition to the statements about uncertainties in Chapter 4, it might be of interest what the error is of a dependent variable $z = f(y_1, \ldots, y_j, \ldots, y_M)$ when the uncertainties of the measured variables y_j are known. This is achieved by propagation of the errors of all y_j.

Let us start with the simplified case that the error $\Delta\mathbf{y} = (\Delta y_1\ \Delta y_2\ \ldots\ \Delta y_M)$ is assumed to be known in order to derive the error in $z = f(\mathbf{y})$. This assumption will be dropped later on and the uncertainty in z will be derived.

The expansion into a Taylor series can be used to express $f(y)$ as linear superposition of power functions [Gel68]

$$\begin{aligned} f(y) &= \sum_{\nu=0}^{\infty} \frac{f^{(\nu)}(y_0)}{\nu!} \cdot (y-y_0)^{\nu} \\ &= f(y_0) + \frac{f^{(1)}(y_0)}{1} \cdot (y-y_0) + \frac{f^{(2)}(y_0)}{2} \cdot (y-y_0)^2 + \\ &\quad \frac{f^{(3)}(y_0)}{6} \cdot (y-y_0)^3 + \ldots \end{aligned}$$

for any function $f(y)$ which is defined in the vicinity of $f(y_0)$ and has infinite continuous derivatives $f^{(\nu)}(y)$. Let the known error in y be $\Delta y = y - y_0$. Then an approximation with only the first derivative $f'(y) = f^{(1)}(y)$ expands $f(y)$ around y_0

$$f(y) \simeq f(y_0) + f'(y_0) \cdot (y - y_0) = f(y_0) + f'(y_0) \cdot \Delta y$$

For higher dimensional functions $\mathbf{y} = (y_1 \cdots y_j \cdots y_M)$ this yields

$$f(\mathbf{y}) \simeq f(\mathbf{y}_0) + \sum_{j=1}^{M} \frac{\partial f(\mathbf{y}_0)}{\partial y_j} \cdot \Delta y_j$$

or

$$\Delta f(\mathbf{y}) = f(\mathbf{y}) - f(\mathbf{y}_0) \simeq \sum_{j=1}^{M} \frac{\partial f(\mathbf{y}_0)}{\partial y_j} \cdot \Delta y_j \,. \tag{7.31}$$

Let us consider the example of Ohm's law

$$R = U/I \,.$$

In an experiment, the voltage U and the current I are measured in order to get some information about the resistor R. The errors might be $\Delta U = U - U_0$ and $\Delta I = I - I_0$. Equation (7.31) reads now as follows

$$\Delta R \simeq \Delta U \cdot \left.\frac{\partial R}{\partial U}\right|_{U_0,I_0} + \Delta I \cdot \left.\frac{\partial R}{\partial I}\right|_{U_0,I_0} = \Delta U \cdot \frac{1}{I_0} + \Delta I \cdot \frac{U_0}{I_0^2}$$

with ΔR as error in R.

Unfortunately, the errors in $\mathbf{y}$ are typically not known, but only their distributions, for instance in terms of the standard uncertainty. Hence, we have to find another description of the error propagation. Let us assume that the most probable value for z is given by

$$\overline{z} = f(\overline{y}_1, \ldots, \overline{y}_j, \ldots, \overline{y}_M)$$

then the uncertainty in the resulting value z is found by considering individual observations $y_{j,1}$, $y_{j,2}$, ...

$$z_i = f(y_{1,i}, \ldots, y_{j,i}, \ldots, y_{M,i}) \ .$$

where $y_{j,i}$ is the ith instance of the jth element of $\mathbf{y}$. The estimated variance of z is

$$\sigma_z^2 = \frac{1}{N-1} \sum_{i=1}^{N} (z_i - \overline{z})^2 \ . \tag{7.32}$$

Analog to equation (7.31), we can express the deviates $z_i - \overline{z}$, i.e. the simple errors, by [Bev92]

$$z_i - \overline{z} \simeq \sum_{j=1}^{M} \frac{\partial f(\mathbf{y}_0)}{\partial y_j} \cdot (y_{j,i} - \overline{y}_j) \tag{7.33}$$

The combination of equations (7.32) and (7.33) yields

$$\begin{aligned}
\sigma_z^2 &\simeq \frac{1}{N-1} \sum_{i=1}^{N} \left[\sum_{j=1}^{M} \frac{\partial f(\mathbf{y}_0)}{\partial y_j} \cdot (y_{j,i} - \overline{y}_j) \right]^2 \\
&\simeq \frac{1}{N-1} \sum_{i=1}^{N} \left[\frac{\partial f(\mathbf{y}_0)}{\partial y_1} \cdot (y_{1,i} - \overline{y}_1) + \frac{\partial f(\mathbf{y}_0)}{\partial y_2} \cdot (y_{2,i} - \overline{y}_2) + \cdots \right]^2 \\
&\simeq \frac{1}{N-1} \sum_{i=1}^{N} \left[\left(\frac{\partial f(\mathbf{y}_0)}{\partial y_1} \right)^2 \cdot (y_{1,i} - \overline{y}_1)^2 + \left(\frac{\partial f(\mathbf{y}_0)}{\partial y_2} \right)^2 \cdot (y_{2,i} - \overline{y}_2)^2 + \right. \\
&\qquad \left. 2 \cdot \frac{\partial f(\mathbf{y}_0)}{\partial y_1} \cdot \frac{\partial f(\mathbf{y}_0)}{\partial y_2} \cdot (y_{1,i} - \overline{y}_1) \cdot (y_{2,i} - \overline{y}_2) + \cdots \right]
\end{aligned} \tag{7.34}$$

A reordering of the terms results to

$$\begin{aligned}
\sigma_z^2 \simeq\ & \left(\frac{\partial f(\mathbf{y}_0)}{\partial y_1} \right)^2 \cdot \frac{1}{N-1} \sum_{i=1}^{N} (y_{1,i} - \overline{y}_1)^2 + \\
& \left(\frac{\partial f(\mathbf{y}_0)}{\partial y_2} \right)^2 \cdot \frac{1}{N-1} \sum_{i=1}^{N} (y_{2,i} - \overline{y}_2)^2 + \cdots + \\
& 2 \cdot \frac{\partial f(\mathbf{y}_0)}{\partial y_1} \cdot \frac{\partial f(\mathbf{y}_0)}{\partial y_2} \cdot \frac{1}{N-1} \sum_{i=1}^{N} (y_{1,i} - \overline{y}_1) \cdot (y_{2,i} - \overline{y}_2) + \cdots
\end{aligned} \tag{7.35}$$

It becomes apparent that the first two sums express the variances of y_1 and y_2, respectively

$$\sigma_{y_1}^2 = \frac{1}{N-1}\sum_{i=1}^{N}(y_{1,i} - \overline{y}_1)^2 \qquad \sigma_{y_2}^2 = \frac{1}{N-1}\sum_{i=1}^{N}(y_{2,i} - \overline{y}_2)^2 \,, \tag{7.36}$$

whereas the third sum corresponds to the covariance of y_1 and y_2

$$\sigma_{y_1,y_2}^2 = \frac{1}{N-1}\sum_{i=1}^{N}(y_{1,i} - \overline{y}_1)\cdot(y_{2,i} - \overline{y}_2) \tag{7.37}$$

Hence, we can write (7.35) as

$$\begin{aligned}\sigma_z^2 \;\simeq\; & \left(\frac{\partial f(\mathbf{y}_0)}{\partial y_1}\right)^2 \cdot \sigma_{y_1}^2 + \left(\frac{\partial f(\mathbf{y}_0)}{\partial y_2}\right)^2 \cdot \sigma_{y_2}^2 + \cdots \\ & + 2\frac{\partial f(\mathbf{y}_0)}{\partial y_1}\frac{\partial f(\mathbf{y}_0)}{\partial y_2} \cdot \sigma_{y_1,y_2}^2 + \cdots \end{aligned} \tag{7.38}$$

what is called *error propagation equation*. In case of uncorrelated deviates $(y_{1,i} - \overline{y}_1)$ and $(y_{2,i} - \overline{y}_2)$, the contribution of the sum in (7.37) will be positive or negative in equal parts, that is, they average out for large N and the covariance vanishes leading to the simplified formula

$$\sigma_z^2 \approx \left(\frac{\partial f(\mathbf{y}_0)}{\partial y_1}\right)^2 \cdot \sigma_{y_1}^2 + \left(\frac{\partial f(\mathbf{y}_0)}{\partial y_2}\right)^2 \cdot \sigma_{y_2}^2 + \cdots \,. \tag{7.39}$$

Regarding the aforementioned example of Ohm's law $R = U/I$, the variance in R is

$$\sigma_{\mathrm{R}}^2 \approx \left(\frac{1}{I_0}\right)^2 \cdot \sigma_{\mathrm{U}}^2 + \left(\frac{U_0}{I_0^2}\right)^2 \cdot \sigma_{\mathrm{I}}^2 \,.$$

7.4 Manual calculation of linear least squares

The general solution given in equation (2.10)

$$\mathbf{a} = \left(\mathbf{J}^{\mathrm{T}} \cdot \mathbf{W} \cdot \mathbf{J}\right)^{-1} \cdot \mathbf{J}^{\mathrm{T}} \cdot \mathbf{W} \cdot \mathbf{y}$$

results in optimal parameters for all linear minimisation problems in terms of least squares. In general, the model parameters can also be calculated 'by hand'. This is the initially used technique in the early years of least-squares history and exemplified in the following for the fitting of a straight line (see also Subsection 1.4.2) and for the determination of optimal filter coefficients (Wiener filter).

7.4.1 Fitting of a straight line using manual derivation

The regression of a straight line is based on the model function

$$y = f(\mathbf{x}|\mathbf{a}) = a_1 + a_2 \cdot x \; .$$

The setup with respect to the minimisation task is the same as before

$$\chi^2 = \sum_{i=1}^{N} w_i \cdot [f(\mathbf{x}_i|\mathbf{a}) - y_i]^2 = \sum_{i=1}^{N} w_i \cdot [(a_1 + a_2 \cdot x_i) - y_i]^2 \longrightarrow \text{Min} \; .$$

Minima are characterised by gradients equal to zero, thus we solve at first the square operator

$$\begin{aligned} \chi^2 &= \sum_{i=1}^{N} w_i \cdot \left[(a_1 + a_2 \cdot x_i)^2 - 2(a_1 + a_2 \cdot x_i)y_i + y_i^2\right] \\ &= \sum_{i=1}^{N} w_i \cdot \left[(a_1^2 + 2a_1a_2x_i + a_2^2x_i^2) - 2a_1y_i - 2a_2x_iy_i + y_i^2\right] \end{aligned}$$

and set the partial derivatives to zero

$$\begin{aligned} \frac{\partial \chi^2}{\partial a_1} &= \sum_{i=1}^{N} w_i \cdot [2a_1 + 2a_2x_i - 2y_i] &= 0 \\ \frac{\partial \chi^2}{\partial a_2} &= \sum_{i=1}^{N} w_i \cdot \left[2a_1x_i + 2a_2x_i^2 - 2x_iy_i\right] &= 0 \; . \end{aligned}$$

Separate summation yields

$$\begin{aligned} \sum_{i=1}^{N} w_i \cdot a_1 + \sum_{i=1}^{N} w_i \cdot a_2x_i - \sum_{i=1}^{N} w_i \cdot y_i &= 0 \\ \sum_{i=1}^{N} w_i \cdot a_1x_i + \sum_{i=1}^{N} w_i \cdot a_2x_i^2 - \sum_{i=1}^{N} w_i \cdot x_iy_i &= 0 \; . \end{aligned}$$

Now, the terms containing elements from **a** have to be separated from others

$$\begin{aligned} a_1 \sum_{i=1}^{N} w_i + a_2 \sum_{i=1}^{N} w_i \cdot x_i &= \sum_{i=1}^{N} w_i \cdot y_i \\ a_1 \sum_{i=1}^{N} w_i \cdot x_i + a_2 \sum_{i=1}^{N} w_i \cdot x_i^2 &= \sum_{i=1}^{N} w_i \cdot x_iy_i \; . \end{aligned}$$

Using matrices we can write

$$\begin{pmatrix} \sum\limits_{i=1}^{N} w_i & \sum\limits_{i=1}^{N} w_i x_i \\ \sum\limits_{i=1}^{N} w_i x_i & \sum\limits_{i=1}^{N} w_i x_i^2 \end{pmatrix} \cdot \begin{pmatrix} a_1 \\ a_2 \end{pmatrix} = \begin{pmatrix} \sum\limits_{i=1}^{N} w_i y_i \\ \sum\limits_{i=1}^{N} w_i x_i y_i \end{pmatrix}$$

and the solution for $\mathbf{a}$ is obviously

$$\begin{pmatrix} a_1 \\ a_2 \end{pmatrix} = \begin{pmatrix} \sum\limits_{i=1}^{N} w_i & \sum\limits_{i=1}^{N} w_i x_i \\ \sum\limits_{i=1}^{N} w_i x_i & \sum\limits_{i=1}^{N} w_i x_i^2 \end{pmatrix}^{-1} \cdot \begin{pmatrix} \sum\limits_{i=1}^{N} w_i y_i \\ \sum\limits_{i=1}^{N} w_i x_i y_i \end{pmatrix} . \tag{7.40}$$

Not surprisingly, this is exact the same result as in equation (2.16). Using the shortcuts

$$S \equiv \sum_{i=1}^{N} w_i , \qquad S_{\mathrm{x}} \equiv \sum_{i=1}^{N} w_i x_i , \qquad S_{\mathrm{xx}} \equiv \sum_{i=1}^{N} w_i x_i^2 ,$$

$$S_{\mathrm{y}} \equiv \sum_{i=1}^{N} w_i y_i , \qquad S_{\mathrm{xy}} \equiv \sum_{i=1}^{N} w_i x_i y_i$$

and applying equation (5.6) for matrix inversion result in

$$\mathbf{a} = \frac{1}{S \cdot S_{\mathrm{xx}} - (S_{\mathrm{x}})^2} \begin{pmatrix} S_{\mathrm{xx}} & -S_{\mathrm{x}} \\ -S_{\mathrm{x}} & S \end{pmatrix} \begin{pmatrix} S_{\mathrm{y}} \\ S_{\mathrm{xy}} \end{pmatrix} . \tag{7.41}$$

i.e., the single parameters can be calculated as follows

$$\begin{aligned} a_1 &= \frac{S_{\mathrm{xx}} \cdot S_{\mathrm{y}} - S_{\mathrm{x}} \cdot S_{\mathrm{xy}}}{S \cdot S_{\mathrm{xx}} - (S_{\mathrm{x}})^2} \\ a_2 &= \frac{S \cdot S_{\mathrm{xy}} - S_{\mathrm{x}} S_{\mathrm{y}}}{S \cdot S_{\mathrm{xx}} - (S_{\mathrm{x}})^2} . \end{aligned}$$

The second derivatives have not to be check for the discrimination between maxima and minima. The found extremum is always a minimum because of $w_i \geq 0$, $[f(\mathbf{x}|\mathbf{a}) - y]^2 \geq 0$ and the typical shape of a parabola.

The first two factors of (7.41) are equal to the covariance matrix which can be used for the evaluation of the uncertainty of the parameter values

$$\begin{aligned}\sigma_{a_1}^2 &= g_{\text{fit}} \cdot \frac{S_{\text{xx}}}{S \cdot S_{\text{xx}} - (S_{\text{x}})^2} \\ &= g_{\text{fit}} \cdot \frac{\sum\limits_{i=1}^{N} w_i x_i^2}{\sum\limits_{i=1}^{N} w_i \cdot \sum\limits_{i=1}^{N} w_i x_i^2 - \left(\sum\limits_{i=1}^{N} w_i x_i\right)^2}\end{aligned}$$

and

$$\begin{aligned}\sigma_{a_2}^2 &= g_{\text{fit}} \cdot \frac{S}{S \cdot S_{\text{xx}} - (S_{\text{x}})^2} \\ &= g_{\text{fit}} \cdot \frac{\sum\limits_{i=1}^{N} w_i}{\sum\limits_{i=1}^{N} w_i \cdot \sum\limits_{i=1}^{N} w_i x_i^2 - \left(\sum\limits_{i=1}^{N} w_i x_i\right)^2} \,.\end{aligned}$$

The goodness-of-fit g_{fit}

$$g_{\text{fit}} = \frac{\chi^2}{N_{\text{free}}} = \frac{1}{N_{\text{free}}} \cdot \sum_{i=1}^{N} w_i \cdot [y_i - f(\mathbf{x}_i|\mathbf{a})]^2 \ ,$$

compensates for the constant factor k introduced by weights $w_i = k/\sigma_i^2$, see Chapter 4, eqs. (4.1) and (4.5).

7.4.2 Wiener filter

The basic idea of Wiener[2] filters had been already introduced in Section 1.4.5, page 19. The impulse response of a digital filter is to be determined aiming at a minimum energy of the error signal $e[n] = d[n] - y[n]$, **Figure 7.4**. The values of a given reference signal $d[n]$ correspond to the observations $d[n] \to y_i$, while the input signal values $x[n] \to x_{i,j}$ are the conditions, and the coefficients of the impulse response are the parameters $g[n] \to a_j$ to be determined. The filter output $y[n]$ is described by the model function

$$y[n] = x[n] * g[n] = \sum_{k=0}^{M-1} g[k] \cdot x[n-k]$$

$$y_i = f(\mathbf{x}|\mathbf{a}) = a_1 \cdot x_{i,1} + a_2 \cdot x_{i,2} + \cdots a_M \cdot x_{i,M} \,.$$

[2] Named after the American mathematician Norbert Wiener (1894 – 1964)

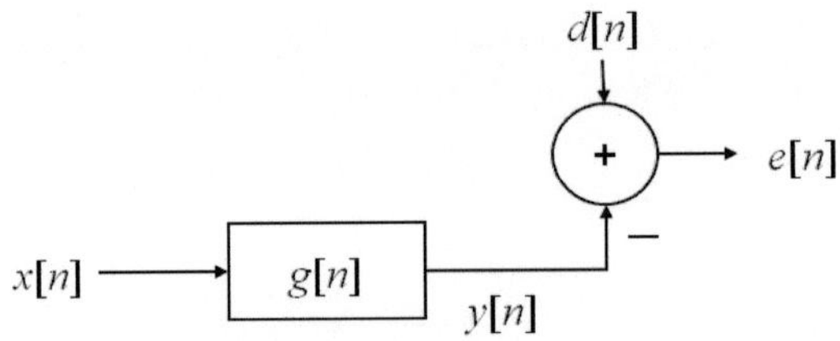

Figure 7.4: Set-up or the optimisation of the impulse response $g[n]$. See text for details.

The goal is to minimise the energy of the squared error (energy of the error signal), thus, it is a least-squares problem:

$$\begin{aligned}
E &= \sum_{n=-\infty}^{\infty} (d[n] - y[n])^2 \Longrightarrow \text{Min.} \\
&= \sum_{n=-\infty}^{\infty} \left(d[n] - \sum_{k=0}^{M-1} g[k] \cdot x[n-k] \right)^2 \Longrightarrow \text{Min.} \\
&= \sum_{n=-\infty}^{\infty} \left[d^2[n] - 2 \cdot d[n] \cdot \sum_{k=0}^{M-1} g[k] \cdot x[n-k] + \left(\sum_{k=0}^{M-1} g[k] \cdot x[n-k] \right)^2 \right] \\
&= \sum_n d^2[n] - 2 \sum_{k=0}^{M-1} g[k] \cdot \sum_n d[n] \cdot x[n-k] + \sum_n \left(\sum_{k=0}^{M-1} g[k] \cdot x[n-k] \right)^2 .
\end{aligned}$$

The observations, i.e. the signal values of the reference signal $d[n]$, are assumed of having constant reliability, hence, the usage of weights is not necessary. We substitute $a_k = g[k]$, determine the partial derivatives $\partial/\partial a_i$ $(i = 0, 1, 2, \ldots, M-1)$ of E and set the derivatives to zero

$$\frac{\partial}{\partial a_i} E = 0 \qquad \forall i$$

$$\begin{aligned}
0 &= \frac{\partial}{\partial a_i} \left\{ -2 \sum_{k=0}^{M-1} a_k \cdot \sum_n d[n] \cdot x[n-k] + \sum_n \left(\sum_{k=0}^{M-1} a_k \cdot x[n-k] \right)^2 \right\} \\
&= -2 \sum_n d[n] \cdot x[n-i] + \sum_n \left[2 \left(\sum_{k=0}^{M-1} a_k \cdot x[n-k] \right) \cdot \frac{\partial}{\partial a_i} \left(\sum_{k=0}^{M-1} a_k \cdot x[n-k] \right) \right] .
\end{aligned}$$

Within the last sum, there is a single term containing a_i. So, the derivative is merely $x[n-i]$:

$$0 = -\sum_n d[n] \cdot x[n-i] + \sum_n \left[\sum_{k=0}^{M-1} a_k \cdot x[n-k] \cdot x[n-i] \right] . \tag{7.42}$$

Equation (7.42) can be transposed to

$$\begin{aligned} \sum_{n=-\infty}^{\infty} d[n] \cdot x[n-i] &= \sum_{k=0}^{M-1} a_k \cdot \sum_{n=-\infty}^{\infty} x[n-k] \cdot x[n-i] \\ &= \sum_{k=0}^{M-1} a_k \cdot \sum_{n=-\infty}^{\infty} x[n] \cdot x[n-(i-k)] . \end{aligned} \tag{7.43}$$

Considering a summation over an infinite range $-\infty < n < \infty$ and exchanging factors we find

$$\begin{aligned} &\sum_n x[n] \cdot x[n-(i-k)] = \sum_n x[n+(i-k)] \cdot x[n] \\ =&\sum_n x[n-(k-i)] \cdot x[n] = \sum_n x[n] \cdot x[n-(k-i)] \\ &\rightsquigarrow (i-k) \sim (k-i) \sim |i-k| . \end{aligned} \tag{7.44}$$

Furthermore, the autocorrelation function of a signal $x[n]$ is defined by

$$r_{xx}[m] = \sum_{n=-\infty}^{\infty} x[n] \cdot x[n-m]$$

and the cross-correlation between $d[n]$ and $x[n]$ by

$$r_{dx}[m] = \sum_{n=-\infty}^{\infty} d[n] \cdot x[n-m]$$

Applied to (7.43) and using (7.44) we get

$$r_{dx}[i] = \sum_{k=0}^{M-1} a_k \cdot r_{xx}[|i-k|] \tag{7.45}$$

or

$$\mathbf{r}_{dx} = \mathbf{R}_{xx} \cdot \mathbf{a}$$

with

$$\mathbf{r}_{dx} = \begin{pmatrix} r_{dx}[0] \\ r_{dx}[1] \\ \vdots \\ r_{dx}[i] \\ \vdots \\ r_{dx}[M-1] \end{pmatrix}, \qquad \mathbf{a} = \begin{pmatrix} a_0 \\ a_1 \\ \vdots \\ a_k \\ \vdots \\ a_{M-1} \end{pmatrix},$$

and

$$\mathbf{R}_{xx} = \begin{pmatrix} r_{xx}[0] & r_{xx}[1] & r_{xx}[2] & \cdots & r_{xx}[M-1] \\ r_{xx}[1] & r_{xx}[0] & r_{xx}[1] & \cdots & r_{xx}[M-2] \\ r_{xx}[2] & r_{xx}[1] & r_{xx}[0] & \cdots & r_{xx}[M-3] \\ \vdots & \vdots & \ddots & \vdots & \\ r_{xx}[M-1] & r_{xx}[M-2] & r_{xx}[M-3] & \cdots & r_{xx}[0] \end{pmatrix}.$$

As the autocorrelation matrix $\mathbf{R}_{xx}$ is symmetric and has Toeplitz[3] form[4], it is always invertible. The filter parameters are computed with

$$\mathbf{a} = \mathbf{R}_{xx}^{-1} \cdot \mathbf{r}_{dx} \,. \tag{7.46}$$

Solving (7.46) based on a limited number of input data is equivalent to the least-squares estimation algorithm using

$$\mathbf{a} = \left(\mathbf{J}^{\mathrm{T}} \cdot \mathbf{J}\right)^{-1} \cdot \mathbf{J}^{\mathrm{T}} \cdot \mathbf{d} \qquad \text{with} \qquad \mathbf{R}_{xx} = \mathbf{J}^{\mathrm{T}} \cdot \mathbf{J}, \qquad \mathbf{r}_{dx} = \mathbf{J}^{\mathrm{T}} \cdot \mathbf{d}\,,$$

and the desired signal (observations) $\mathbf{d} = (d[0] \ \cdots \ d[n] \ \cdots \ d[N-1])^{\mathrm{T}}$.

7.5 Combined treatment of different model functions

The joint processing of different model functions becomes important (i) when these functions depend on at least one common parameter and/or (ii) the observations are not scalars y_i but vectors $\mathbf{y}_i$ and weights have to be assigned to these vectors instead to their inseparable elements.

Both of these reasons are valid in the example of image alignment described in Subsection 2.6.7. Let us assume that we have taken two snap-shots from the

[3] Named after the German mathematician Otto Toeplitz (1881-1940).

[4] A Toeplitz matrix is a matrix in which each descending diagonal from left to right contains constant values.

same scene. The position of the camera, however, has slightly changed between the two shots in terms of translation in parallel to the image plane and rotation around the optical axis. Consequently, the second image is a translated and slightly rotated version of the first image.

If it is possible to identify pairs of corresponding points (landmarks), the parameters of the camera movement can be estimated and the images can be aligned to each other.

The vectors $\mathbf{u}_i = (u_i \; v_i)^{\mathrm{T}}$ are the landmark coordinates in the first image (conditions) and the vectors $\mathbf{x}_i = (x_i \; y_i)^{\mathrm{T}}$ are the corresponding coordinates in a second image (observations), which were found, for instance, by a matching algorithm. The observations are points (x_i, y_i) in two-dimensional space

$$\begin{pmatrix} x_i \\ y_i \end{pmatrix} = \begin{pmatrix} a_1 \\ a_2 \end{pmatrix} + \begin{pmatrix} \cos(a_3) & -\sin(a_3) \\ \sin(a_3) & \cos(a_3) \end{pmatrix} \cdot \begin{pmatrix} u_i \\ v_i \end{pmatrix} ,$$

depending on conditions (u_i, v_i) which are coordinates as well. The parameter vector $(a_1 \; a_2)^{\mathrm{T}}$ describes the translational component, while the matrix

$$\begin{pmatrix} \cos(a_3) & -\sin(a_3) \\ \sin(a_3) & \cos(a_3) \end{pmatrix}$$

describes the rotation.

Actually, there are two model functions involved with $\mathbf{u} = (u \; v)^{\mathrm{T}}$, namely

$$f_{\mathrm{x}}(\mathbf{u}|\mathbf{a}) = \quad x \quad = a_1 + \cos(a_3) \cdot u - \sin(a_3) \cdot v$$

and

$$f_{\mathrm{y}}(\mathbf{u}|\mathbf{a}) = \quad y \quad = a_2 + \sin(a_3) \cdot u + \cos(a_3) \cdot v$$

which are linked via a_3. Based on an extension of the maximum-likelihood principle (see Section 6.2) towards two-dimensional normal distributions, the minimisation problem can be formulated as

$$\begin{aligned} \chi^2 &= \frac{1}{2}\sum_{i=1}^{N} w_i \cdot \left\{[x_i - f_{\mathrm{x}}(\mathbf{u}_i|\mathbf{a})]^2 + [y_i - f_{\mathrm{y}}(\mathbf{u}_i|\mathbf{a})]^2\right\} \longrightarrow \text{Min} \\ &= \frac{1}{2}\sum_{i=1}^{N} w_i \cdot \left\{[x_i - f_{\mathrm{x}}(\mathbf{u}_i|\mathbf{a})]^2\right\} + \frac{1}{2}\sum_{i=1}^{N} w_i \cdot \left\{[y_i - f_{\mathrm{y}}(\mathbf{u}_i|\mathbf{a})]^2\right\} \longrightarrow \text{Min} \end{aligned}$$

With respect to equation (2.6) for updating the model parameters, it is somewhat unclear in this setup how the Jacobian matrix and the vector of residuals, eqs.

(2.3) and (2.4), should be filled. Instead, the direct use of Gauss-Newton method, eq. (6.20),

$$\mathbf{\Delta a} = -\mathbf{H}^{-1} \cdot \mathbf{g}$$

is proposed to solve this problem. The elements of $\mathbf{g}$ are equal to the derivatives of χ^2

$$g_j = \frac{\partial \chi^2(\mathbf{a})}{\partial a_j} ,$$

i.e.,

$$\begin{aligned} g_j = -\sum_{i=1}^{N} w_i \cdot & \left\{ \frac{\partial f_{\mathrm{x}}(\mathbf{u}_i|\mathbf{a})}{\partial a_j} \cdot [x_i - f_{\mathrm{x}}(\mathbf{u}_i|\mathbf{a})] \right\} - \\ \sum_{i=1}^{N} w_i \cdot & \left\{ \frac{\partial f_{\mathrm{y}}(\mathbf{u}_i|\mathbf{a})}{\partial a_j} \cdot [y_i - f_{\mathrm{y}}(\mathbf{u}_i|\mathbf{a})] \right\} , \end{aligned} \tag{7.47}$$

with j comprising all parameters from both model functions. Analogously to eq. (6.21) this can be written in matrix notation

$$\mathbf{g} = -\mathbf{J}_{\mathrm{x}}^{\mathrm{T}} \cdot \mathbf{W} \cdot \mathbf{r}_{\mathrm{x}} - \mathbf{J}_{\mathrm{y}}^{\mathrm{T}} \cdot \mathbf{W} \cdot \mathbf{r}_{\mathrm{y}} , \tag{7.48}$$

where the indices 'x' and 'y' refer to the model functions $f_{\mathrm{x}}(\mathbf{u}_i|\mathbf{a})$ and $f_{\mathrm{y}}(\mathbf{u}_i|\mathbf{a})$, respectively. $\mathbf{W}$ is jointly used for both functions.

According to eq. (6.19), the elements of the Hessian matrix $\mathbf{H}$ are defined as

$$H_{jk} = \frac{\partial^2 \chi^2(\mathbf{a})}{\partial a_j \partial a_k} = \frac{\partial g_j}{\partial a_k}$$

leading to

$$\begin{aligned} H_{jk} = & \sum_{i=1}^{N} w_i \cdot \left\{ \frac{\partial f_{\mathrm{x}}(\mathbf{u}_i|\mathbf{a})}{\partial a_j} \cdot \frac{\partial f_{\mathrm{x}}(\mathbf{u}_i|\mathbf{a})}{\partial a_k} - \frac{\partial^2 f_{\mathrm{x}}(\mathbf{u}_i|\mathbf{a})}{\partial a_j \partial a_k} \cdot [x_i - f_{\mathrm{x}}(\mathbf{u}_i|\mathbf{a})] \right\} + \\ & \sum_{i=1}^{N} w_i \cdot \left\{ \frac{\partial f_{\mathrm{y}}(\mathbf{u}_i|\mathbf{a})}{\partial a_j} \cdot \frac{\partial f_{\mathrm{y}}(\mathbf{u}_i|\mathbf{a})}{\partial a_k} - \frac{\partial^2 f_{\mathrm{y}}(\mathbf{u}_i|\mathbf{a})}{\partial a_j \partial a_k} \cdot [y_i - f_{\mathrm{y}}(\mathbf{u}_i|\mathbf{a})] \right\} \end{aligned} \tag{7.49}$$

and

$$\mathbf{H} = \mathbf{Q}_{\mathrm{x}} + \mathbf{J}_{\mathrm{x}}^{\mathrm{T}} \cdot \mathbf{W} \cdot \mathbf{J}_{\mathrm{x}} + \mathbf{Q}_{\mathrm{y}} + \mathbf{J}_{\mathrm{y}}^{\mathrm{T}} \cdot \mathbf{W} \cdot \mathbf{J}_{\mathrm{y}} . \tag{7.50}$$

There are only two differences compared to the derivations in Section 6.4.2: (i) the sizes of $\mathbf{g}$ and $\mathbf{H}$ are based on the total number of parameters of both model functions and (ii) both, $\mathbf{g}$ and $\mathbf{H}$, are result of the summation of two components.

The next subsections will explain the combined treatment of model functions based on examples.

7.5.1 Example 1: Coordinate transformation

The concrete contents of **g** and **H** in the case of the aforementioned example of image alignment via coordinate transformation have to be derived based on the minimisation of

$$\chi^2 = \frac{1}{2}\sum_{i=1}^{N} w_i \cdot \Big\{[x_i - (a_1 + \cos(a_3)\cdot u_i - \sin(a_3)\cdot v_i)]^2 + [y_i - (a_2 + \sin(a_3)\cdot u_i + \cos(a_3)\cdot v_i)]^2\Big\} \longrightarrow \text{Min} .$$

Let us define

$$\begin{aligned}\beta_1 &= x_i - [a_1 + \cos(a_3)\cdot u_i - \sin(a_3)\cdot v_i]\\ \beta_2 &= y_i - [a_2 + \sin(a_3)\cdot u_i + \cos(a_3)\cdot v_i] \ ,\end{aligned}$$

then the elements of **g** can be written as

$$\begin{aligned}g_1 &= \frac{\partial\chi^2}{\partial a_1} = -\sum_{i=1}^{N} w_i\cdot\beta_1 \ ,\\ g_2 &= \frac{\partial\chi^2}{\partial a_2} = -\sum_{i=1}^{N} w_i\cdot\beta_2 \ ,\\ g_3 &= \frac{\partial\chi^2}{\partial a_3} = \sum_{i=1}^{N} w_i\cdot\{\beta_1\cdot[u_i\cdot\sin(a_3) + v_i\cdot\cos(a_3)]\\ &\qquad + \beta_2\cdot[-u_i\cdot\cos(a_3) + v_i\cdot\sin(a_3)]\}\\ &= \sum_{i=1}^{N} w_i\cdot\{[(x_i - a_1)\cdot u_i + (y_i - a_2)\cdot v_i]\cdot\sin(a_3) + \\ &\qquad [(x_i - a_1)\cdot v_i - (y_i - a_2)\cdot u_i]\cdot\cos(a_3)\} \ ,\end{aligned}$$

and the elements of **H** are

$$h_{11} = \frac{\partial^2\chi^2}{\partial^2 a_1} = \sum_{i=1}^{N} w_i \ , \qquad h_{22} = \frac{\partial^2\chi^2}{\partial^2 a_2} = \sum_{i=1}^{N} w_i$$

$$h_{12} = h_{21} = \frac{\partial^2\chi^2}{\partial a_1\partial a_2} = \frac{\partial^2\chi^2}{\partial a_2\partial a_1} = 0$$

$$h_{13} = h_{31} = \frac{\partial^2\chi^2}{\partial a_1\partial a_3} = \frac{\partial^2\chi^2}{\partial a_3\partial a_1} = \sum_{i=1}^{N} w_i\cdot[-u_i\cdot\sin(a_3) - v_i\cdot\cos(a_3)] \ ,$$

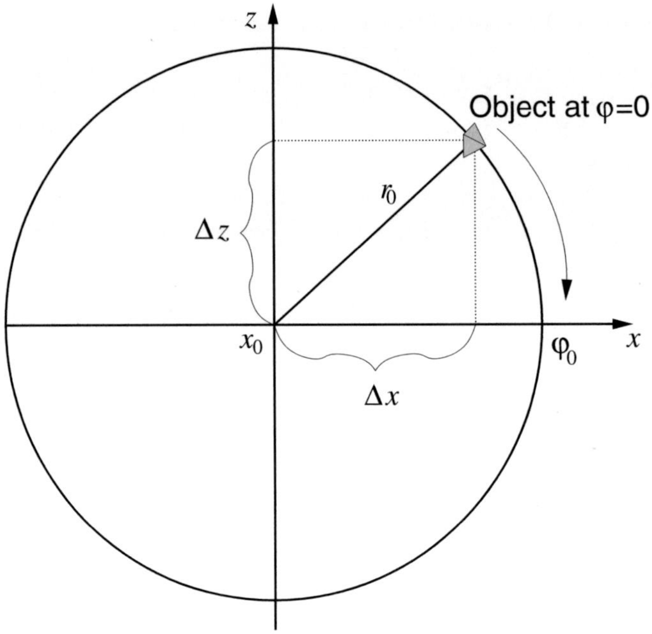

Figure 7.5: Circular movement of an object around the rotation axis (y axis) perpendicular to the x and z axes. The direction of the projection is parallel to the z axis.

$$\begin{aligned} h_{23} = h_{32} &= \frac{\partial^2 \chi^2}{\partial a_2 \partial a_3} = \frac{\partial^2 \chi^2}{\partial a_3 \partial a_2} \\ &= \sum_{i=1}^{N} w_i \cdot [u_i \cdot \cos(a_3) - v_i \cdot \sin(a_3)] \ , \end{aligned}$$

$$\begin{aligned} h_{33} = \frac{\partial^2 \chi^2}{\partial^2 a_3} = \sum_{i=1}^{N} w_i \cdot \{ &[(x_i - a_1) \cdot u_i + (y_i - a_2) \cdot v_i] \cdot \cos(a_3) - \\ &[(x_i - a_1) \cdot v_i - (y_i - a_2) \cdot u_i] \cdot \sin(a_3)\} \ . \end{aligned}$$

7.5.2 Example 2: Circular movement

The modelling of circular object movements is a similar scenario (**Figure 7.5**). Let us imagine that only the projection of the object onto the x-y plane can be observed by a camera and let us assume in addition that the camera is aligned such that the object moves along a horizontal line in the image, i.e., the y coordinate is constant, whereas the x coordinate follows the model function

$$x = x_0 + r_0 \cdot \cos(\varphi - \varphi_0),$$

which can be turned into a linear function (with respect to $\mathbf{a} = (a_1\ a_2\ a_3)^{\mathrm{T}}$)

$$x = a_1 + a_2 \cdot \cos(\varphi) + a_3 \cdot \sin(\varphi)$$

with

$$a_1 = x_0, \quad a_2 = r_0 \cdot \cos(\varphi_0), \quad \text{and} \quad a_3 = r_0 \cdot \sin(\varphi_0)\ ,$$

according to the expositions in Section 1.3.

If the observed coordinates (x, y) have to be regarded as an inseparable unit, i.e., the x value cannot be measured correctly if the observation of y is wrong and vice versa, then the combined treatment should be applied. The minimisation problem for this example is

$$\begin{aligned}
\chi^2 &= \frac{1}{2}\sum_{i=1}^{N} w_i \cdot \left\{[x_i - f_x(\mathbf{u}_i|\mathbf{a})]^2 + [y_i - f_y(\mathbf{u}_i|\mathbf{a})]^2\right\} \longrightarrow \text{Min} \\
&= \frac{1}{2}\sum_{i=1}^{N} w_i \cdot \left\{[x_i - (a_1 + a_2 \cdot \cos(\varphi_i) + a_3 \cdot \sin(\varphi_i))]^2 \right. \\
&\qquad\qquad\qquad\qquad \left. + [y_i - a_4]^2\right\} \longrightarrow \text{Min}\ .
\end{aligned}$$

Using eqs. (7.47) and (7.49), the elements of $\mathbf{g}$ and $\mathbf{H}$ are

$$\begin{aligned}
\frac{\partial\chi^2}{\partial a_1} &= -\sum_{i=1}^{N} w_i \cdot [x_i - (a_1 + a_2 \cdot \cos(\varphi_i) + a_3 \cdot \sin(\varphi_i))]\ , \\
\frac{\partial\chi^2}{\partial a_2} &= -\sum_{i=1}^{N} w_i \cdot [x_i - (a_1 + a_2 \cdot \cos(\varphi_i) + a_3 \cdot \sin(\varphi_i))] \cdot \cos(\varphi_i)\ , \\
\frac{\partial\chi^2}{\partial a_3} &= -\sum_{i=1}^{N} w_i \cdot [x_i - (a_1 + a_2 \cdot \cos(\varphi_i) + a_3 \cdot \sin(\varphi_i))] \cdot \sin(\varphi_i)\ , \\
\frac{\partial\chi^2}{\partial a_4} &= -\sum_{i=1}^{N} w_i \cdot [y_i - a_4]\ ,
\end{aligned}$$

and

$$\frac{\partial^2 \chi^2}{\partial^2 a_1} = \sum_{i=1}^{N} w_i = S \,, \qquad \frac{\partial^2 \chi^2}{\partial^2 a_2} = \sum_{i=1}^{N} w_i \cdot \cos(\varphi_i) \cdot \cos(\varphi_i) = S_{\mathrm{cc}} \,,$$

$$\frac{\partial^2 \chi^2}{\partial^2 a_4} = \sum_{i=1}^{N} w_i = S \,, \qquad \frac{\partial^2 \chi^2}{\partial^2 a_3} = \sum_{i=1}^{N} w_i \cdot \sin(\varphi_i) \cdot \sin(\varphi_i) = S_{\mathrm{ss}} \,,$$

$$\begin{aligned}
\frac{\partial^2 \chi^2}{\partial a_1 \partial a_2} = \frac{\partial^2 \chi^2}{\partial a_2 \partial a_1} &= \sum_{i=1}^{N} w_i \cdot \cos(\varphi_i) = S_{\mathrm{c}} \,, \\
\frac{\partial^2 \chi^2}{\partial a_1 \partial a_3} = \frac{\partial^2 \chi^2}{\partial a_3 \partial a_1} &= \sum_{i=1}^{N} w_i \cdot \sin(\varphi_i) = S_{\mathrm{s}} \,, \\
\frac{\partial^2 \chi^2}{\partial a_1 \partial a_4} = \frac{\partial^2 \chi^2}{\partial a_4 \partial a_1} &= 0 \,, \\
\frac{\partial^2 \chi^2}{\partial a_2 \partial a_3} = \frac{\partial^2 \chi^2}{\partial a_3 \partial a_2} &= \sum_{i=1}^{N} w_i \cdot \cos(\varphi_i) \cdot \sin(\varphi_i) = S_{\mathrm{cs}} \,, \\
\frac{\partial^2 \chi^2}{\partial a_2 \partial a_4} = \frac{\partial^2 \chi^2}{\partial a_4 \partial a_2} &= 0 \,, \\
\frac{\partial^2 \chi^2}{\partial a_3 \partial a_4} = \frac{\partial^2 \chi^2}{\partial a_4 \partial a_3} &= 0 \,.
\end{aligned}$$

The entire Hessian matrix is

$$\mathbf{H} = \begin{pmatrix} S & S_{\mathrm{c}} & S_{\mathrm{s}} & 0 \\ S_{\mathrm{c}} & S_{\mathrm{cc}} & S_{\mathrm{cs}} & 0 \\ S_{\mathrm{s}} & S_{\mathrm{cs}} & S_{\mathrm{ss}} & 0 \\ 0 & 0 & 0 & S \end{pmatrix} = \begin{pmatrix} & & & 0 \\ & [\mathbf{H}_{\mathrm{x}}] & & 0 \\ & & & 0 \\ 0 & 0 & 0 & S \end{pmatrix} .$$

Its inverse is equal to

$$\mathbf{H}^{-1} = \begin{pmatrix} & & & 0 \\ & [(\mathbf{H}_{\mathrm{x}})^{-1}] & & 0 \\ & & & 0 \\ 0 & 0 & 0 & 1/S \end{pmatrix} .$$

For equal weights $w_i = 1.0$ in this example, the combined treatment leads to exactly the same parameters a_j as the separate fitting would. If, however, the observations x_i and y_i do have the same (but in i variable) uncertainty per definition, then the application of combined fitting is mandatory.

7.6 Weighted least squares with correlated data

Chapter 2 has introduced a diagonal matrix $\mathbf{W}$ containing one weight w_i per observation y_i. Fitting a constant, for example, which is nothing more than averaging all observations, can be done in a weighted fashion. Furthermore, it has been stated that the weight w_i should be the reciprocal of σ_i^2, when σ_i is the uncertainty of the measurement y_i. That also means, σ_i^2 is the variance of the error $e = y_\text{t} - y_i$, with y_t being the true value.

Now let us imagine that it is very early in the morning and we would like to guess what today's temperature will be at noon. We can rely on two sources: weather forecasting application software (app) and our neighbour. The neighbour regularly watches the sky and has inherited his knowledge from his father, who himself learned from his father etc. Thus, the neighbour is really an expert in predicting the local weather. We would like to combine the two temperature-forecast values creating a single average value. Based on earlier predictions, we know that the uncertainty of the neighbours forecast is $\sigma_\text{N} = 1.20°\text{C}$, while the app is less reliable: $\sigma_\text{A} = 1.41°\text{C}$. Let the today's temperature forecasts be $T_\text{N} = 17°\text{C}$ (neighbour) and $T_\text{A} = 15°\text{C}$ (app).

The remaining question is: what is the best guess of the temperature?[5] Here are some possible answers:

1. Anything between 15°C and 17°C.
2. Something between 15°C and 17°C. The exact value depends on details that were not disclosed in the statement of the problem.
3. 16°C, the linear average of the two predictions.
4. 17°C, the result with the lowest uncertainty.
5. Something between 10.77°C and 20.60°C, which is a guess that covers ± 3 times the standard deviations (uncertainties) of the predictions for both methods.
6. No answer is possible unless more information is provided regarding the estimators 'neighbour' and 'app'.

[5] This example is inspired by [zPWA].

7. The best estimate is

$$T^* = \frac{w_N \cdot T_N + w_A \cdot T_A}{w_N + w_A} = \frac{\left(\frac{1}{\sigma_N}\right)^2 \cdot T_N + \left(\frac{1}{\sigma_A}\right)^2 \cdot T_A}{\left(\frac{1}{\sigma_N}\right)^2 + \left(\frac{1}{\sigma_A}\right)^2}$$
$$= \frac{\left(\frac{1}{1.20^\circ\mathrm{C}}\right)^2 \cdot 17^\circ\mathrm{C} + \left(\frac{1}{1.41^\circ\mathrm{C}}\right)^2 \cdot 15^\circ\mathrm{C}}{\left(\frac{1}{1.20^\circ\mathrm{C}}\right)^2 + \left(\frac{1}{1.41^\circ\mathrm{C}}\right)^2} = 16.16^\circ\mathrm{C}\,,$$

which is the weighted mean of the two values, with weights equal to the inverse error variances.

Which is the correct answer? Intuition leads people to typically oscillate between answer 4, 5, and 7 with equilibrium at 4 or 5, because this problem would not have been stated if 7 were correct.

However, only answer 6 is correct. It is not possible to find the correct average of the two estimates unless the amount of correlation between the errors of the two estimators is known. It is not even guaranteed that the best average lies between the two temperature predictions!

The problem is as follows: If predictions (or measurements in general) are positively correlated, they tend to lie on the same side of the true value. In other words: the prediction errors have the same sign in most of the cases. Let T_t be the true temperature and the neighbours guess $T_N < T_t$. Hence, the prediction error $e = T_t - T_N$ is positive. If the app forecast is correlated with the neighbour's prediction, it probably also has a positive error leading to $T_A < T_t$. At this point we can already see that the true value lies outside the range of the two predictions. The question now is: "Which of the two forecasts is closer to the true value?" Since T_N has a lower uncertainty, it should be closer on average. In summary, if the predictions are positively correlated, the best guess can be larger than $T_N = 17^\circ$C!

Equation (2.2) on page 26 is only applicable in the special case that the observational errors are not correlated. Then the weights matrix $\mathbf{W}$ is diagonal and reduced to the inverse error variances. The more general case considers the full covariance matrix $\mathbf{C} = (c_{ik})$ of the measurement errors. The matrix elements are either computed based on the values of N error signals $\{e_i[n]\}$ $(i = 1, 2, \ldots, N)$, which result from observations from repeated experiments with different but fixed conditions $\mathbf{x}_i$ or they stem from other numerical models providing information on uncertainties. The elements are

$$c_{ik} = c_{ki} = \mathbf{C}[i,k] = \frac{1}{K-1} \sum_{n=1}^{K} (e_i[n] - \overline{e_i}) \cdot (e_k[n] - \overline{e_k})\,, \tag{7.51}$$

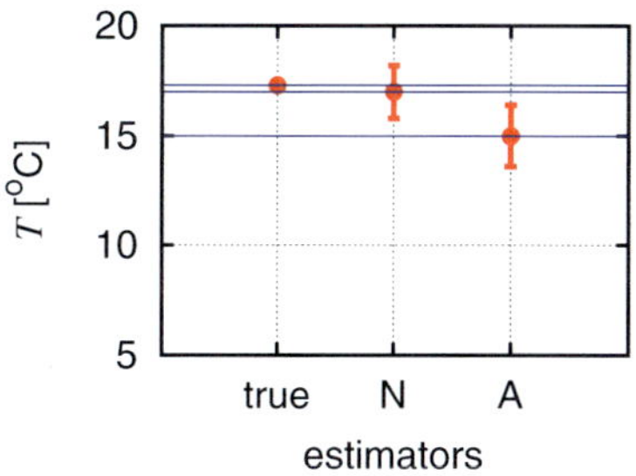

Figure 7.6: Possible arrangement of true temperature and estimate values including uncertainty in case of highly correlated estimation errors (N... prediction of neighbour, A... forecast of the app)

with

$$\overline{e_i} = \frac{1}{K} \sum_{n=1}^{K} e_i[n] \; , \tag{7.52}$$

and with K being the number of repetitions, while $e_i[n]$ is the error in observation $y_i[n]$ of the nth repetition of the experiment with conditions $\mathbf{x}_i$ and $e_k[n]$ is the measurement error of the nth repetition of the experiment under conditions $\mathbf{x}_k$.

If the errors in the two experiments are correlated, the minimisation procedure must take the elements of the inverse covariance matrix into account

$$\chi^2 = \sum_{i,k=1}^{N} [y_i - f(\mathbf{x}_i|\mathbf{a})] \cdot \mathbf{C}^{-1}[i,k] \cdot [y_k - f(\mathbf{x}_k|\mathbf{a})] \longrightarrow \text{Min} \; . \tag{7.53}$$

Let us return to the example of temperature forecast. If the predictions of both estimators (the neighbour and the app) have the tendency to contemporaneously overestimate or underestimate the true temperature (e.g., simply because there are unconsidered factors influencing the weather), then the predictions or, more precisely, the prediction errors are correlated.

Computing the covariances requires recordings of many true temperatures and estimates from the past. Since there are only two estimators in the weather example, the covariance comprises four elements. Two of them have already been provided above: $\sigma_{\text{N}}^2 = (1.20°\text{C})^2$ and $\sigma_{\text{A}}^2 = (1.41°\text{C})^2$. The other two are equal $\sigma_{\text{N,A}}^2 = \sigma_{\text{A,N}}^2$ and are assumed to be $(1.21°\text{C})^2$.

The model function is simply $f(\mathbf{x}_i|\mathbf{a}) = a_1$ and the general set-up according to equation (2.10) requires

$$\mathbf{J} = \begin{pmatrix} 1 \\ 1 \end{pmatrix}, \quad \mathbf{W} = \begin{pmatrix} 1/\sigma_{\mathrm{N}}^2 & 0 \\ 0 & 1/\sigma_{\mathrm{A}}^2 \end{pmatrix}, \quad \text{and} \quad \mathbf{y} = \begin{pmatrix} 17 \\ 15 \end{pmatrix}$$

leading to

$$a_1 = \left(\mathbf{J}^{\mathrm{T}} \cdot \mathbf{W} \cdot \mathbf{J}\right)^{-1} \cdot \mathbf{J}^{\mathrm{T}} \cdot \mathbf{W} \cdot \mathbf{y} = 16.16\ .$$

In the case of correlated data, this changes to

$$a_1 = \left(\mathbf{J}^{\mathrm{T}} \cdot \mathbf{C} \cdot \mathbf{J}\right)^{-1} \cdot \mathbf{J}^{\mathrm{T}} \cdot \mathbf{C} \cdot \mathbf{y} = 17.10\ .$$

It is self-evident that the best combination of both estimations also can be derived directly from equation (7.53)

$$\chi^2 = \sum_{i,k=1}^{2} (y_i - a_1) \cdot \begin{pmatrix} c_{11} & c_{21} \\ c_{12} & c_{22} \end{pmatrix}^{-1} \cdot (y_k - a_1) \longrightarrow \text{Min}\ .$$

Based on matrix inversion (see eq.(5.6) on page 143)

$$\begin{pmatrix} c_{11} & c_{21} \\ c_{12} & c_{22} \end{pmatrix}^{-1} = \begin{pmatrix} d_{11} & d_{21} \\ d_{12} & d_{22} \end{pmatrix},$$

we can write

$$\begin{aligned}
\chi^2 &= \sum_{i,k=1}^{2} (y_i - a_1) \cdot \begin{pmatrix} d_{11} & d_{21} \\ d_{12} & d_{22} \end{pmatrix} \cdot (y_k - a_1) \longrightarrow \text{Min} \\
&= (y_1 - a_1) \cdot d_{11} \cdot (y_1 - a_1) + (y_2 - a_1) \cdot d_{22} \cdot (y_2 - a_1) + \\
&\quad 2 \cdot (y_1 - a_1) \cdot d_{12} \cdot (y_2 - a_1) \\
&= d_{11} \cdot (y_1^2 - 2 \cdot a_1 \cdot y_1 + a_1^2) + d_{22} \cdot (y_2^2 - 2 \cdot a_1 \cdot y_2 + a_1^2) + \\
&\quad 2 \cdot d_{12} \cdot (y_1 \cdot y_2 - (y_1 + y_2) \cdot a_1 + a_1^2) \\
&= (d_{11} + d_{22} + 2 \cdot d_{12}) \cdot a_1^2 - \\
&\quad 2 \cdot (d_{11} \cdot y_1 + d_{22} \cdot y_2 + d_{12} \cdot (y_1 + y_2)) \cdot a_1 + \\
&\quad d_{11} \cdot y_1^2 + d_{22} \cdot y_2^2 + 2 \cdot d_{12} \cdot y_1 \cdot y_2\ .
\end{aligned}$$

This can be solved via

$$0 = \frac{\partial \chi^2}{\partial a_1} = 2 \cdot (d_{11} + d_{22} + 2 \cdot d_{12}) \cdot a_1 - 2 \cdot (d_{11} \cdot y_1 + d_{22} \cdot y_2 + d_{12} \cdot (y_1 + y_2))$$

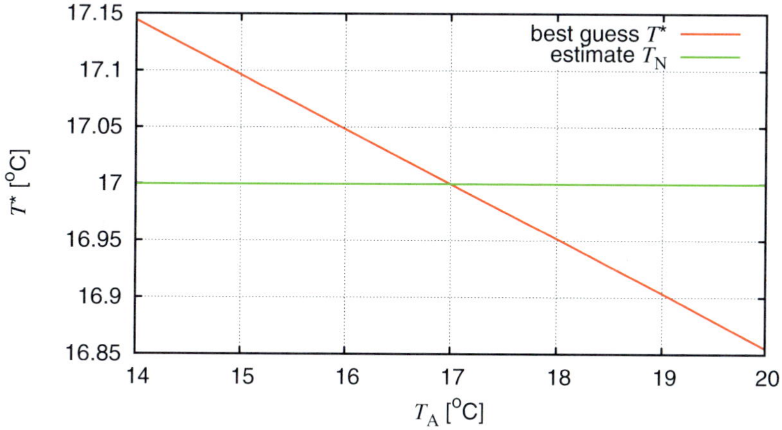

Figure 7.7: Most likely temperature (best guess) depending on the app estimate T_A

and with $y_1 = T_N, y_2 = T_A$

$$\begin{aligned} a_1 &= \frac{d_{11} \cdot y_1 + d_{22} \cdot y_2 + d_{12} \cdot (y_1 + y_2)}{d_{11} + d_{22} + 2 \cdot d_{12}} \\ &= \frac{2.7640 \cdot 17 + 2.0020 \cdot 15 - 2.0355 \cdot (17 + 15)}{2.7640 + 2.0020 - 2 \cdot 2.0355} = 17.10 \,, \end{aligned} \tag{7.54}$$

which leads, of course, to the same result.

Now, it is of interest, how the best guess changes if the parameters are different. Based on equation (7.54), **Figure 7.7** shows the influence of the estimate of the app, while all other variables are kept constant. It can clearly be seen that the neighbours estimate $T_N = 17°\text{C}$ is always between the best guess and the prediction of the app.

What is the influence of the covariance $\sigma^2_{N,A}$? **Figure 7.8** provides the answer. When there is little correlated data (i.e., $\sigma_{N,A} < \sigma_N$), the best guess lies between the two estimates T_N and T_A. Beyond this point, the best combination quickly departs from T_N, the prediction with lower uncertainty, going to infinity. The reason is the pole in equation (7.54) at $0 = d_{11} + d_{22} + 2 \cdot d_{12}$. It seems that we should make some considerations about the possible range of $\sigma^2_{N,A} = \sigma^2_{12}$.

The covariance σ^2_{ik} $(i \neq k)$ is equal to zero if the errors in the two predictions are uncorrelated. For positive correlation it is positive, for negative correlation it is negative. However, what is the upper bound of the absolute value $|\sigma^2_{ik}|$? Here,

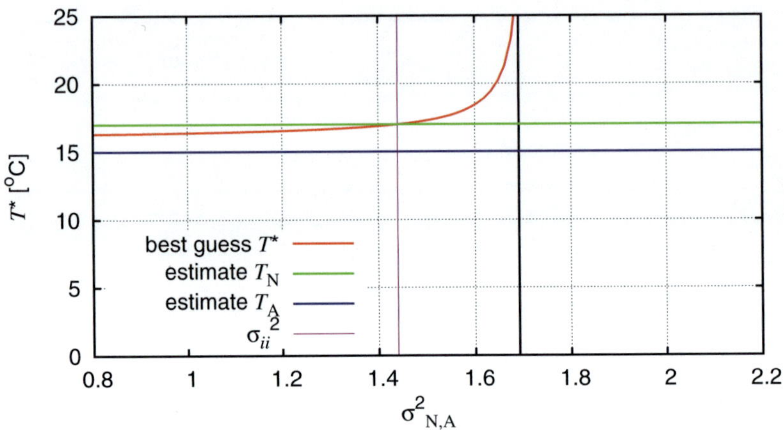

Figure 7.8: Most likely temperature (best guess) depending on the correlation between both estimators, expressed by the covariance $\sigma^2_{\mathrm{N,A}}$; σ^2_{ii} is the variance of the more reliable estimate; see text for more details

Cauchy's inequality

$$\left(\sum_{n=1}^{K} a_n \cdot b_n\right)^2 \leq \left(\sum_{n=1}^{K} a_n^2\right) \cdot \left(\sum_{n=1}^{K} b_n^2\right) \tag{7.55}$$

comes into play, at which equality holds for $\mathbf{a} = c \cdot \mathbf{b}$, (c=constant). With the substitutions $a_n = (e_i[n] - \overline{e_i})$ and $b_n = (e_k[n] - \overline{e_k})$ we get

$$\left(\sum_{n=1}^{K}(e_i[n] - \overline{e_i}) \cdot (e_k[n] - \overline{e_k})\right)^2 \leq \left(\sum_{n=1}^{K}(e_i[n] - \overline{e_i})^2\right) \cdot \left(\sum_{n=1}^{K}(e_k[n] - \overline{e_k})^2\right)$$

$$\left|\frac{1}{K} \cdot \sum_{n=1}^{K}(e_i[n] - \overline{e_i}) \cdot (e_k[n] - \overline{e_k})\right| \leq \sqrt{\left(\frac{1}{K} \cdot \sum_{n=1}^{K}(e_i[n] - \overline{e_i})^2\right) \cdot \left(\frac{1}{K} \cdot \sum_{n=1}^{K}(e_k[n] - \overline{e_k})^2\right)}\,,$$

which is the upper bound

$$|\sigma^2_{ik}| \leq \sqrt{\sigma^2_{ii}\sigma^2_{kk}}\,. \tag{7.56}$$

This value is marked in Figure 7.8 with the vertical black line. Larger values are impossible, but the range close to and including equality in equation (7.56) is also unsuitable for optimisation using (7.53), because the inversion of the covariance matrix

$$\mathbf{C}^{-1} = \begin{pmatrix} \sigma_{11}^2 & \sigma_{12}^2 \\ \sigma_{12}^2 & \sigma_{22}^2 \end{pmatrix}^{-1} = \frac{1}{\sigma_{11}^2 \cdot \sigma_{22}^2 - \sigma_{12}^2 \cdot \sigma_{12}^2} \cdot \begin{pmatrix} \sigma_{22}^2 & -\sigma_{12}^2 \\ -\sigma_{12}^2 & \sigma_{11}^2 \end{pmatrix}$$

becomes unstable or is not possible in the case of $\left|\sigma_{12}^2\right| = \sqrt{\sigma_{11}^2 \sigma_{22}^2}$.

Now, it is time to make some final remarks. The indispensable prerequisite for taking the correlation between measurements into account is the knowledge of this correlation. This requires the execution of the experiment several times for each set of conditions $\mathbf{x}_i$ leading to an estimated correlation. If the estimated correlation is very high, the corresponding covariance can approach its theoretical maximum (eq.(7.56)) and the best guess becomes unreasonably high. This effect might also be caused by numerical issues (inaccuracy of estimated values, rounding etc.). Besides these considerations, one should keep in mind that the method of least squares assumes normally distributed data, a fact, which does not apply in any application. In the framework of predicting values of image signals, for example, the high (and varying) correlation between the errors of different predictors makes the method according to eq.(7.53) impracticable and another (empirical) solution had to be found [Str15]. As a minimum requirement, one should implement a safe guard limiting the best guess to values which are not too far outside the range of the measured values. The treatment of correlated quantities when estimating a weighted mean can also be found in [Cox06a].

7.7 Total least squares

Introducing the principle of data fitting in Chapter 1, one of the first assumption was that the condition vector $\mathbf{x}$ is known exactly. It may sometimes happen, however, that the conditions also have errors associated with them that are not negligible with respect to the errors in the observations. If, for instance, current and voltage have to be measured in order to determine the characteristic curve of the resistance of an electrical component, it can be expected that both values are subject to random errors. These cases are referred to as ‘errors in variables’ or ‘measurement error models’ and typically the method of ordinary least squares is not appropriate (compare Section 2.1). Instead of minimising the squared distances between the observations and the model function, i.e. $[y_i - f(\mathbf{x}_i|\mathbf{a})]^2$, the shortest distance between the measured point $(\mathbf{x}_i, y_i)$ and the model function in point $(\mathbf{x}_{i,\text{TLS}}, f(\mathbf{x}_{i,\text{TLS}}|\mathbf{a}))$ should be taken into account (see **Figure 7.9**). Since

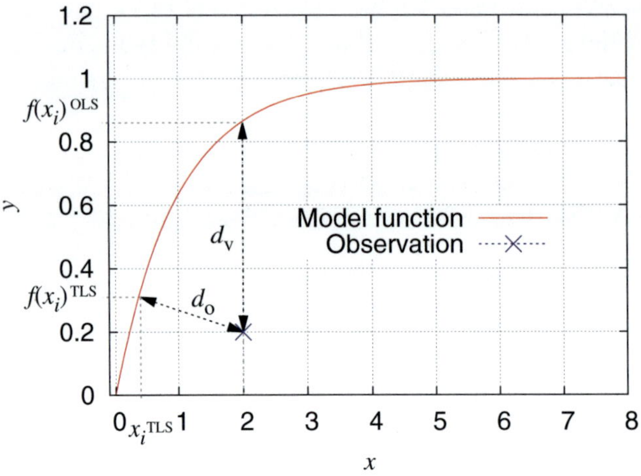

Figure 7.9: Distances between an observed point and model function: vertical distance $d_\text{v} = y_i - f(x_i)^{\text{OLS}}$ used in ordinary least-squares fitting (OLS), orthogonal distance $d_\text{o} = \sqrt{(y_i - f(x_i)^{\text{TLS}})^2 + (x_i - x_i^{\text{TLS}})^2}$ used in total least-squares fitting (TLS)

the distances between observation and model function are determined orthogonally to the model graph, the total-least-squares fitting is also called *orthogonal least-squares fitting* or *orthogonal distance regression* (ODR).

7.7.1 Orthogonal fitting of a circle

For the case of fitting a circle, it is obvious that ideally the sum of the squared distance between the data points and the graph should be minimum (**Figure 7.10**). Looking for the best circle centre (a_1, a_2) and radius a_3, the minimisation criterion consequently is

$$\sum_i \left[\sqrt{(x_{1,i} - a_1)^2 + (x_{2,i} - a_2)^2} - a_3\right]^2 \longrightarrow \text{Min} .$$

In comparison to the definition of $\chi^2(\mathbf{a})$ (equation (2.2) on page 26), we can define

$$f(\mathbf{x}_i|\mathbf{a}) = \sqrt{(x_{1,i} - a_1)^2 + (x_{2,i} - a_2)^2} - a_3$$

and

$$y_i \equiv 0 .$$

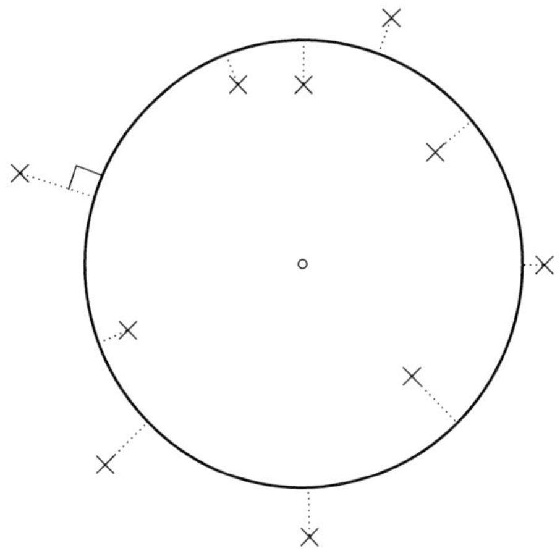

Figure 7.10: Orthogonal distances between observed points and graph of a circle

Hence, the underlying model function is

$$y_i = 0 = \sqrt{(x_{1,i} - a_1)^2 + (x_{2,i} - a_2)^2} - a_3 + \varepsilon_i \;, \tag{7.57}$$

which is obviously different from the model function (2.22) used in Subsection 2.6.10. Unfortunately, (7.57) is a real nonlinear function with respect to the vector of model parameters $\mathbf{a} = (a_1\ a_2\ a_3)^{\mathrm{T}}$ and has to be solved iteratively.

With

$$f = \sqrt{(x_1 - a_1)^2 + (x_2 - a_2)^2} - a_3$$

the partial derivatives are

$$\begin{aligned} \frac{\partial f_i}{\partial a_1} &= \frac{a_1 - x_{1,i}}{\sqrt{(x_{1,i} - a_1)^2 + (x_{2,i} - a_2)^2}} \\ \frac{\partial f_i}{\partial a_2} &= \frac{a_2 - x_{1,i}}{\sqrt{(x_{1,i} - a_1)^2 + (x_{2,i} - a_2)^2}} \\ \frac{\partial f_i}{\partial a_3} &= -1 \,. \end{aligned}$$

The computation of an initial guess for vector **a** was already explained in Subsection 2.6.10 (page 43). Applying this total least-squares approach to the problem discussed in Subsection 3.5.9 (page 104), the parameters change to $a_1 = 14.9609$, $a_2 = 15.3466$, $a_3 = 9.7200$, and the mean squared distance between observed data points and the graph of the approximated circle reduces from 0.7893 to 0.7851.

7.7.2 General approach

While in the example of circle fitting described above the TLS solution could be derived directly, the general case must be treated in a much more complicated manner.

If the condition vector $\mathbf{x}_i = (x_1 \; \cdots \; x_k \; \cdots \; x_K)_i^{\mathrm{T}}$ has an associated error of $\underline{\delta}_i = (\delta_1 \; \cdots \; \delta_k \; \cdots \; \delta_K)_i^{\mathrm{T}}$, and the vector elements δ_k are normally distributed ($\delta_k \sim \mathcal{N}(0, \sigma_\delta^2)$), then the method of least squares results in the maximum-likelihood estimate for the parameters in $\mathbf{a}$ (compare also Section 6.2). Extending the derivations of [Bog87] towards multidimensional condition vectors ($x_i \mapsto \mathbf{x}_i$) the fitting problem given in equation (1.2) has to be changed to

$$y_i = f(\mathbf{x}_i + \underline{\delta}_i | \mathbf{a}) + \varepsilon_i = f(\mathbf{x}_i | \mathbf{a})^{\mathrm{TLS}} + \varepsilon_i \,. \tag{7.58}$$

The orthogonal distance regression searches for those parameters which cause the sum of the squared orthogonal distances r^2 from the observation points $(\mathbf{x}_i, y_i)$ to the model function $f(\mathbf{x}_i|\mathbf{a})$ to be minimised. Let $\delta_{k,i} = x_{k,i}^{\mathrm{TLS}} - x_{k,i}$ be the error assigned to the kth condition associated to the ith observation, then, the orthogonal distance is expressed by

$$\begin{aligned} r_i^2 &= \left[y_i - f(\mathbf{x}_i|\mathbf{a})^{\mathrm{TLS}}\right]^2 + \left[x_{1,i}^{\mathrm{TLS}} - x_{1,i}\right]^2 + \cdots \\ &\quad + \left[x_{k,i}^{\mathrm{TLS}} - x_{k,i}\right]^2 + \cdots + \left[x_{K,i}^{\mathrm{TLS}} - x_{K,i}\right]^2 \\ &= \varepsilon_i^2 + \sum_{k=1}^{K} \delta_{k,i}^2 \,. \end{aligned}$$

In order to minimise the distance r^2 we must sum over all i

$$\sum_{i=1}^{N} r_i^2 = \sum_{i=1}^{N} \left\{ \varepsilon_i^2 + \sum_{k=1}^{K} \delta_{k,i}^2 \,. \right\} \longrightarrow \text{Min}$$

Introducing weights and eliminating ε_i using equation (7.58), the minimisation problem is finally

$$\chi^2 = \frac{1}{2} \sum_{i=1}^{N} w_i \cdot \left\{ [y_i - f(\mathbf{x}_i + \underline{\delta}_i|\mathbf{a})]^2 + \sum_{k=1}^{K} v_{k,i} \cdot \delta_{k,i}^2 \right\} \longrightarrow \text{Min}, \tag{7.59}$$

with $w_i = 1/\sigma_{\varepsilon_i}^2$ and $v_{k,i} = \sigma_{\varepsilon_i}^2 / \sigma_{\delta_{k,i}}^2$.

In comparison to the ordinary least-squares (OLS) approach, $N \cdot K$ new unknown parameters come into play, namely $\delta_{k,i}$ ($i = 1, 2, \ldots, N;\; k = 1, 2, \ldots, K$). [Bog87] proposes to split (7.59) into sets of equations, similar to what we have used in Section 2.6.7 to combine the two model functions of the coordinate transform. The problem remains over-determined, since the generalisation to multi-dimensional conditions leads also to $K \cdot N$ new functions $g(\mathbf{x}|\mathbf{a}, \underline{\delta})$

$$
\begin{aligned}
g_i(\mathbf{x}_i|\mathbf{a}, \underline{\delta}_i) &= w_i \cdot [y_i - f(\mathbf{x}_i + \underline{\delta}_i|\mathbf{a})]^2 & i = 1, \dots, N \\
g_{i+N}(\mathbf{x}_i|\mathbf{a}, \underline{\delta}_i) &= w_i \cdot v_{1,i} \cdot \delta_{1,i}^2 & i = 1, \dots, N \\
&\vdots \\
g_{i+k\cdot N}(\mathbf{x}_i|\mathbf{a}, \underline{\delta}_i) &= w_i \cdot v_{k,i} \cdot \delta_{k,i}^2 & i = 1, \dots, N \\
&\vdots \\
g_{i+K\cdot N}(\mathbf{x}_i|\mathbf{a}, \underline{\delta}_i) &= w_i \cdot v_{K,i} \cdot \delta_{K,i}^2 & i = 1, \dots, N
\end{aligned}
\tag{7.60}
$$

and we have to minimise

$$\chi^2 = \frac{1}{2} \sum_{i=1}^{(K+1)\cdot N} \{g_i(\mathbf{x}_i|\mathbf{a}, \underline{\delta}_i)\} \longrightarrow \text{Min}$$

which is an ordinary least-squares problem with M parameters a_j, $K \cdot N$ parameters $\delta_{k,i}$ and $(K+1) \cdot N$ observations.

Example: Straight-line fitting using TLS

Using the model function

$$y = a_1 + a_2 \cdot x \quad \mapsto \quad y_i = a_1 + a_2 \cdot (x_i + \delta_i) + \varepsilon_i \ ,$$

the condition vector $\mathbf{x}_i$ reduces to a scalar x_i and its error $\underline{\delta}_i$ reduces to δ_i. There are N new parameters δ_i and, in total, two separate models (compare eq. 7.60) can be set up as

$$
\begin{aligned}
g_i(\mathbf{x}_i|\mathbf{a}, \underline{\delta}_i) &= w_i \cdot [y_i - (a_1 + a_2 \cdot (x + \delta_i))]^2 & i = 1, \dots, N \\
g_{i+N}(\mathbf{x}_i|\mathbf{a}, \underline{\delta}_i) &= w_i \cdot v_i \cdot \delta_i^2 & i = 1, \dots, N
\end{aligned}
$$

It becomes apparent that, due to the parameters δ_i, the fitting task has turned into a nonlinear problem. The required vector of residuals and the Jacobian matrix are

$$\mathbf{r} = \begin{pmatrix} y_1 - (a_1 + a_2 \cdot (x_1 + \delta_1)) \\ y_2 - (a_1 + a_2 \cdot (x_2 + \delta_2)) \\ \vdots \\ y_N - (a_1 + a_2 \cdot (x_N + \delta_N)) \\ 0 - \delta_1 \\ 0 - \delta_2 \\ \vdots \\ 0 - \delta_N \end{pmatrix}$$

and

$$\mathbf{J} = \left(\begin{array}{cc|cccc} 1 & x_1+\delta_1 & a_2 & 0 & \cdots & 0 \\ 1 & x_2+\delta_2 & 0 & a_2 & \cdots & 0 \\ \vdots & \vdots & \vdots & \vdots & \ddots & \vdots \\ 1 & x_N+\delta_N & 0 & 0 & \cdots & a_2 \\ \hline 0 & 0 & 1 & 0 & \cdots & 0 \\ 0 & 0 & 0 & 1 & \cdots & 0 \\ \vdots & \vdots & \vdots & \vdots & \ddots & \vdots \\ 0 & 0 & 0 & 0 & \cdots & 1 \end{array}\right) .$$

$\mathbf{J}$ comprises four blocks. The top left block corresponds to the Jacobian matrix of the OLS problem; the other three are result of the TLS derivations. The two blocks on the right-hand side are diagonal sub-matrices, because each row in $\mathbf{J}$ corresponds to a certain i. Row $i = 1$ of $\mathbf{J}$ contains the partial derivatives of the equation

$$y_1 = a_1 + a_2 \cdot (x_1 + \delta_1) + \varepsilon_1$$

which are

$$\frac{\partial y_1}{\partial a_1} = 1 \quad \frac{\partial y_1}{\partial a_2} = x_1 + \delta_1 \quad \frac{\partial y_1}{\partial \delta_1} = a_2 \, .$$

All other partial derivatives are zero

$$\frac{\partial y_1}{\partial \delta_j} = 0 \qquad \text{for} \qquad j > 1 \, .$$

The next row of $\mathbf{J}$ contains

$$\begin{aligned} \partial y_2/\partial a_1 &= 1 \\ \partial y_2/\partial a_2 &= x_2 + \delta_2 \\ \partial y_2/\partial \delta_1 &= 0 \qquad\qquad \delta_1 \text{ does not exist for } i = 2 \\ \partial y_2/\partial \delta_2 &= a_2 \, , \end{aligned}$$

and obviously it holds in general

$$\frac{\partial y_i}{\partial \delta_j} = 0 \qquad \text{if} \qquad j \neq i \, .$$

The bottom right block of $\mathbf{J}$ is diagonal for same reasons.
□□□

The major disadvantage of this approach is the large normal matrix $\mathbf{J}^{\mathrm{T}} \cdot \mathbf{W} \cdot \mathbf{J}$ whose size increases from $M \times M$ for OLS to $(M+N) \times (M+N)$ for TLS with one-dimensional conditions, i.e. $K = 1$. However, the special structure of $\mathbf{J}$ consisting of a zero block and two diagonal blocks can be exploited to reduce the amount of work to solve the TLS problem in about the same time as the corresponding OLS problem [Bog87].

7.8 Test questions

7.1 In case of varying parameters, there is no constant parameter vector $\mathbf{a}$. Is optimisation still possible? If yes, how is it? If not, why?

7.2 Derive the formula for the adaptation of model parameters based on the minimisation of the squared error $[y_i - f(\mathbf{x}_i|\mathbf{a})]^2$ (LMS filter).

7.3 What are the three typical operations in an evolutionary approach to optimisation?

7.4 If the gradients of the error surface are not known, how is it still possible to drive the optimisation towards the correct solution?

7.5 What is the fundamental idea of Chauvenet's criterion?

7.6 What are the drawbacks of outlier detection using Chauvenet's criterion?

7.7 If an experiment depends on two measurement variables, both subject to errors, what is the resulting error variance of the experiment?

7.8 How must data fitting be performed, if the result is based on two different model functions?

7.9 Assuming correlation between observation errors,how should the optimisation procedure be modified? Which problems have to be considered?

7.10 What is the basic idea of TLS?

7.11 How can total least-squares problems be solved? How does the Jacobian matrix look like for $f(x|\mathbf{a}) = a_1 + a_2 \cdot x + a_3 \cdot x^2$?

Appendix A

Comparison of Approaches to Outlier Detection

Based on Monte Carlo simulations, the cluster-based outlier detection (ClubOD) (see Subsection 3.4.2) is tested with different data distributions and compared with the method of standardised residuals (z-score). It is shown that the cluster-based approach identifies outliers more reliably, even for a normal data distribution, and the advantages are discussed in detail.

A.1 Normally distributed data

It has been investigated whether the removal of observations according to the $\pm\kappa_{\rm O}\cdot\hat{\sigma}_{\rm y}$ rule of the standardised residuals method (eq. 3.15 on page 58) is in fact critical, and whether the described cluster-based approach leads to more reliable results. For different numbers N of data points, 10^5 sets of observations y_i, drawn from a normal distribution ($\sigma_{\rm y} = 1; \overline{y} = 0$), have been generated and analysed individually.

A.1.1 Data sets without outliers

The method of standardised residuals has been tested in three modes: with a constant value of $\kappa_{\rm O} = 3.5$ and with two different adaptive values according to Chauvenet's approach, eq. (7.30), respectively. The cluster-based outlier detection (ClubOD) has been applied as described in subsection 3.4.2.

The average percentage of data sets having at least one data point declared as being an outlier has been recorded (**Figure A.1**). When using a constant value of $\kappa_{\rm O}$, the chance of classifying data points as outlier naturally increases with increasing N, since the tails of the Gaussian bell become more and more filled. The curve corresponding to $\nu_0 = 0.5$ does not seem to converge to the value of 50%. The reason for this lies in counting only the number of sets with outliers without considering the number of outliers per set. If one counts sets with two

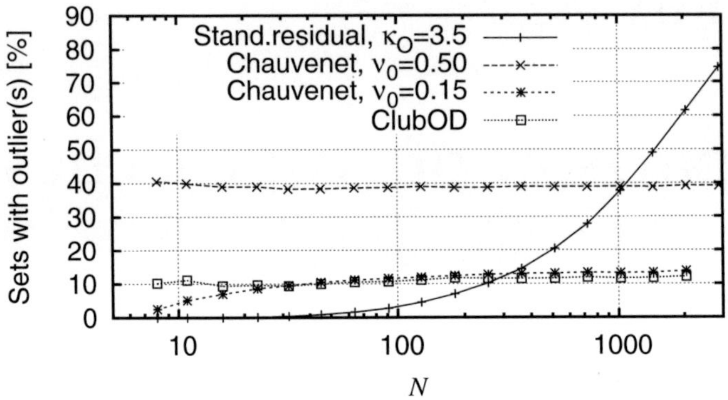

Figure A.1: Normally distributed observations: percentage of data sets with at least one observation classified as outlier.

outliers twice, sets with three outliers three times, and so on, the result will in fact converge towards 50%. According to the chosen thresholds κ_1 for the cluster-based method (ClubOD), each set shows on average 0.15 outliers leading to about 10% – 12% sets containing potential outliers.

The results of Figure A.1 reveal one major problem. Even though all values have been drawn from the same distribution, some of them have been classified as outliers by definition. The question is, does the removal of these falsely classified observations harm the data analysis, i.e. the estimation of the true value of y? In order to answer, the simulation mentioned above has also compared the mean value $\hat{y}$ of the entire set with the mean value $\hat{y}'$ of the reduced set, i.e. the set after removal of putative outliers. Theoretically, $\hat{y}$ should be equal to zero, due to the parameters of the normal distribution used. It has been found that the removal improves the estimate of the mean value of y in less than fifty percent of all sets containing one or more observations classified as being contaminant (**Figure A.2**). There are no significant differences between all three cases, despite the different values of κ_O or the different methods.

It follows that the removal of putative outliers actually yields poorer results when sets of normally distributed data are free of contaminants, whereas ClubOD shows less degradation than the method of standardised residuals for $N < 30$.

However, as the removal does not always negatively influence the estimate of $\hat{y}$, there is a chance that this relation will change in favour of removal where the presence of outliers can be expected.

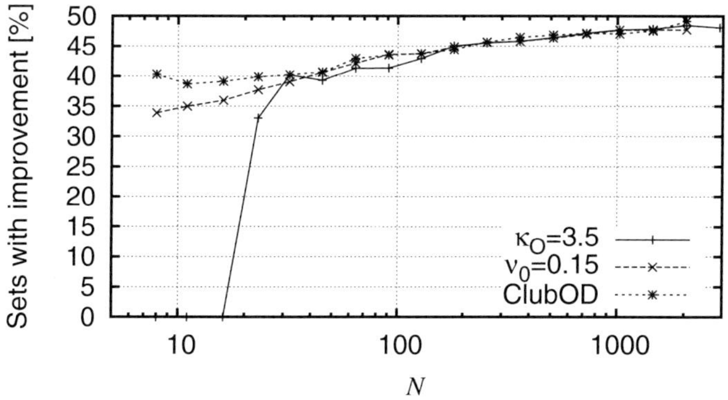

Figure A.2: Normally distributed observations: percentage of data sets with better estimates of y after the removal of observations classified as outliers.

A.1.2 Data sets containing outliers

In order to investigate the effects of real outliers, the simulations have been run again with one, two or three of the original observations substituted by some values drawn from a normal distribution with other parameters ($\sigma_{\mathrm{y}} = 1, \overline{y} = 4.0$). **Figure A.3** shows the results in comparison to the outlier-free case. Naturally, the percentage of data sets with putative contaminants has increased. It is not 100% because the inserted outliers may have values similar to the other observations, and are not detected in these cases. Furthermore, the chance of detecting outliers with the method of standardised residuals decreases for small N with an increasing number of inserted contaminants, because the estimate of σ_{y} is strongly influenced by the outliers and it becomes less likely that the remaining 'good' observations will form a distinct unit.

The proposed cluster-based approach proves advantageous when detecting outliers in small data sets, because it is not dependent on the estimation of σ_{y}. In larger data sets, observations are more frequently located in the tails of the Gaussian distribution, closing the gap between the bulk of good observations and outliers. Consequently, fewer outliers are detected on average. This behaviour is also beneficial, as we have to ask ourselves whether the outliers deliberately included can still be regarded as contaminant when similar values are also common for true data points.

Figure A.4 clearly shows that, as soon as outliers are present, the removal of outliers is statistically advantageous, especially for small data sets. It should

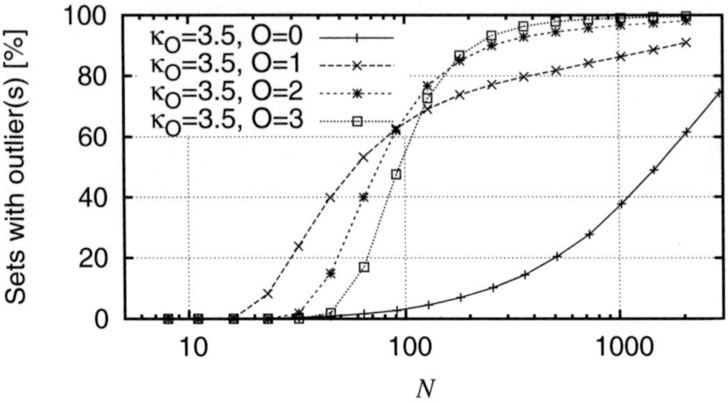

a)

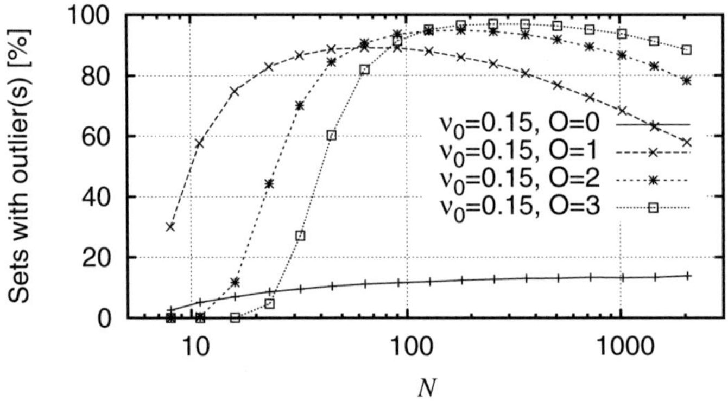

b)

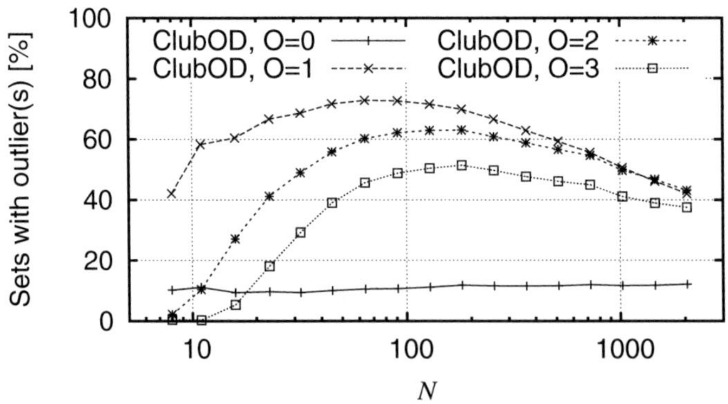

c)

Figure A.3: Normally distributed data: percentage of data sets with at least one observation classified as an outlier after insertion of one, two or three contaminants; a) $\kappa_\text{O} = 3.5$, b) Chauvenet's criterion $\nu_0 = 0.15$, c) cluster criterion.

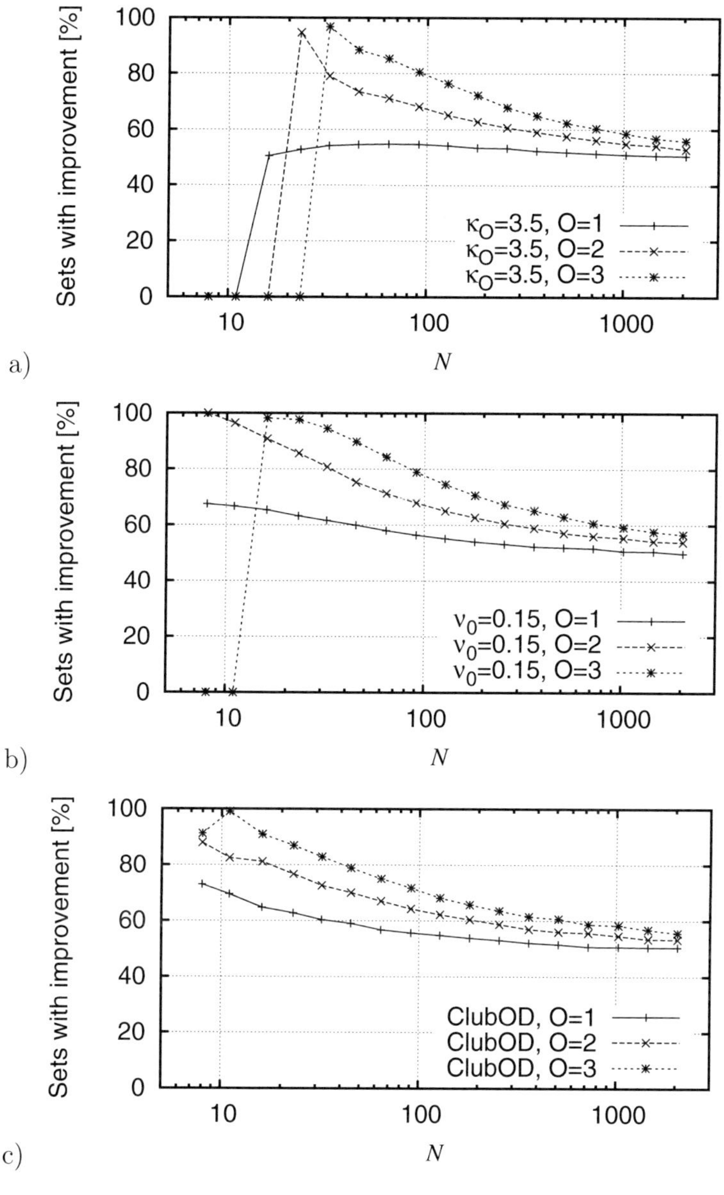

Figure A.4: Normally distributed data: percentage of data sets with better estimates of y after the removal of observations classified as outliers. Data sets with insertion of one, two or three contaminants; a) $\kappa_{\mathrm{O}} = 3.5$, b) Chauvenet's criterion $\nu_0 = 0.15$, c) cluster criterion.

also be noted that the amount of improvement of the estimated value $\hat{y}$ is on average higher than its degradation (**Figure A.5**). The changes are given as absolute values. With increasing N, the influence of outliers on the estimation of y decreases, and so do the changes.

A.2 Non-Gaussian distribution

Albeit originally developed for normally distributed deviates, the cluster-based method also works well for other distributions. As a matter of course, the method of standardised residuals will fail in these cases. Two examples are discussed here.

A.2.1 Laplace distribution

The Laplace distribution is a two-sided exponential distribution

$$f(\Delta) = \frac{1}{2 \cdot b} \cdot \exp\left(-\frac{|\Delta - \mu|}{b}\right) ,$$

which was investigated with $\mu = 0$ and $b = 1$. It turns out that the estimated standard deviation $\hat{\sigma}_{\mathrm{y}}$ is not suitable anymore as basis for discrimination of good observations and outliers. It causes the $\kappa_{\mathrm{O}} \cdot \hat{\sigma}_{\mathrm{y}}$ criterion to reject far too many observations with increasing N (**Figure A.6**, in comparison with Fig.A.3, O=0). The cluster-based criterion, however, only shows a somewhat increased tendency to declare observations as contaminant. On average, only about 1.5 samples from the end of the tail are declared as contaminant per data set. That means, it is resistant to long tails in the distribution of deviates.

Most interestingly, the removal of putative outliers improves the estimated value on average. The proposed cluster-based method has here the highest percentage of improved data sets (**Figure A.7**).

A.2.2 Uniformly distributed data

When applying both outlier detection schemes to uniformly distributed and outlier-free data ($-1 \leq y_i \leq +1$), almost no outliers are detected and the few cases of wrong classification mostly decrease the quality of the estimated $\hat{y}$. Using a fixed threshold $\kappa_{\mathrm{O}} = 3.5$, none of the observations is classified as an outlier.

After the insertion of outliers, uniformly distributed in the range of $2 \ldots 4$, the proportion of sets classified as containing at least one outlier significantly increases (**Figure A.9** a). Only in cases of small data sets, do the inserted outliers

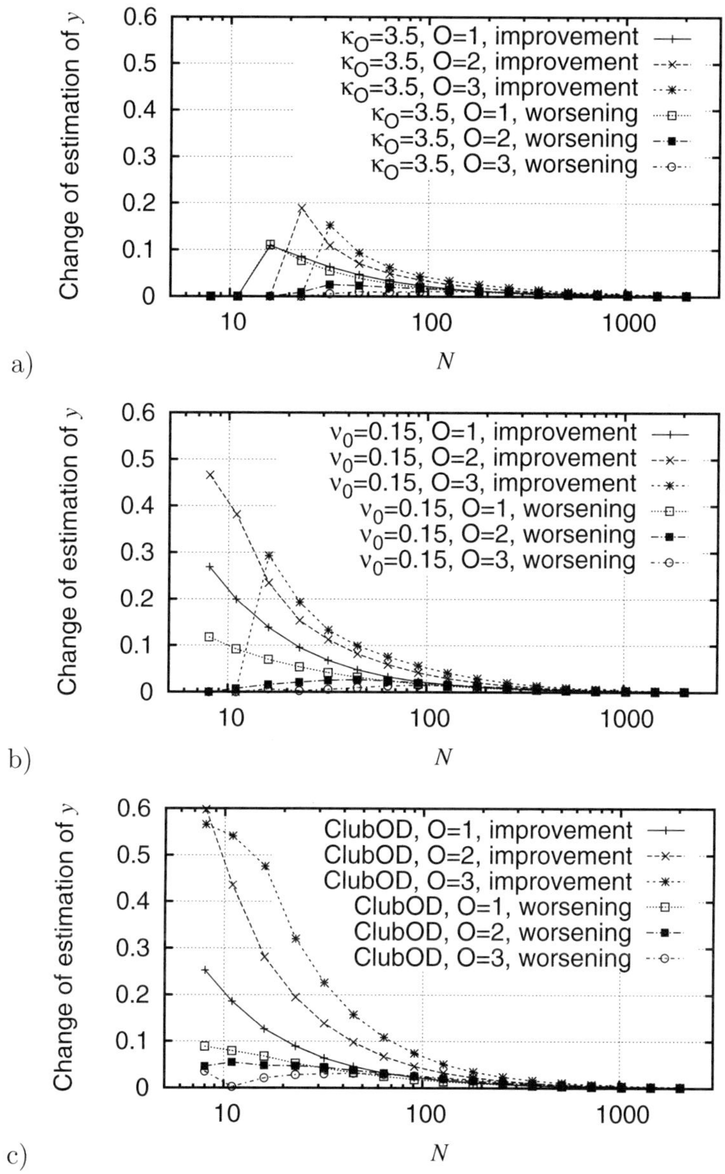

Figure A.5: Normally distributed data: quantitative change of estimated $\hat{y}$ after the removal of observations classified as outliers. Data sets with insertion of one, two or three contaminants; a) $\kappa_O = 3.5$, b) Chauvenet's criterion $\nu_0 = 0.15$, c) cluster criterion.

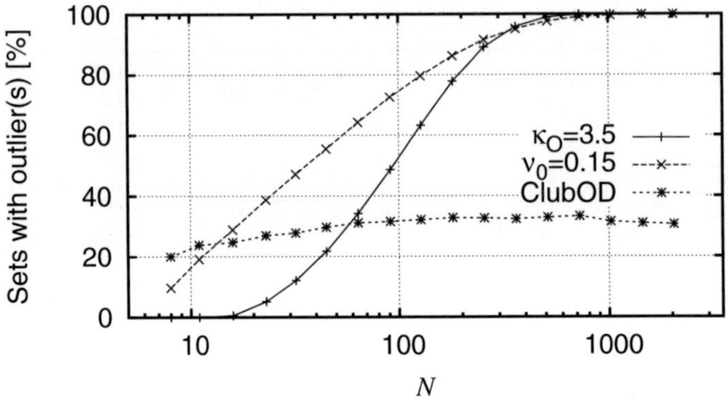

Figure A.6: Laplace distribution: percentage of data sets with at least one observation classified as an outlier.

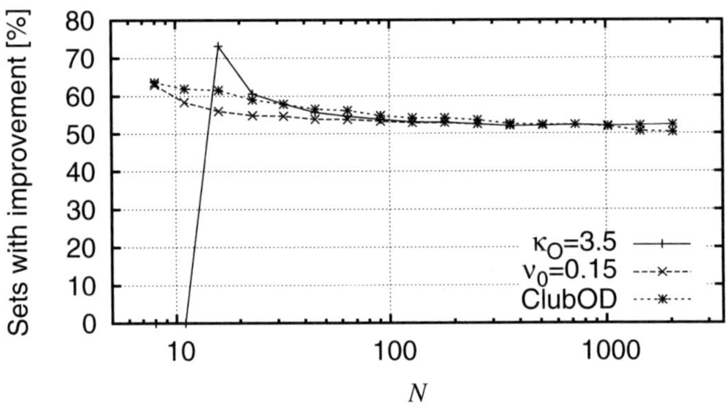

Figure A.7: Laplace distribution: percentage of data sets with better estimates of y after the removal of observations classified as outliers.

influence the determination of σ_Δ (standardised residuals) or the computation of d_{glob} (cluster-based approach) so strongly that the outliers are likely to become part of the cluster of 'good' observations. In terms of improvement after removal of putative contaminants, there is no significant difference between the approaches. That is why **Figure A.9** b) only shows the results based on a single inserted outlier.

It might be of interest that, if the data contains two outliers, the proposed approach typically finds both, whereas the method of standardised residuals often detects only one (**Figure A.10**).

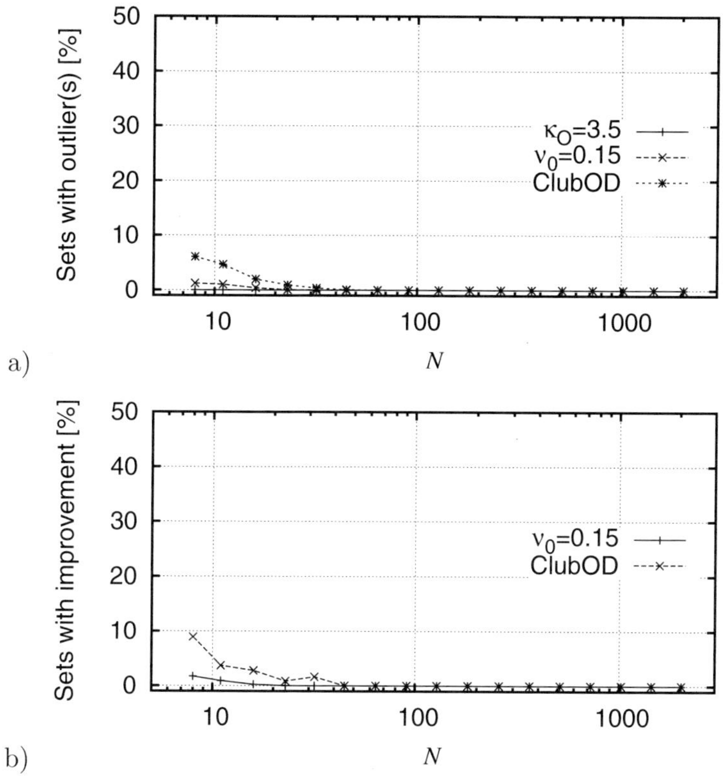

Figure A.8: Uniformly distributed observations: a) percentage of data sets with at least one observation classified as outlier; b) percentage of data sets with better estimates of y after the removal of observations classified as outliers.

A.3 Discussion

The Monte-Carlo simulations presented above underline numerically that the removal of observations decreases the estimation accuracy if normal distribution is assumed and the data set contains no outliers. This result can be generalised for any unbounded distribution. If any arbitrarily large value is an element of the distribution, none of the observations may be regarded as an outlier, regardless of the criterion that is used. The situation changes as soon the presence of outliers can be assumed. In the majority of cases, the removal of potential outliers improves the data analysis, and the improvements are higher than the effect, by which the estimation becomes worse.

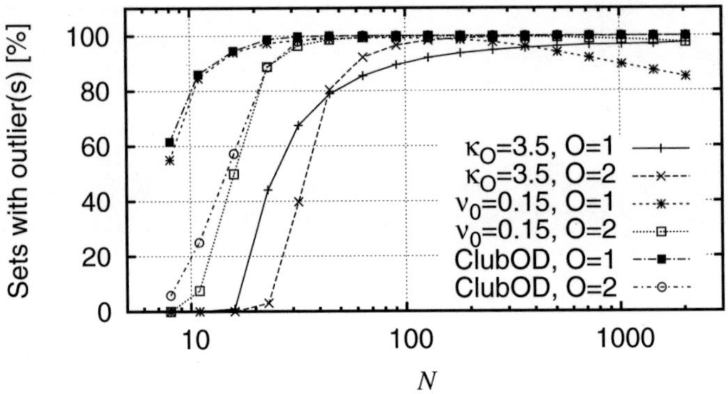

a)

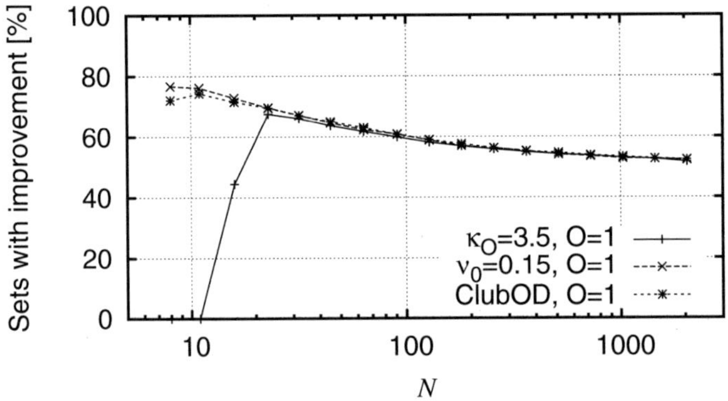

b)

Figure A.9: Uniformly distributed observations: a) percentage of data sets with at least one observation classified as an outlier; b) percentage of data sets with better estimates of y after the removal of observations classified as outliers.

The simulations have also revealed that the method of standardised residuals sometimes places the threshold λ_O in between similar observations, which contradicts the intuitive decision as to the definition of outliers. In contrast, the cluster-based method considers the distances between deviates, and only removes those points that are in fact remote from the cluster of true observations. In addition, the cluster-based method is able to adapt itself to the possibly sparse distribution of deviates.

The benefits from the cluster-based approach become especially apparent, when the deviates are not normally distributed. In the case of Laplace distribution, the method of standardised residuals declares far too many observations as contaminant, whereas the proposed method is only slightly affected. In case of uniformly

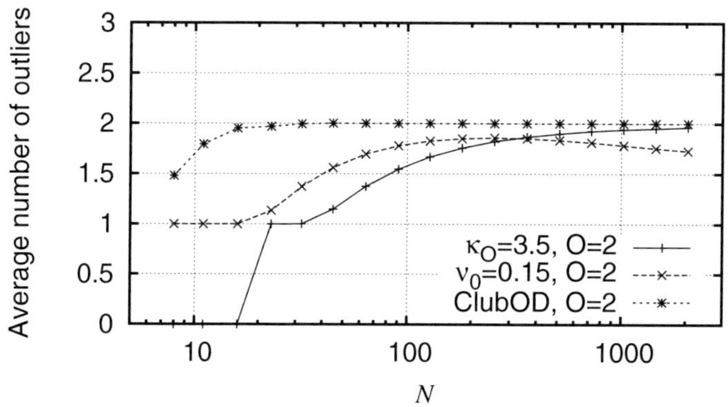

Figure A.10: Uniformly distributed observations: average number of detected outliers per data set, with two outliers inserted on purpose.

distributed data, only the proposed approach is able to detect true outliers with sufficient reliability, because it finds multiple outliers, while the method of standardised residual often only finds one out of two inserted outliers, for example.

The statistical relevance of results has been obtained by large scale tests based on simulated data. Presenting results of a particular data set from a concrete application would not prove or disprove the effectiveness of the approaches.

The inherent principle of the cluster-based method is generally compatible to any distribution of deviates, as soon as the deviates of outliers are more distant to others than the deviates of true data points, making it a very versatile method.

Appendix B

Implementation

B.1 Functionality

The source code provided along with this book[1] addresses the tasks of linear as well as nonlinear data fitting using the method of weighted least-squares approximation as described in this book. It is by no means a software package for general purposes, but implements the discussed equations more or less in a one-to-one mapping. In other words: the implementation is neither optimised for speed nor for performance, but enables a maximum of understanding.

The software contains a selection of linear and nonlinear model function that were discussed in the previous chapters. The choice of arbitrary functions is not supported, however, with some basic programming skills in C, the interested reader should be able to add model functions of his or her needs.

The reader should be aware that there are other free software packages for least-squares optimisation available. Here only the 'fit' command of the gnuplot program [Gnupl] is mentioned. It enables the definition of arbitrary nonlinear functions. However, the conditions may appear in maximal three dimensions.

Please note that the packages known to the author do support neither the automated weighting nor outlier detection.

The next section explains the usage of the software. Special problems of implementation are discussed afterwards.

B.2 Manual

The program can be called with a number options, most of them will discussed below in detail. **Figure B.1** and **Figure B.2** show the program output in case that a wrong option was given on the command line. Most options require an additional parameter. This is indicated by a following string starting with

[1] see 'OnlinePlus' area of the publishers website

```
-i %s  ... input data file (compulsory)
-o %s  ... output file (compulsory)
-m  %d ... model function (compulsory)
       0 ... constant
       1 ... f(x|a) = a1 + SUM_j=2^M a_j * x_(j-1)
       2 ... f(x|a) = a1 + a2 * cos( x) + a3 * sin( x) [in degrees]
       3 ... f(x|a) = a1 + x[n-1] + a2 * x[n-1] + a3 * x[n-3]
       5 ... f(x|a) = a1 + a2 * cos( x - a3) [in degrees]
       6 ... f(x|a) = a1 + a2 * exp( a3 * x)
       7 ... f(x|a) = log( a1 * x)
       8 ... f(x|a) = a1 * exp( (x-a2)^2 * a3) +
                      a4 * exp( (x-a5)^2 * a6)
       9 ... f(x|a) = a1 * exp( a2 * x)
      10 ... ln(f(x|a)) = ln(a1) + a2 * x
      11 ... f(x|a) = a1 * exp( -|x|^a2 * a3)
      12 ... f(x|a) = a1 + a2 * x + a3 * cos( x - a4) [in radians]
      13 ... f(x|a) = a1 * exp( (x-a2)^2 * a3) +
      16 ... f(x|a) = a1 + a2 * x + a3 * x^2
      17 ... f(x|a) = a1 + a2 * x + a3 * x^2 + a4 * x^3
      18 ... f(x|a) = sum_{j=1}^M aj * x^(j-1) (linear)
      20 ... f(x|a) = a1 + a2*x1 + a3*x1^2 + a4*x2 +a5*x2^2
      21 ... f1(x|a) = a1 + cos(a3) * x1 - sin(a3) * x2 [in radians]
             f2(x|a) = a2 + sin(a3) * x1 + cos(a3) * x2 [in radians]
      22 ... f(x|a) = a1 + a2*cos(a3*x-a4)          [in radians]
      23 ... f(x|a) = a1 + a2*cos(a3*x-a4) + a5*cos(2*a3*x-a6)
      24 ... f(x|a) = 0 = (x1 - a1)^2 + (x2 - a2)^2 - a3*a3 (circle)
      25 ... f(x|a) = x1^2 + x2^2 = a1*x1 + a2*x2 - a3
                                (circle, linear)
      26 ... f(x|a) = 0 = (sqrt[(x1 - a1)^2 + (x2 - a2)^2] - a3)^2
                                           (circle, TLS)
      30 ... neural network 3x3x1, feed forward
      31 ... neural network 3x2x1
      32 ... neural network 1x2x1
      33 ... neural network 2x2x1
      34 ... neural network 1x3x1
      40 ... f(x|a) =(a1 + a2*x + a3*x*x + a4*x*x*x) /
                     (1 + a5*x + a6*x*x + a7*x*x*x)      NIST_THURBER
      41 ... f(x|a) = a1 * (x**2 + a2*x) /
                     (x*x + a3*x + a4)                   NIST_MGH09
      42 ... f(x|a) = a1 / (1 + exp(a2 - a3*x))          NIST_Rat42
      43 ... f(x|a) = a1 / [1 + exp(a2 - a3*x)]^(1/a4)   NIST_Rat43
      44 ... f(x|a) = a1 / a2 * exp(-0.5*((x -a3)/ a2)^2) NIST_Eckerle4
      45 ... f(x|a) = a1 * exp( a2 / (x+a3))             NIST_MGH10
      46 ... f(x|a) = a1 * (x+a2)^(-1/a3)                NIST_Bennett5
      47 ... f(x|a) = a1 *(1 - exp( -a2 * x)             NIST_BoxBOD
```

Figure B.1: Legal options of the fitting program (part 1)

```
-a  %d ... inversion algorithm (default: 1)
        0 - cofactor method
        1 - singular value decomposition
        2 - LU decomposition
-a[j]  %f ... provides initial value for  a_j (j=1,2,..,9)
-b  %d ... observations per bin, for '-w 2' (default: 50)
-c     ... enable scaling of conditions
-cc %s ... comma-separated list of column(s) containing
            conditions x (default: 1,2,...)
-co %d ... column containing observations y (default: 2)
-cw %d ... column containing weights
-f     ... forget weights after outlier removal
-H     ... enable true Hessian matrix
-I %d   ... maximum number of iterations (default: 2000)
-M %d   ... number of parameters (for '-m 1' only)
-n     ... force usage of numerical derivation
-L     ... use Gaus-Newton instead of Levenberg-Marquardt
-s     ... disable special SVD function for solving linear model
-t %f  ... target value for chisq (maximum error)
            default: iteration until convergence
-w %d   ... weighting  (default: 0)
        0 ... no weighting
        1 ... based on deviates
        2 ... binning
-x %d   ... outlier removal (default: 0)
        0 ... no outlier removal
        1 ... z-score + Chauvenet's criterion
        2 ... cluster criterion  (ClubOD)
        3 ... M-score + Chauvenet's criterion
        4 ... RANSAC
```

Figure B.2: Legal options of the fitting program (part 2)

character '%'. If the next letter is 'd', then an integer number is expected, 'f' stands for a floating-point number, and 's' indicates that a string is required. There are only three mandatory options '-i', '-o', and '-m'. All other options are assigned a default value.

B.2.1 Input and output

Option '-i' expects the file name of an ASCII text file containing the data to be processed. It must be organised in columns. Comment lines (lines starting with '#') and lines containing only white spaces (space, tabulator, line feed, carriage return) are ignored (**Figure B.3**).

Please ensure that all lines are finished with the *newline* (or *line feed*) character '\n' (ASCII code 10). Some text editors insert just single *carriage return* characters '\r' (ASCII code 13). These are not recognised by the C function `fgets()`,

```
# This is comment line

# condition1 condition2   observation
  1            1.0          10.3
  1            2.3          11.5
  2            1            12.3
  2            2            14.4
# end of data
```

Figure B.3: Input example with conditions in columns one and two and observations in column three; neither option '-cc' nor '-co' are required

which is used for scanning the text. Wrong interpretation of the input data would be the consequence.

Based on the chosen model function (**option '-m'**), the program generally knows how many conditions have to be red. There are two exceptions. Model function '-m 1' activates multivariate regression

$$f(\mathbf{x}|\mathbf{a}) = a_1 + \sum_{j=2}^{M} a_j \cdot x_{j-1} \,.$$

The number of parameters (number of dimensions plus 1) must be given, in addition, via **option '-M'**. The same holds true for fitting a polynomial function ('-m 18' and '-m 19')

$$f(\mathbf{x}|\mathbf{a}) = \sum_{j=1}^{M} a_j \cdot x^{j-1} \,.$$

By default, the first condition $x_{1,i}$ is red from the first column, the second condition $x_{2,i}$ (if required) from the second column and so on. If the conditions are located differently, **option '-cc'** can be used to specify the corresponding columns. Let the two conditions $x_{1,i}$ and $x_{2,i}$ be located in columns two and three, for example, then '-cc 2' indicates the column of the first condition. The next condition is automatically taken from the next column. If, however, the conditions are not sorted in ascending order, option '-cc' requires a comma-separated list of columns.

When the number of conditions is *cond_dim*, the observations y_i are expected in the column number *cond_dim* + 1, otherwise the column must be specified using the **option '-co'**.

```
# this is file "dummy.txt"
# containing dummy observation and condition values
     # also comment followed by an empty line

# this is file "dummy.txt"
# containing dummy observation and condition values
  # also comment

# obs cond2  cond1  obs_exp  weights
  1    2      3      3        1
  3    2      4      2        2
  4    2      5      1.5      4

# comment
  5    3      3      1        2
  7    3      4      0.8      1
  7    3      5      0.75     1
```

Figure B.4: Input example with conditions in unexpected order; option '-cc' and '-co' are required

In the example of Figure B.3, the columns must not specified by neither option '-cc' nor '-co'. **Figure B.4** shows an input-file example with a different arrangement of columns. The columns deviate from the default settings and the corresponding **options '-cc 3,2'** and **'-co 1'** must be provided when calling the fitting program.

With the the option '-cc 0', the program assumes that no special condition values are provided and sequential numbers starting from $1, 2, \ldots$ are used as conditions, that means the condition is merely the number of the experiment.

The new version of the fitting program supports the provision of weights associated with each observation. Via option'-cw' the column containing the weights can be specified.

The program writes the data-fitting results into a text file determined by **option '-o'**. Dependent on the chosen options, the content of this file can vary a lot. The output was designed to provide the user with a maximum of feedback about the convergence process. Especially, when using weights estimation in combination with nonlinear model functions, this file can be become rather long, because of the nested iterations. In order to make these files suitable as input for gnuplot ([Gnupl]), all non-empty lines start with '#' apart from the final values.

Example 1

The command-line options "-i dummy.txt -o dummy_cc0_m0.txt -m 0 -cc 0" tell the program that the data has to be red from the text file "dummy.txt" (Figure B.4) and that none of the columns is used as condition values. '-m 0' indicates the fitting of a constant value, i.e. model function $y = a_1$. Despite the absence of conditions, the dimensionality of the condition space is set to *cond_dim* = 1 for algorithmic reasons. Hence, the observations are expected in column *cond_dim* + 1 = 2. The output of the fitting process is written to the text file "dummy_cc0_m0.txt", see **Figure B.5**. The output file starts with general information about the fitting process, i.e., which data file was used, which model function was approximated, how many observations were used, and how many model parameters had to be determined. This is followed by information about the iterated estimation of weights. Since weighting was not applied in this example, there is only a single run with number '# 0'. There is some feedback of the least-squares algorithm in each iteration. In the case of Figure B.5, there is some information about the matrix inversion via singular value decomposition.

After the fitting process, the results are output (χ^2, goodness-of-fit, number of outliers, standard uncertainty of observations σ_{y} and other). Of highest importance are probably the uncertainties of estimated parameters which are additionally given in percent. According to Chapter 4, this value should always be inspected. An uncertainty above 10% can indicate suspicious fitting results and the program outputs a warning also on the console.

The output finishes with a list of used data, the approximated observations ('fitted') and information about the estimated weights w_i, the estimated uncertainty σ_i of each observation y_i, the global uncertainty σ_{y}, and the difference between the approximated and measured observations $(y_i - \hat{y}_i)$ with $\hat{y}_i$ according to equation (3.4) on page 52.

Figure B.6 shows the comparison between observations and fitted model function.

Example 2

The command-line options "-i dummy.txt -o dummy_cc23_co1_m1_M3.txt -cc 2,3 -co 1 -m 1 -M 3" tell the program that the data must be red from the text file "dummy.txt" and has to be fitted against a plane. The input data is from the same input file as before (Fig. B.4). **Figure B.7** shows the corresponding output file. It contains similar information as Figure B.5. There are, however, two columns of conditions now, which are interpreted as raster points on a grid,

```
# main ===============================
# use data file: dummy.txt
# constant function
# Number of observations: 6
# Number of parameters   : 1
# condition columns: 0
# observations column: 2
#
# mode of weighting:  no weighting used
# mode of outlier detection:  no outlier detection
#
# fitting a linear system
# use special SVD based algoritm
#
#===================================
# main: approximation with final weights #: 0
#   -- ls - start  -----------------------------
#
#   singular values (thresh = 1E-012)
#   s1=2.4494897427832,
#   -- ls - end  -----------------------------
#
# main
# Parameters:  a1=2.500000,
#
# | chisq: 1.500000
# | mean of |deviates|: 0.500000
# | variance of |deviates|: 0.000000
# | goodnes of fit: 0.300000
# | number of outliers: 0
#
# evaluation of results
#  number of outliers: 0
#  parameters: a1=2.500000
#  chi square.................: 1.5
#  goodness of fit...........: 0.3
#  uncertainty in observations: 0.547722557505
#
#  (co)variance of parameters:
#     0.166666667

#  (co)variance of parameters multiplied with goodness of fit
#         0.05

#  resulting uncertainty of parameters
#     0.223606798
#         8.9443%
#
# Final_Parameters a1= 2.5
#
#  cond1        observed      fitted   weight uncertainty glob.uncertainty difference
   1.000000     2.000000     2.500000     1.000000     1.000000  0.547723    -0.500000
   2.000000     2.000000     2.500000     1.000000     1.000000  0.547723    -0.500000
   3.000000     2.000000     2.500000     1.000000     1.000000  0.547723    -0.500000
   4.000000     3.000000     2.500000     1.000000     1.000000  0.547723    +0.500000
   5.000000     3.000000     2.500000     1.000000     1.000000  0.547723    +0.500000
   6.000000     3.000000     2.500000     1.000000     1.000000  0.547723    +0.500000
```

Figure B.5: Example 1: Output example for fitting a constant with '-cc 0'

while the observation is the value of elevation. The uncertainty of the parameters is indicated as suspicious. Obviously, the model function does not fit the data very well.

In order to provide a proper input to gnuplot [Gnupl], rows with different condition one (cond1) are separated by an empty line. This enables to draw a plane mesh as can be seen in **Figure B.8**.

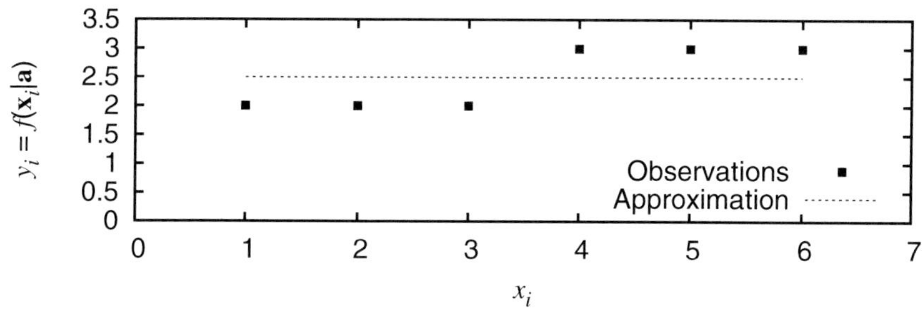

Figure B.6: Example 1: Comparison between observations and fitted model function

Example 3

The third example is to present a case of nonlinear data fitting. With the command-line options "-i dummy.txt -o dummy_cc0_co4_m9.txt -cc 0 -co 4 -m 9" the output in **Figure B.9** and **Figure B.10** is produced. As the parameters have to be estimated in an iterative manner, the generated output is much longer. In total, thirteen iterations are necessary to reach the convergence criterion. Figure B.9 shows only the results of iteration #1 and #11 – #13. The difference compared to the linear fitting starts with the output of the initially estimated parameter values. As can be seen in comparison to the final value, the initial value of a_1 is not very good. The reason lies in the assumption of the corresponding routine that the condition values start close to zero, while in the example the first condition is equal to one. Nevertheless, the algorithm finally approaches the global minimum.

In each iteration of least-squares approximation, the updates Δa_j, the resulting (intermediate) parameter estimates, and the value of χ^2 are given. As expected, χ^2 is decreasing in the course of iteration.

In case that the update does not lead to a decreasing χ^2 in a particular round, the update is rejected (for example in iteration #11) and the parameter μ of the Levenberg-Marquardt method is increased in order to give the gradient-descent approach a higher weight (compare Subsection 6.4.4). The result of the fitting process is visualised in **Figure B.11**.

B.2.2 Initialisation of model parameters

As pointed out in Section 2.3, it is required to initialise all parameters a_j with a certain value, if the model function is nonlinear. Typically, the corresponding

```
# main ================================
# use data file: dummy.txt
# linear in x, order 2
# Number of observations: 6
# Number of parameters  : 3
# condition columns: 2 3
# observations column: 1
#
# mode of weighting:  no weighting used
# mode of outlier detection:  no outlier detection
# fitting a linear system
# use special SVD based algoritm
#
#====================================
# main: approximation with final weights #: 0
#  -- ls - start  -----------------------------
#
#  singular values (thresh = 1E-012)
#  s1=11.944357370871, s2=0.33724838998668, s3=1.4894933770784,
#  -- ls - end  ------------------------------
#
# main
# Parameters:  a1=-7.933333, a2=3.266667, a3=1.075000,
#
# | chisq: 0.204167
# | mean of |deviates|: 0.152778
# | variance of |deviates|: 0.010687
# | goodnes of fit: 0.068056
# | number of outliers: 0
#
# evaluation of results
#  number of outliers: 0
#  parameters: a1=-7.933333 a2=3.266667 a3=1.075000
#  chi square.................: 0.204166666667
#  goodness of fit............: 0.0680555555556
#  uncertainty in observations: 0.260874597375
#
#  (co)variance of parameters:
#      8.33333333     -1.66666667              -1
#     -1.66666667     0.666666667 -3.60822483E-016
#              -1 -3.60822483E-016           0.25

#  (co)variance of parameters multiplied with goodness of fit
#      0.56713     -0.113426    -0.0680556
#    -0.113426     0.0453704 -2.4556E-017
#   -0.0680556 -2.4556E-017     0.0170139

#  resulting uncertainty of parameters
#      0.753080095       0.213003217       0.130437299
#           9.4926%           6.5205%          12.134%
# One parameter has relative high uncertainty !
#
# Final_Parameters a1= -7.93333333333 a2= 3.26666666667 a3= 1.075
#  cond1          cond2          observed       fitted    weight uncertainty glob.uncertainty difference
    2.000000      3.000000      1.700000      1.825000      1.000000       1.000000  0.260875    -0.125000
    2.000000      4.000000      3.000000      2.900000      1.000000       1.000000  0.260875    +0.100000
    2.000000      5.000000      4.000000      3.975000      1.000000       1.000000  0.260875    +0.025000

    3.000000      3.000000      5.000000      5.091667      1.000000       1.000000  0.260875    -0.091667
    3.000000      4.000000      6.500000      6.166667      1.000000       1.000000  0.260875    +0.333333
    3.000000      5.000000      7.000000      7.241667      1.000000       1.000000  0.260875    -0.241667
```

Figure B.7: Example 2: Output example for fitting a plane with equal weights

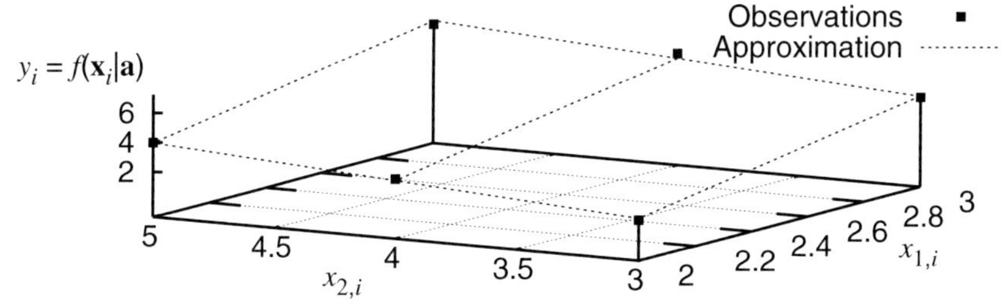

Figure B.8: Example 2: Comparison between observations and fitted model function

routines guess starting values from the observations themselves. In case that theses guesses are not sufficient and, for example, the algorithm does not find the global minimum, it is possible to provide initial parameter when starting the program using the **options '-a1', '-a2', . . . , '-a9'**.

B.2.3 Processing control

There are some options controlling the kind of processing. **Option '-I'** allows for changing the maximum number of iterations (default is 2000). **Option '-s'** switches off the special SVD-based algorithm for the fitting of linear problems. Instead of equation (6.5), now equation (6.11) is used. In the case of nonlinear fitting, **Option '-L'** switches from the Levenberg-Marquardt approach, equation (6.25), to the Gauss-Newton method, equation (6.22).

B.2.4 Weights and outliers

Example 1

Option '-w' enables the estimation of weights for each single observation. The choice of '-w 1' triggers the estimation based on the deviates (see Subsection 3.3.2). Based on the parameters "-i dummy.txt -o dummy_cc0_co1_m1_M2_w1.txt -co 1 -cc 0 -m 1 -M 2 -w 1", the programs output reflects this iterative estimation process (**Figure B.12** and **Figure B.13**)

The program starts with ordinary least-squares approximation, i.e. equal weights. The sorted deviates are inspected and the weights accordingly derived. After 12 cycles, weights estimations has converged and two out of the six observations

```
# main ===============================
# use data file: dummy.txt
# exponential 2
# Number of observations: 6
# Number of parameters  : 2
# condition columns: 0
# observations column: 4
#
# algorithm for inversion: SVD
# mode of weighting:  no weighting used
# mode of outlier detection:  no outlier detection
# fitting a nonlinear system
# use Levenberg-Marquardt method
# initial Parameters
# a1=1.660000000, a2=-0.271084337,
#
#==================================
# main: approximation with final weights #: 0
#  -- ls - start  ------------------------------
#  Iteration of least squares: 1
#  Updates:  da1=2.3233817125245, da2=-0.10885501320102,
#
#  Parameters:  a1=3.98338171252, a2=-0.37993935055,
#
#  chisq: 0.32117299135648  mu_fac: 1.000000e-004
#
#  Iteration of least squares: 2
#  Updates:  da1=0.069147985340399, da2=0.056935233145382,
#
#  Parameters:  a1=4.05252969786, a2=-0.323004117405,
#
#  chisq: 0.061775130101859  mu_fac: 5.000000e-005
    .
    .
    .
#  Iteration of least squares: 11
#  Updates:  da1=1.6863979690383E-010, da2=-2.1206452195462E-011,
#
#  rejection, chisq=0.059597741261129
#
#  Parameters:  a1=4.05787643972, a2=-0.328323111241,
#
#  chisq: 0.059597741261129  mu_fac: 3.164063e-005
#
#  Iteration of least squares: 12
#  Updates:  da1=1.6771781155617E-010, da2=-2.1142623269165E-011,
#
#  Parameters:  a1=4.05787643989, a2=-0.328323111262,
#
#  chisq: 0.059597741261129  mu_fac: 1.582031e-005
#
#  Iteration of least squares: 13
#  Updates:  da1=1.6346232385932E-012, da2=-1.4711255482683E-013,
#
#  Parameters:  a1=4.05787643989, a2=-0.328323111263,
#
#  chisq: 0.059597741261129  mu_fac: 7.910156e-006
#
#  convergence after 13 iterations
#  -- ls - end  ------------------------------
#
# main
# Parameters:  a1=4.057876, a2=-0.328323,
#
# | chisq: 0.059598
# | mean of |deviates|: 0.081178
# | variance of |deviates|: 0.003343
# | goodnes of fit: 0.014899
# | number of outliers: 0
    .
    .
    .
```

Figure B.9: Example 3: Output example for fitting a nonlinear model function (part 1)

```
  .
  .
  .
#
# evaluation of results
#  number of outliers: 0
#  parameters: a1=4.057876 a2=-0.328323
#  chi square.................: 0.0595977412611
#  goodness of fit............: 0.0148994353153
#  uncertainty in observations: 0.122063243097
#
#  (co)variance of parameters:
#      3.35192706     -0.302687448
#    -0.302687448     0.0380923652

#  (co)variance of parameters multiplied with goodness of fit
#    0.0499418  -0.00450987
#  -0.00450987  0.000567555

#  resulting uncertainty of parameters
#     0.223476666     0.0238234072
#         5.5072%          7.2561%
#
# Final_Parameters a1= 4.05787643989 a2= -0.328323111263
#  cond1          observed       fitted    weight uncertainty glob.uncertainty difference
   1.000000       3.000000       2.922200    1.000000        1.000000  0.122063    +0.077800
   2.000000       2.000000       2.104365    1.000000        1.000000  0.122063    -0.104365
   3.000000       1.500000       1.515417    1.000000        1.000000  0.122063    -0.015417
   4.000000       1.000000       1.091297    1.000000        1.000000  0.122063    -0.091297
   5.000000       0.800000       0.785876    1.000000        1.000000  0.122063    +0.014124
   6.000000       0.750000       0.565933    1.000000        1.000000  0.122063    +0.184067
```

Figure B.10: Example 3: Output example for fitting a nonlinear model function (part 2)

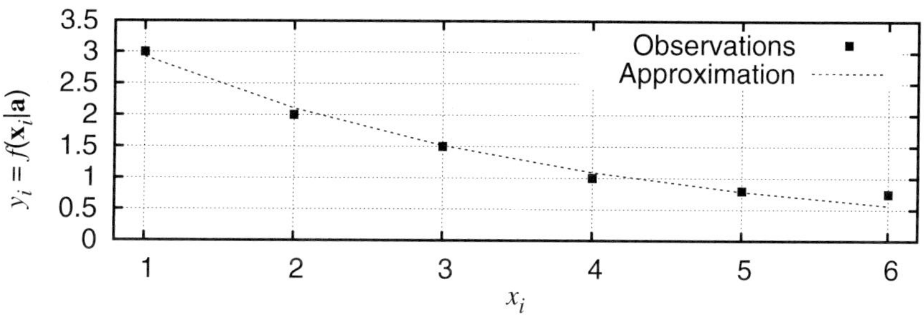

Figure B.11: Example 3: Comparison between observations and fitted model function

are strongly down-weighted, as they are much more distant to the fitted straight line than the others. The thirteenth cycle is used to fit the data with the final weights. Please note that the minimisation target χ^2 must not decrease with changing weights.

Figure B.14 shows that the approximated line is dominated by the four observations owing high weights.

```
# main ================================
# use data file: dummy.txt
# linear in x, order 1
# Number of observations: 6
# Number of parameters  : 2
# condition columns: 0
# observations column: 1
#
# mode of weighting:  estimate weights based on deviates
# mode of outlier detection:  no outlier detection
# fitting a linear system
# use special SVD based algorithm
#====================================
# main: weights iteration #: 0
#  -- ls - start  ------------------------------
#  singular values (thresh = 1E-012)
#  s1=1.0463419638985, s2=9.7931184254345,
#  -- ls - end  ------------------------------
#
# main
# Parameters:  a1=0.733333, a2=1.085714,
#
# | chisq: 0.204762
# | mean of |deviates|: 0.147619
# | variance of |deviates|: 0.012336
# | goodnes of fit: 0.051190
# | number of outliers: 0
#
# enter weight estimation
# -- est_weights1 - start ----------------------
#  Number of observations: 6
#  max_deviate: 0.338095
#  lambda_L = 0.119048
#
#  dev_sort[i],  weights[ idx[i] ]
#   0  9.523809524e-003      70.56000000
#   1  7.619047619e-002      70.56000000
#   2  9.523809524e-002      70.56000000
#   3  1.190476190e-001      70.56000000
#   4  2.476190476e-001      16.30917160
#   5  3.380952381e-001       8.74826423
# -- est_weights1 - end  -------------------------
#
# main
#  i       observ          calc       deviates    weights
#  0       1.70000       1.81905       0.11905      70.560000
#  1       3.00000       2.90476       0.09524      70.560000
#  2       4.00000       3.99048       0.00952      70.560000
#  3       5.00000       5.07619       0.07619      70.560000
#  4       6.50000       6.16190       0.33810       8.748264
#  5       7.00000       7.24762       0.24762      16.309172
#====================================
# main: weights iteration #: 1
#  -- ls - start  ------------------------------
#  singular values (thresh = 1E-012)
#  s1=7.5396235253955, s2=56.330172166194,
#  -- ls - end  ------------------------------
#
# main
# Parameters:  a1=0.724590, a2=1.080075,
#
# | chisq: 3.856002
# | mean of |deviates|: 0.146679
# | variance of |deviates|: 0.013516
# | goodnes of fit: 0.964000
# | number of outliers: 0
#
# enter weight estimation
# -- est_weights1 - start ----------------------
#  Number of observations: 6
#  max_deviate: 0.375037
#  lambda_L = 0.115261
#
    :
```

Figure B.12: Example for fitting with weights estimation based on deviates (part 1)

```
:
# enter weight estimation
# -- est_weights1 - start ----------------------
#  Number of observations: 6
#  max_deviate: 0.499166
#  lambda_L = 0.024958
#
#  dev_sort[i]          weights[ idx[i] ]
#   0  2.410498630e-006    1605.35067674
#   1  8.387731592e-004    1605.35067674
#   2  1.670314822e-003    1605.35067674
#   3  1.675135820e-003    1605.35067674
#   4  2.974885015e-001      11.29951055
#   5  4.991660478e-001       4.01337669
# -- est_weights1 - end  -------------------------
#
# main
#  i       observ        calc      deviates    weights
#  0       1.70000      1.99749     0.29749      11.299511
#  1       3.00000      2.99832     0.00168    1605.350677
#  2       4.00000      3.99916     0.00084    1605.350677
#  3       5.00000      5.00000     0.00000    1605.350677
#  4       6.50000      6.00083     0.49917       4.013377
#  5       7.00000      7.00167     0.00167    1605.350677
#
# convergence of weights
#
#===================================
# main: approximation with final weights #: 12
#   -- ls - start  -----------------------------
#
#   singular values (thresh = 1E-012)
#   s1=28.754092057198, s2=331.76700514052,
#   -- ls - end  ------------------------------
#
# main
# Parameters:  a1=0.996652, a2=1.000836,
#
# | chisq: 2.010113
# | mean of |deviates|: 0.133474
# | variance of |deviates|: 0.038464
# | goodnes of fit: 0.502528
# | number of outliers: 0
#
# evaluation of results
#  number of outliers: 0
#  parameters: a1=0.996652 a2=1.000836
#  chi square.................: 2.010113
#  goodness of fit............: 0.502528
#  uncertainty in observations: 0.021643
#
#  (co)variance of parameters:
#   0.00114784155 -0.000264948098
# -0.000264948098 7.07291791E-005
#  (co)variance of parameters multiplied with goodness of fit
#  0.000576823 -0.000133144
# -0.000133144 3.55434E-005

#  resulting uncertainty of parameters
#    0.0240171358   0.00596182952
#         2.4098%         0.59568%
#
# Final_Parameters a1= 0.996652139801 a2= 1.00083636245
#  cond1         observed       fitted    weight uncertainty glob.uncertainty difference
   1.000000      1.700000      1.997489    11.299511      0.297489  0.021643   -0.297489
   2.000000      3.000000      2.998325 1605.350677      0.024958  0.021643   +0.001675
   3.000000      4.000000      3.999161 1605.350677      0.024958  0.021643   +0.000839
   4.000000      5.000000      4.999998 1605.350677      0.024958  0.021643   +0.000002
   5.000000      6.500000      6.000834    4.013377      0.499166  0.021643   +0.499166
   6.000000      7.000000      7.001670 1605.350677      0.024958  0.021643   -0.001670
```

Figure B.13: Example for fitting with weights estimation based on deviates (part 2)

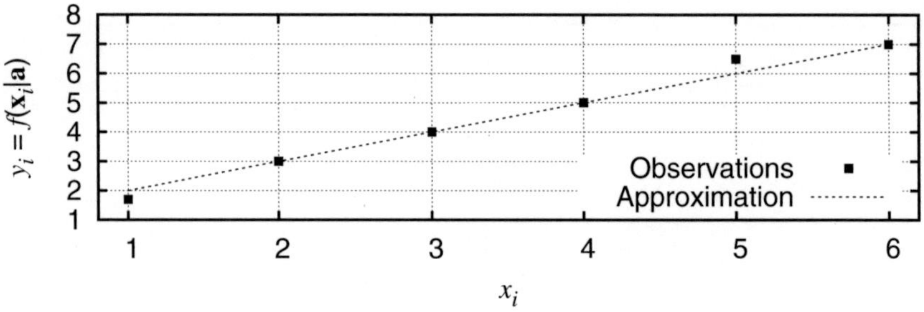

Figure B.14: Example 1: Comparison between observations and fitted model function when using estimation of weights

Example 2

Using the same options as in the previous example but including **option '-x'** enables the detection and removal of outlying observations. Independent of the detection scheme, the two down-weighted observations are identified as being contaminant and the fitting of the remaining four values yields a perfect straight-line fit.

The outlier detection is started directly after the convergence of the weights. **Figure B.15** shows the end of the programs output.

κ_1 is determined based on the number of observations, while κ_2 is set to a fixed value of 2. The routine for outlier detection outputs a table with sorted absolute deviates $\Lambda_s[n]$, the distances between the deviates $d[n]$, the global distance $d_{\text{glob}}[n]$, and the quotient $q[n] = d[n]/d_{\text{glob}}[n]$ (see Subsection 3.4.2). If $q[n] > \kappa_1$, then also the local distance $d_{\text{loc}}[n]$ and the quotient $r[n] = d[n]/d_{\text{loc}}[n]$ are computed. The final breakdown point is marked with a double-hash character. The table starts with $n = 4$ in this example, since sixty percent of the observations are assumed to be non-outliers. The weights of the contaminant data points are set to zero. The last cycle performs the least-squares approximation using the remaining observations and the actual weights.

Example 3

If sufficient observations under equal or similar conditions are available, the standard uncertainty of observations can be reliably determined using the binning method (Subsection 3.3.1). **Figure B.16** contains the program output for such an example. In total, 500 data points were fitted to a polynomial function. The

```
        .
        .
        .
#
# convergence of weights
# -- outlier_detection2 - start ----------------------
#  Number of observations: 6
#  kappa1: 7.300000
#  kappa2: 2.000000
#
#   n  deviate    d[n]        d_glob       d[n]/d_glob    d_loc   d[n]/d_loc
# ---------------------------------------------------------------------------
#   4  2.975e-001  2.958e-001  5.00e-004    591.09  6.865e-006  43088.449324 * ##
#   5  4.992e-001  2.017e-001  1.02e-001      1.99
#
#  outliers detected !   lambda_0 = 0.297489
#
#   dev_sort[i]          dist_all[i]       weights[ idx[i] ]
#   0  2.410498624e-006 0.000000000e+000    1605.35067674
#   1  8.387731592e-004 8.363626605e-004    1605.35067674
#   2  1.670314822e-003 8.315416633e-004    1605.35067674
#   3  1.675135820e-003 4.820997248e-006    1605.35067674
#   4  2.974885015e-001 2.958133657e-001       0.00000000
#   5  4.991660478e-001 2.016775463e-001       0.00000000
# -- outlier_detection2 - end  ----------------------
#
#===================================
# main: approximation with final weights #: 12
#  -- ls - start  -----------------------------
#
#  singular values (thresh = 1E-012)
#  s1=28.642351244568, s2=331.5853018607,
#  -- ls - end  -----------------------------
#
# main
# Parameters:  a1=1.000000, a2=1.000000,
#
# | chisq: 0.000000
# | mean of |deviates|: 0.000000
# | variance of |deviates|: 0.000000
# | goodnes of fit: 0.000000
# | number of outliers: 2
#
# evaluation of results
#  number of outliers: 2
#  parameters: a1=1.000000 a2=1.000000
#  chi square.................: 2.5644567358E-026
#  goodness of fit............: 1.2822283679E-026
#  uncertainty in observations: 3.45721314435E-015
#
#  (co)variance of parameters:
#   0.00115684559 -0.000266964368
# -0.000266964368 7.11904981E-005

#  (co)variance of parameters multiplied with goodness of fit
# 1.48334E-029 -3.42309E-030
# -3.42309E-030 9.12825E-031

#  resulting uncertainty of parameters
# 3.85141563E-015 9.55418632E-016
#      3.85E-013%       9.55E-014%
#
# Final_Parameters a1= 1.000000000 a2= 1.000000000
#  cond1        observed      fitted   weight uncertainty glob.uncertainty difference
   1.000000     1.700000     2.000000    0.000000 9999.000000 0.000000   -0.300000
   2.000000     3.000000     3.000000 1605.350677     0.024958 0.000000   -0.000000
   3.000000     4.000000     4.000000 1605.350677     0.024958 0.000000   +0.000000
   4.000000     5.000000     5.000000 1605.350677     0.024958 0.000000   +0.000000
   5.000000     6.500000     6.000000    0.000000 9999.000000 0.000000   +0.500000
   6.000000     7.000000     7.000000 1605.350677     0.024958 0.000000   +0.000000
```

Figure B.15: Output example for fitting with outlier removal after estimation of weights

```
# main ================================
# use data file: polynom_biased.xy
# polynomial of 2nd order
# Number of observations: 500
# Number of parameters  : 3
# condition columns: 1
# observations column: 3
# fitting a linear system
# use special SVD based algoritm
#  -- est_weights2 - start ------------------------------
#  Number of bins: 10
#  bin number_of_observations   a1          a2          last_cond variance
#     0         50        9.9279e+000   2.3206e+000   1.3817e-001   8.412e-002
#     1         50        9.1731e+000   7.2364e+000   2.8010e-001   9.925e-002
#     2         50        7.7854e+000   1.1755e+001   5.0366e-001   5.764e-002
#     3         50        5.0093e+000   1.7649e+001   6.4755e-001   1.124e-001
#     4         50       -2.1232e+000   2.8296e+001   9.4245e-001   1.738e-001
#     5         50       -1.0234e+001   3.7062e+001   1.1285e+000   1.877e+000
#     6         50       -8.1908e+000   3.5651e+001   1.2778e+000   8.256e+000
#     7         50       -1.7926e+001   4.5995e+001   1.5758e+000   4.079e+001
#     8         50        1.3498e+001   3.1553e+001   1.7752e+000   1.295e+002
#     9         50       -1.0895e+002   1.0205e+002   1.9717e+000   2.689e+002
#  -- est_weights2 - end  ------------------------------
#====================================
# main: approximation with final weights #: 0
#  -- ls - start  ------------------------------
#  singular values (thresh = 1E-012)
#  s1=56.53362686549, s2=15.010473603365, s3=2.9327109772612,
#  -- ls - end  ------------------------------
# main
# Parameters:  a1=10.205032, a2=-2.092809, a3=18.615686,
# | chisq: 567.991503
# | mean of |deviates|: 3.613647
# | variance of |deviates|: 47.361795
# | goodnes of fit: 1.142840
# | number of outliers: 0
#
# evaluation of results
#  number of outliers: 0
#  parameters: a1=10.205032 a2=-2.092809 a3=18.615686
#  chi square.................: 567.991503173
#  goodness of fit...........: 1.14284004663
#  uncertainty in observations: 1.06903697159
#
#  (co)variance of parameters:
#   0.00218900526  -0.00918148845   0.00795165394
#  -0.00918148845   0.0539568789   -0.0553097955
#   0.00795165394  -0.0553097955    0.064873586

#  (co)variance of parameters multiplied with goodness of fit
#   0.00250168     -0.010493    0.00908747
#    -0.010493     0.0616641   -0.0632102
#   0.00908747    -0.0632102    0.0741401

#  resulting uncertainty of parameters
#    0.0500168259       0.248322536       0.272286856
#            0.49%             11.9%             1.46%
# 1 parameter(s) have relative high uncertainty !
#
# Final_Parameters a1= 10.2050318876 a2= -2.09280850099 a3= 18.615685773
#  cond1         observed       fitted    weight uncertainty glob.uncertainty difference
    0.013620    10.002783     10.179981    11.888098      0.290031  1.069037    -0.177198
    0.013805     9.292838     10.179688    11.888098      0.290031  1.069037    -0.886850
    0.015035    10.121643     10.177775    11.888098      0.290031  1.069037    -0.056132
    0.015083     9.922627     10.177701    11.888098      0.290031  1.069037    -0.255074
    0.022013     9.813321     10.167984    11.888098      0.290031  1.069037    -0.354663
    :
    1.958257    76.087557     77.493657     0.003719     16.397803  1.069037    -1.406100
    1.960152   107.943153     77.627920     0.003719     16.397803  1.069037   +30.315233
    1.960207    94.607138     77.631819     0.003719     16.397803  1.069037   +16.975319
    1.960227    85.994650     77.633237     0.003719     16.397803  1.069037    +8.361413
    1.971709    99.422305     78.449640     0.003719     16.397803  1.069037   +20.972665
```

Figure B.16: Example for fitting with estimation of weights via binning

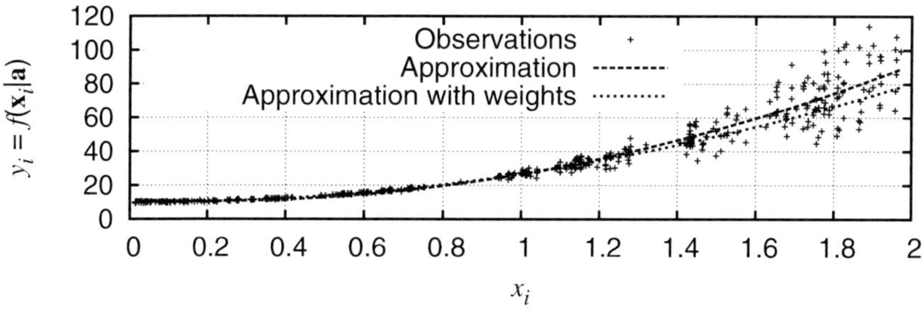

Figure B.17: Example 3: Comparison between observations and fitted model function when using estimation of weights

uncertainty of observation increases with the condition value and, in addition, they are influence by a positive bias.

First, observations from fifty consecutive conditions were fitted to straight lines, shown at the beginning of the output text file in Figure B.16. The deviation from this fit was taken as uncertainty for all of these fifty data points. The least-squares approximation then utilises the corresponding weights. The success is underlined by a goodness-of-fit value of $g_{\text{fit}} = 1.14$. However, the parameter a_2 shows a relative high uncertainty of about 12%. This indicates an inappropriate model function. The model function used for the fitting was $f(x|\mathbf{a}) = a_1 + a_2 \cdot x + a_3 \cdot x^2$, while the true process was $f(x|\mathbf{a}) = a_1 + a_3 \cdot x^2$, i.e. a_2 should be equal to zero.

Figure B.17 shows the approximated curves with and without the estimation of weights.

B.3 General organisation of source code

The source code is organised in several '*.c' and '*.h' files.
The file `fitting.c` contains the main function. It

- reads the options from the command line,
- initialises the Jacobian matrix,
- initialises the model parameters, if a nonlinear model function was chosen,
- iterates the weights estimation,
- calls the least-squares routine with appropriate parameters,
- starts the outlier detection and removal after convergence of the weights estimation,
- re-runs the least-squares routine, if at least one outlier was found, and
- evaluates and outputs the results.

All functions not having a special header file have a prototype declaration in `prototypes.h`.

The least-squares fitting as such is encoded in file `ls.c`. All operations according to equation (2.6)

$$\mathbf{\Delta a} = \left(\mathbf{J}^{\mathrm{T}} \cdot \mathbf{W} \cdot \mathbf{J}\right)^{-1} \cdot \mathbf{J}^{\mathrm{T}} \cdot \mathbf{W} \cdot \mathbf{r} \,. \tag{B.1}$$

are implemented here. Dependent on the chosen option '-a', one out of three different methods for the matrix inversion is used, i.e. cofactor method, LU decomposition, or singular value decomposition. The cofactor method, however, is not computed recursively as described in Subsection 5.3.1, but uses formulae which are precomputed up to a matrix size of 5×5. Hence, it is only applicable for problems with less than six model parameters.

In the case of nonlinear model functions, the Levenberg-Marquardt method (page 168, eq. (6.25))

$$\mathbf{\Delta a} = \left(\mathbf{J}^{\mathrm{T}} \cdot \mathbf{W} \cdot \mathbf{J} + \mu \cdot \mathbf{I}\right)^{-1} \cdot \mathbf{J}^{\mathrm{T}} \cdot \mathbf{W} \cdot \mathbf{r} \;\;. \tag{B.2}$$

is applied and the fitting is iterated until convergence. The convergence criterion is identical to eq. (2.8) with $\epsilon = 10^{-10}$ (defined in `macros.h`), i.e. when the correction terms Δa_j for all parameters fall under a given threshold, the iteration is stopped. Only, if the provided target value for χ^2 (option '-t') is not reached, the parameters are randomised in order to escape a possible local minimum, see also Section B.4.4. If χ^2 has dropped down to zero, the algorithm stops as well. The maximum number of repeated iterations, however, is limited to 2000, if not differently specified via option '-I'.

Linear fitting is performed via equation (6.11)

$$\mathbf{a} = \mathbf{V} \cdot \mathbf{S}^{-1} \cdot \mathbf{U}^{\mathrm{T}} \cdot \mathbf{\Sigma} \cdot \begin{pmatrix} y_1 & y_2 & \cdots & y_N \end{pmatrix}^{\mathrm{T}} \,. \tag{B.3}$$

by default. The matrices $\mathbf{V}$, $\mathbf{S}$, and $\mathbf{U}$ are results of a singular value decomposition (see Subsection 5.3.4). $\mathbf{\Sigma}$ is a diagonal matrix containing the reciprocal uncertainties of the observations y_i (see Section 6.3). This method is implemented in `solve_lin.c`, which is called from `ls.c`.

There is another fitting routine `ls_straighline.c` for the fitting of straight lines only. It is used for the piece-wise approximation of observations in the case of weights estimation via binning (Section 3.3.1) and uses eq. (7.41) from Section 7.4 with equal weights $w_i = 1.0 \;\forall\; i$.

The methods for weights estimation are encoded in file `est_weights.c` according to the explanations in Chapter 3. The file `outlier_detection.c` contains the algorithms for outlier detection.

`functions.c` contains all implemented model functions and/or their derivatives, respectively. They will be discussed below. The header file `functions.h` provides the corresponding prototypes of the functions.

As described in Section 2.3, the model parameters have to be initialised with reasonable values in the case of nonlinear model functions. This is done in special initialisation functions in the files `init_collection.c` and `init_gauss2.c` and will be described exemplarily below. The functions from the NIST data base (see Section B.7) are encoded in separate files `functions_NIST.c`, `functions_NIST.h`, and `init_NIST.c`.

The matrix inversion via singular value decomposition is encoded in `svd_inversion.c`, which calls `singvaldec.c`. Other matrix operations can be found in `decomp_LU.c` and `matrix_utils.c`. The latter also contains special functions for the allocation of memory. It is accompanied with a header file `matrix_utils.h` for prototyping purposes.

`get_option.c` and `get_option.h` are responsible for the command-line parsing. All options start with a dash (minus sign) followed by one or more characters. The lists of available and compulsory options must be given in `usage.c`, which also provides help in cases of wrong command-line options.

`errmsg.c` and `errmsg.h` support a simple kind of error handling. The sorting of arrays of different types is possible using the functions in `heap_sort.c`.

Finally, there are two other header files, `macros.h` and `defines.h`, containing macros and definitions required in several functions.

B.4 Model functions

The file `functions.c` contains all necessary information about the implemented model functions. Linear functions require simply the first partial derivatives with respect to the single model parameters. In contrast to this, fitting nonlinear functions additionally needs the model function as such.

B.4.1 Numerical differentiation

In case of complicated nonlinear functions it might be simpler to use numerical differentiation instead of analytical determination of the partial first derivatives. This can be achieved by inspecting the functional course close to the actual parameter vector $\mathbf{a}$ using a distance δ_j

$$\frac{\partial f}{\partial a_j} \simeq \frac{f(\mathbf{a}+\delta_j\mathbf{e}_j) - f(\mathbf{a}-\delta_j\mathbf{e}_j)}{2\cdot\delta_j} \tag{B.4}$$

where $\mathbf{e}_j$ is the unit vector along a_j, i.e. as for instance $\mathbf{e}_1 = (1\,0\,0\,\ldots)^{\mathrm{T}}$ for a_1. δ_j should be only a portion of the parameter value. The implementation (function `f_deriv()`) uses $\delta_j = a_j/C$ with $C = 10^6$ yielding a variation of relatively small amounts. If $a_j = 0$, then δ_j is set to the arbitrary value of 0.001.

The drawback of the numerical differentiation is its computational complexity, since the model function has to be called twice with different parameter settings. In addition, the results of nonlinear approximation are somewhat less accurate than the results based on analytical derivations (see Section B.7).

B.4.2 Handling of multi-dimensional conditions

The treatment of model functions with a single condition per observation is rather straightforward, since both, conditions x_i and observations y_i, can be stored in one-dimensional arrays, '`cond[]`' and '`obs[]`'. The joint handling of these model functions with those having more than one condition requires the re-use of these arrays for multi-dimensional conditions. The implemented code writes the conditions $\mathbf{x}_i$ in alternating order into the array `cond[]`. If there are two conditions per observations, for example, the length of `cond[]` must be doubled. All functions accessing this array must increase the array index accordingly. This concerns, for example, the files `fitting.c` and `init_collection.c` (functions: `flin_deriv()`, `frotation()`, `fcircle()`). **Figure B.18** shows function `frotation()`, where the distinction between the interleaved conditions is made using the modulo-2 operation '`i%2`'.

B.4.3 Limitation of parameter space

Section 2.3 had already discussed the problem of restricted ranges of allowed parameter values. The exponential function

$$y_i = a_1 \cdot \exp(a_2 \cdot x_i)$$

is one example, where the domain of definition has to be limited. The corresponding function is implemented in `fexpon2_deriv()` in file `functions.c`. If it is known that the observations describe a decaying process, the parameter a_2 must be negative. In case that this parameter becomes non-negative in the course of optimisation, it is set virtually to zero by changing the partial derivatives for a_1 and a_2 according to

$$\frac{\partial f}{\partial a_1} = \begin{cases} \exp(a_2 \cdot x_i) & \text{if } a_2 < 0 \\ 1 & \text{otherwise} \end{cases}$$

```
/*------------------------------------------------------------
 * frotation()
 *     x = f1(u|a) = a1 + cos(a3) * u - sin(a3) * v
 *     y = f2(u|a) = a2 + sin(a3) * u + cos(a3) * v
 *------------------------------------------------------------*/
double
frotation( int i, double *cond, double *a)
{
  if (i%2 == 0)
  {
    /* equation for x */
    return a[0] + cos(a[2]) * cond[i] - sin(a[2]) * cond[i+1];
  }
  else
  {
    /* equation for y */
    return a[1] + sin(a[2]) * cond[i-1] + cos(a[2]) * cond[i];
  }
}
```

Figure B.18: Source code example with interleaved access to conditions of a two-dimensional function

and

$$\frac{\partial f}{\partial a_2} = \begin{cases} x_i \cdot a_1 \cdot \exp(a_2 \cdot x_i) & \text{if } a_2 < 0 \\ x_i \cdot a_1 & \text{otherwise} \end{cases} .$$

In doing so, the error surface is modified virtually and the optimisation does not walk into this direction.

If, however, the model function is only defined in a certain range as, for example, the logarithmic function

$$f(x|a_1) = \log(a_1 \cdot x) = \log(a_1) + \log(x) \qquad a_1 > 0 ,$$

then the parameter a_1 should be fixed before evaluating the function or its derivation.

B.4.4 Initialisation of parameters

The proper initialisation can reduce the number of required iterations and might be even essential, if the error surface has many local minima. This problem was already addressed more in detail in Section 2.3. The responsible functions can be found in `init_collection.c` and `init_gauss2.c`. Two examples are discussed in the following.

Example 1: Fitting of a circle (`init_collection.c`)
Using the model function

$$(y_1 - a_1)^2 + (y_2 - a_2)^2 = r^2 \ ,$$

three parameters must be determined. Under the assumption that the sample points are evenly distributed on the circle, the centre coordinates (a_1, a_2) can be estimated by

$$a_{1\text{init}} = \frac{1}{N} \cdot \sum_{i=1}^{N} y_{1i} \qquad \text{and} \qquad a_{2\text{init}} = \frac{1}{N} \cdot \sum_{i=1}^{N} y_{2i} \ .$$

The estimated radius is the averaged distance between all observed points and the estimated circle centre

$$r_{\text{init}} = \frac{1}{N} \cdot \sum_{i=1}^{N} \sqrt{(y_{1i} - a_{1\text{init}})^2 + (y_{2i} - a_{2\text{init}})^2} \ .$$

□□□

Example 2: Neural network (`init_collection.c`)
Initially all parameters are set to uniformly distributed random values in the range of $[-1; 1]$, if not specified differently on the command line. In order to keep the input of the hidden neurons also within $[-1; 1]$, the maximum and minimum condition values are determined and the parameter values accordingly normalised, see `init_NN3x3x1()` for example.
□□□

The provided source code supports the definition of a target value for $\chi^2_{\max}$ (option '-t'), which can be used to escape local minima. If the actual value of χ^2 is higher than $\chi^2_{\max}$ after convergence or a maximum number of iterations, the parameters are re-initialised (`ls.c`, lines 541ff. and lines 614ff., respectively). However, calling the initialisation routines again only makes sense if they have a random component yielding different parameters in **a** for each call.

B.5 Special algorithms

B.5.1 LU decomposition

The LU decomposition described in Section 5.3.3 is performed using Crout's algorithm, see [Pre92] for details. The result is a single matrix, which contains both, the λ's and the υ's

$$\begin{pmatrix} \upsilon_{11} & \upsilon_{12} & \cdots & \upsilon_{1M} \\ \lambda_{21} & \upsilon_{22} & \cdots & \upsilon_{2M} \\ \vdots & \vdots & \ddots & \vdots \\ \lambda_{M1} & \lambda_{M2} & \cdots & \upsilon_{MM} \end{pmatrix} . \tag{B.5}$$

B.5.2 Singular value decomposition

For a detailed description of the singular value decomposition see for instance [Bjo96]. The implementation in `singvaldec()` is based on a FORTRAN source code published on the internet [SVD]. The translation to C also takes into account that the indices are starting from zero in C and not from one as in the FORTRAN language.

B.5.3 Sorting

The functions for weights estimation require the sorting of values at different stages. The file `heap_sort.c` contains two routines for the sorting based on the heap-sort algorithm. The function `heap_sort_d()` sorts the values in the input vector array `values[]`. A second routine `heap_sort_d_()` sorts in addition an index array. This is helpful, if the values must stay in their order and the sorted access is solely performed via the sorted index array. In this case, the value array passed to the heap-sort function must be a copy of the original vector array in order to keep the original values unsorted.

B.6 Possible optimisations

It must be emphasised again that the provided source code is neither optimised for speed nor it is using elegant implementations, but is designed to be readable in connection with the text of this book.

Besides optimisations with respect to the speed of calculations, the user might be interested to have more influence on the performance of the data fitting. For instance the convergence criterion is hard-coded, but could be provided by the user via an additional program option. The same holds true for desired parameter ranges in terms of minimum and/or maximum values for all a_j.

The combined fitting of two model functions as described in Section 7.5 is not implemented yet. The total least-squares approach (Section 7.7) works only for the special case of fitting a circle.

The implementation of the neural networks based on the sigmoid function might be not optimal. The convergence process is highly dependent on the starting parameter values. Future implementations should consider to use radial basis functions as activation function for the neurons.

Suggestions by the reader towards the future improvement of the program are welcome!

B.7 Performance Test

The National Institute of Standardization and Technology (NIST)[2] maintains a website containing statistical reference datasets [Strd].

In response to industrial concerns about the numerical accuracy of computations from statistical software, the Statistical Engineering and Mathematical and Computational Sciences Divisions of NIST's Information Technology Laboratory are providing datasets with certified computational results for a variety of statistical methods enabling the objective evaluation of statistical software.

Several datasets addressing linear as well as nonlinear problems are listed; either stemming from real world measurements or prepared by hand in order to stress the fitting algorithms. The datasets are labelled according to the level of difficulty as 'Lower', 'Average', or 'Higher'. All datasets are accompanied with the underlying model function $y = f(\mathbf{x}|\mathbf{a})$, certified model parameters a_j, which lead to the true global minimum, and their standard uncertainties σ_{a_j}. Producing correct results for all datasets in this collection will provide some degree of assurance, in the sense that a certain software package provides correct results for datasets known to yield incorrect results for some software. The certified values are 'best-available' solutions, obtained using 128-bit precision and confirmed by at least two different algorithms and software packages using analytic derivatives.

This section will evaluate the implemented algorithms using only the data with 'higher' difficulty. The NIST website states that "a good nonlinear least squares procedure should be able to duplicate the certified results to at least 4 or 5 digits".

The developed fitting software has been tested on a Pentium® 4 CPU (32 bit). Since the NIST datasets do not contain weights, $\sigma_i = 1.0$ is used for all tests.

[2] Founded in 1901, NIST is a non-regulatory federal agency within the U.S. Department of Commerce. NIST's mission is to promote U.S. innovation and industrial competitiveness by advancing measurement science, standards, and technology in ways that enhance economic security and improve our quality of life.

Table B.1: Results of fitting difficult linear model functions: number of correct digits in mantissae of estimated model parameters

Dataset	Model type	M	Default	'-s -a 1'	'-s -a 2'	'-c'	'-s -a 1 -c'
Filip	polynomial	11	2	failed	failed	7	0
Longley	multilinear	7	9	failed	9	11	7
Wampler 1	polynomial	6	10	3	5	10	6
Wampler 2	polynomial	6	11	5	9	11	9
Wampler 3	polynomial	6	7	3	5	10	7
Wampler 4	polynomial	6	5	3	5	8	7
Wampler 5	polynomial	6	3	3	5	6	6

B.7.1 Fitting linear systems

The NIST data base contains seven linear problems of higher difficulty, two observed sets (Filip and Longley) and five artificially generated (Wampler1-5).

The default setup of the developed fitting software for solving linear problems is to apply the special SVD-based approach according to equation (B.3)

$$\mathbf{a} = \mathbf{V} \cdot \mathbf{S}^{-1} \cdot \mathbf{U}^{\mathrm{T}} \cdot \mathbf{\Sigma} \cdot \begin{pmatrix} y_1 & y_2 & \cdots & y_N \end{pmatrix}^{\mathrm{T}} ,$$

where $\mathbf{V}$, $\mathbf{S}$, and $\mathbf{U}$ are the matrices derived from the Jacobian matrix via singular value decomposition (see Subsections 5.3.4 and 6.3.2). The matrix $\mathbf{\Sigma}$ is a diagonal matrix containing the reciprocal uncertainties of the observations y_i (see Section 6.3).

Alternatively, option '-s' switches to equation (B.1)

$$\mathbf{a} = \left(\mathbf{J}^{\mathrm{T}} \cdot \mathbf{W} \cdot \mathbf{J}\right)^{-1} \cdot \mathbf{J}^{\mathrm{T}} \cdot \mathbf{W} \cdot \mathbf{y} ,$$

where an explicit inversion of the normal matrix $\mathbf{N} = \mathbf{J}^{\mathrm{T}}\mathbf{W}\mathbf{J}$ is performed. The algorithm for inversion can be chosen via option '-a'. However, '-a 0' is only applicable for model functions with five parameters in maximum. The option '-c' enables the scaling of the condition values (see Subsection 6.3.3).

Table B.1 indicates the number of correct digits in the mantissae of estimated parameter values. The numbers are approximate results, since the number of identical digits can vary for different a_j.

It can be clearly seen that the special SVD approach (column 'Default') outperforms the standard setup (option '-s') for all datasets. Polynomials of high

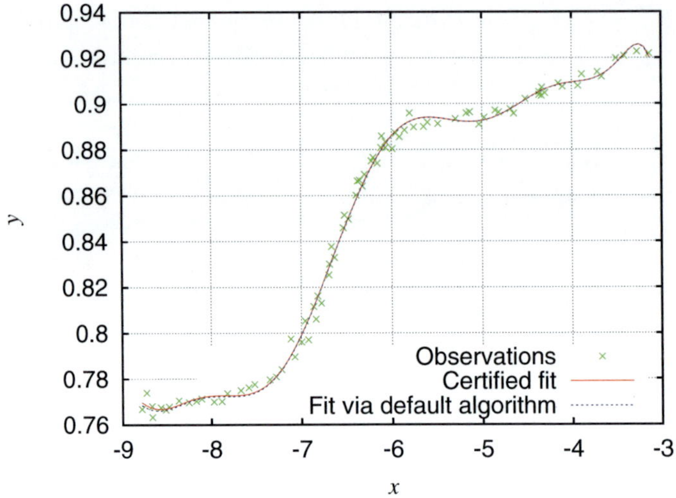

Figure B.19: Filip dataset and its approximation

order, as the Filip set, cannot be fitted via the normal matrix at 32-bit precision without special treating of the data. Even the special SVD approach yields only two correct digits. However, the visual quality of fitting is already rather good (**Figure B.19**).

When using option '-s' for the Filip data, the program outputs an error message, because at least one of the computed model parameter variance is negative, which indicates numerical instability. The same happens for the Longley dataset when using the options '-s -a 1'.

The Wampler datasets where designed to stress the fitting algorithms under test. The observations deviate more or less from a polynomial of fifth order. The model parameters are always equal to one or equal to 10^{-b} ($b \in \mathbb{N}$) in the case of the Wampler2 set. Even though the results of the designed fitting program deviate somewhat from this exact solution, the plots are indistinguishable as can be seen for the case of Wampler4 (**Figure B.20**).

Interestingly, when using the standard approach, the inversion via LU decomposition (option '-a 2') yields better results for all tested datasets than the inversion via singular value decomposition (option '-a 1').

The normalisation of the condition values by their absolute maximum, before the least-squares approximation is applied (option '-c', see Subsection 6.3.3), distinctly increases the numerical accuracy. In addition, the standard approach (option '-s -a 1 -c') does not fail, neither in application to the Filip data nor to the Longley data. In general, the accuracy of the estimated parameters is increased

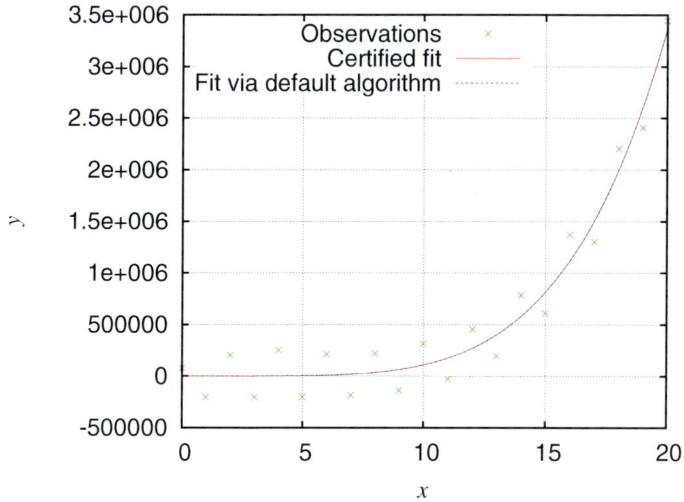

Figure B.20: Wampler4 dataset and fitted curves

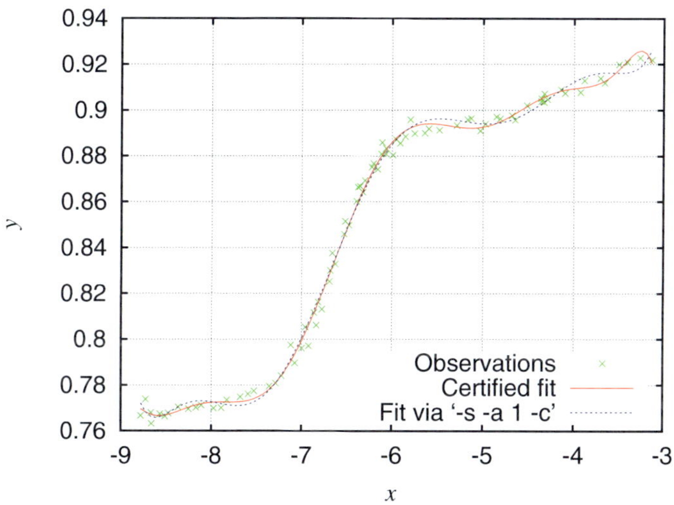

Figure B.21: Filip dataset fitted with options '-s -a 1 -c'

compared to column '-s -a 1'. Only the estimated parameters of the Filip data differ strongly from the true values. Nevertheless, the fitted curve is still close to the true one (**Figure B.21**).

B.7.2 Fitting nonlinear systems

The robustness and reliability of nonlinear least squares software depends on the algorithm used and how it is implemented, as well as on the characteristics of the actual problem being solved. As already pointed out in Section 2.3, nonlinear least squares solvers are particularly sensitive to the starting values provided for a given problem. For this reason, the NIST website provides three sets of starting values for each problem: the first is relatively far from the final solution; the second relatively close; and the third is the actual certified solution.

Table B.2 list the results of nonlinear fitting based on equation (B.2)

$$\mathbf{\Delta a} = \left(\mathbf{J}^{\mathrm{T}} \cdot \mathbf{W} \cdot \mathbf{J} + \mu \cdot \mathbf{I}\right)^{-1} \cdot \mathbf{J}^{\mathrm{T}} \cdot \mathbf{W} \cdot \mathbf{r} \ .$$

The table contains at least two lines for each dataset corresponding to the two sets of initial parameter values, which are provided along with the data. The abbreviation 'st. 1' stands for the parameter set, which is far from the final solution. The lines 'st. 2' list the results stemming from initial parameter values close to the certified solution. The datasets BoxBOD and Eckerle4 have a third entry. As these model functions have a more or less simple structure, it is possible to guess suitable starting values directly from the observations.

Each column contains two values. The first one is the number of digits which are identical to the certified values; the second value in parenthesis is the number of iterations till convergence of the algorithm. Cases without convergence to a minimum are indicated by 'no conv.'.

As can be concluded from the column 'default', the developed software is very robust and reliable. There is only a single case (MGH10 start set 1) where the algorithm could not converge. In all other cases, the fitting program could determine the model parameter with sufficient accuracy.

Even though start set 1 defines initial parameter values, which are more distant from the global minimum than the values from start set 2, the number of iterations is not always higher. For the datasets Thurber and Bennett5, the algorithm converges faster, when starting from the remote point.

Five additional tests have been made in order to test single features of the fitting algorithm:

- '-n' ... usage of numerical derivation instead of analytical derivation,

Table B.2: Results of fitting difficult nonlinear model functions: number of correct digits in mantissae of estimated model parameters

Dataset	Model	M	Default	'-n '	'-a 0 '	'-a 2'	'-L'	'-L -a 2'
MGH09 st. 1	rational	4	8 (233)	8 (231)	8 (230)	9 (232)	0 (29)	failed
MGH09 st. 2	rational	4	9 (44)	9 (47)	8 (47)	8 (46)	0 (118)	8 (46)
Thurber st. 1	rational	7	8 (72)	8 (73)	/	9 (73)	0 (66)	0 (46)
Thurber st. 2	rational	7	10 (83)	9 (83)	/	8 (82)	11 (66)	11 (66)
BoxBOD autom.	expon.	2	10 (22)	10 (23)	10 (22)	10 (22)	11 (20)	11 (20)
BoxBOD st. 1	expon.	2	8 (27)	failed/1/	8 (27)	8 (27)	11 (21)	failed/2/
BoxBOD st. 2	expon.	2	11 (24)	8 (26)	11 (24)	11 (24)	11 (19)	11 (19)
Rat42 st. 1	expon.	3	11 (28)	8 (29)	10 (30)	11 (28)	failed/1/	failed/1/
Rat42 st. 2	expon.	3	10 (23)	10 (23)	10 (23)	10 (23)	11 (10)	10 (10)
Rat43 st. 1	expon.	4	9 (34)	8 (35)	9 (34)	10 (33)	0 (5)	failed/1/
Rat43 st. 2	expon.	4	10 (23)	9 (26)	9 (25)	7 (24)	10 (17)	10 (17)
MGH10 st. 1	expon.	3	0 (no conv.)	0 (no conv.)	0 (no conv.)	0 (no conv.)	0 (9)	failed/3/
MGH10 st. 2	expon.	3	7 (420)	7 (419)	7 (418)	7 (420)	11 (18)	11(18)
Eckerle4 autom.	expon.	3	11 (11)	11 (8)	11 (11)	11 (11)	11 (8)	11 (8)
Eckerle4 st.1	expon.	3	11 (71)	9 (75)	11 (42)	11 (71)	failed/4/	failed/2/
Eckerle4 st.2	expon.	3	11 (11)	11 (10)	11 (11)	11 (11)	11 (10)	11 (10
Bennett5 st.1	misc.	3	5 (1103)	5 (1129)	5 (1134)	5 (1127)	11 (45)	11 (40)
Bennett5 st.2	misc.	3	6 (1754)	6 (1752)	6 (1754)	6 (1753)	11 (16)	11 (34)

- '-a 0' ... usage of direct inversion of the normal matrix instead of using the SVD,
- '-a 2' ... usage of LU decomposition for inversion of the normal matrix,
- '-L' ... usage of Gauss-Newton method instead of Levenberg-Marquardt,
- '-L -a 2' ... combination of Gauss-Newton method and LU decomposition.

Column '-n' proves that the implemented numerical derivation is accurate enough in general. Only fitting the BoxBOD dataset starting from the remote point fails. The failures are labelled in the table and the numbers have following meaning

- failed/1/ ... Result too large (CPU message),
- failed/2/ ... error message from LU decomposition "max. element is zero",
- failed/3/ ... error message from SVD "degraded matrix"[3],
- failed/4/ ... error message from SVD "singular matrix",

Column '-a 0' and '-a 2' show that there is only little advantage of SVD over other methods for matrix inversion. More interesting is the behaviour if the optimisation is performed without the Levenberg-Marquardt term $\mu \cdot \mathbf{I}$. The Gauss-Newtom method fails in two cases and does not converge to the desired minimum in five other cases. Then, again, it leads to more accurate parameters values for the Bennett5 and MGH10 (start set 2) datasets using a fraction of iterations. If the Gauss-Newton method is accompanied with the LU decomposition for inversion of the normal matrix, the algorithm tends to fail instead to converge to a non-desired target.

The **Figures** B.22–B.25 show the fitting results (default approach) for selected datasets. The corresponding model functions are

- BoxBOD: $f(x|\mathbf{a}) = a_1 \cdot (1 - \mathrm{e}^{-a_2 \cdot x})$
- Thurber: $f(x|\mathbf{a}) = \dfrac{a_1 + a_2 \cdot x + a_3 \cdot x^2 + a_4 \cdot x^3}{1 + a_5 \cdot x + a_6 \cdot x^2 + a_7 \cdot x^3}$
- MGH10: $f(x|\mathbf{a}) = a_1 \cdot \mathrm{e}^{a_2/(x+a_3)}$
- Bennett5: $f(x|\mathbf{a}) = a_1 \cdot (a_2 + x)^{-1/a_3}$

[3] The failure message for MGH10 st.1 '-L -a 2' does not stem from the matrix inversion within the iterations, because the LU decomposition is used here. The SVD is required for the determination of the covariance matrix after a minimum was found.

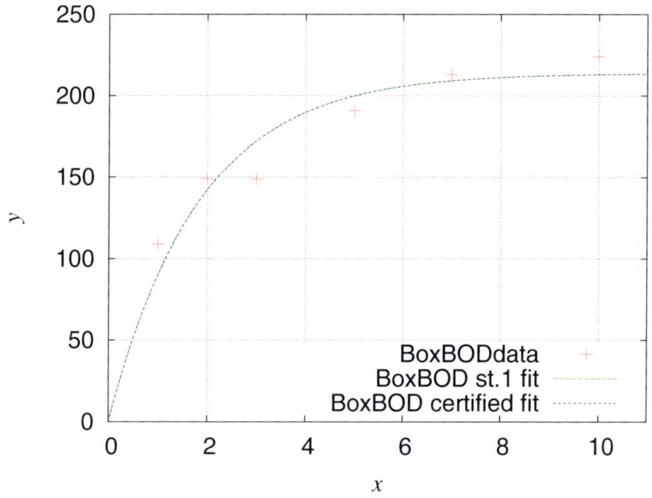

Figure B.22: BoxBOD dataset and fitted curve

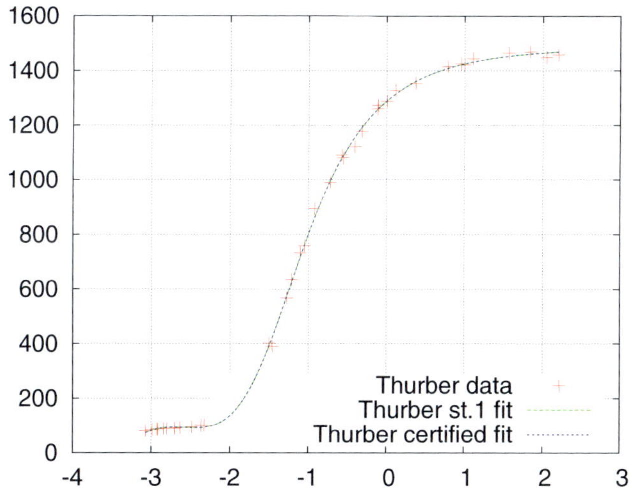

Figure B.23: Thurber dataset and fitted curve

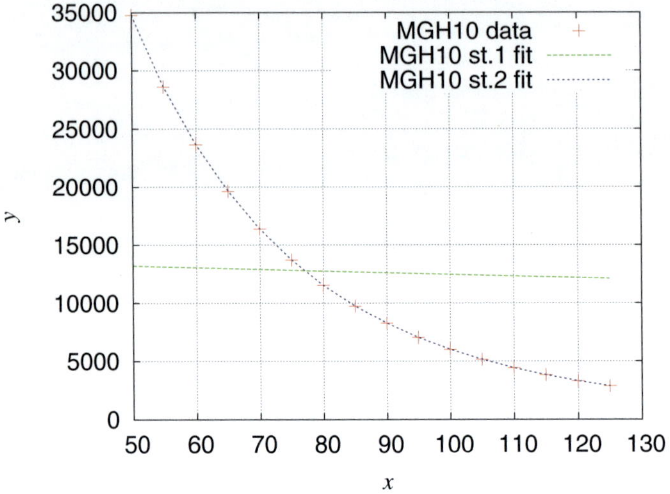

Figure B.24: MGH10 dataset and fitted curve

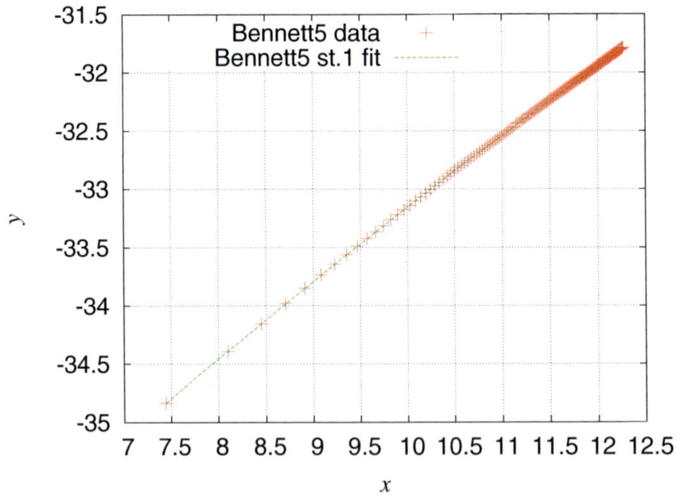

Figure B.25: Bennett5 dataset and fitted curve

List of symbols

$\in$	is element of
$\forall$	for all
$\approx$	approximately equal
$\|g\|$	absolute value of g
$\|\|\mathbf{g}\|\|$	(Euclidean) norm of vector $\mathbf{g}$; $\|\|\mathbf{g}\|\| = \sqrt{\mathbf{g}^{\mathrm{T}} \cdot \mathbf{g}}$
a_j	model parameter with $j = 1, 2, \ldots, M$
$\alpha_{l,k}$	weight of a neural network between l^{th} input unit and k^{th} hidden neuron
β_k	weight of a neural network between k^{th} hidden neuron and output neuron
Δa_j	change of the model parameter a_j
$\mathbf{a}$	vector of model function parameters $(a_1 \ \ldots \ a_j \ \ldots \ a_M)$
$\mathbf{\Delta a}$	vector of model function parameters changes
$\hat{\mathbf{a}}$	vector of estimated model parameters
$\mathbf{A} = (a_{ij})$	matrix with elements a_{ij}
$\mathbf{A}^{\mathrm{T}} = (a_{ji})$	transposed matrix of $\mathbf{A}$
b_j	model parameter with $j = 1, 2, \ldots, M$
$\mathbf{C}$	covariance matrix
$\mathbf{C}_a$	covariance matrix of model parameters $\mathbf{a}$
χ^2	value to be minimised in the course of least-squares approximation
Δ_i	deviate $y_i - \hat{y}_i$ (also: residual)
e	error $e = t - x$ between target value t and current signal value x
$e[n]$	error $e[n] = d[n] - x[n]$ between desired value $d[n]$ and current signal value $x[n]$
E	energy of error e
ε_i	error in y_i
$f(\mathbf{x}\|\mathbf{a})$	model function
$f(x)$	function of x
$f'(x)$	first derivative of function $f(x)$
$f^{(\nu)}(x)$	ν^{th} derivative of function $f(x)$
$\Phi(\cdot)$	activation function of a neuron
g_{fit}	goodness of fit
$\mathbf{g}$	vector of gradients $\mathbf{g} = \left(\frac{\partial \chi^2(\mathbf{a})}{\partial a_1} \ \frac{\partial \chi^2(\mathbf{a})}{\partial a_2} \ \ldots \ \frac{\partial \chi^2(\mathbf{a})}{\partial a_M} \right)^{\mathrm{T}}$
$\mathbf{H}$	Hessian matrix $\mathbf{H} = \left(\frac{\partial^2 \chi^2(\mathbf{a})}{\partial a_j \partial a_k} \right)$
I	current

$\mathbf{I}$	identity matrix
i	number of an observation $i = 1, 2, \ldots, N$
$\mathbf{J}$	Jacobian matrix
$\mathbf{L}$	lower triangular matrix
λ_{L}	threshold for discrimination between observations with equal weights and observations with adaptively estimated weights
λ_{O}	threshold for discrimination between good observations and outliers
Λ_i	absolute deviate $\lvert y_i - \hat{y}_i \rvert = \lvert \Delta_i \rvert$
μ	mean value of a (Gaussian) distribution; learning rate in the context of neural networks
M	number of model parameters
N	number of observations
N_{free}	number of degrees of freedom, $N_{\mathrm{free}} = N - M$
$\mathcal{N}(\mu, \sigma^2)$	Gaussian distribution
$\mathbf{N}$	normal matrix
$\mathbb{N}$	set of natural numbers
r	radius
$\mathbf{r}$	vector of residuals $(y_i - f(\mathbf{x}_i \vert \mathbf{a}))$, $i = 1, 2, \ldots, N$
P	probability
$\prod_{i=0}^{p} b_i$	product $b_0 \cdot b_1 \cdot b_2 \cdot \ldots \cdot b_p$
R	resistor
$\mathbf{S}$	matrix of singular values
σ	standard deviation of Gaussian distribution
σ_{ε}^2	variance of normally distributed noise
$\sigma_{a_j}^2$	variance of estimated parameter a_j
σ_i	standard uncertainty of of the parent distribution of y_i
σ_{y}	standard uncertainty of y
$\hat{\sigma}_{\mathrm{y}}$	estimated standard uncertainty of y
$\mathbf{\Sigma}$	diagonal matrix of reciprocal uncertainties
$\sum_{i=0}^{p} b_i$	sum $b_0 + b_1 + b_2 + \cdots + b_p$
U	voltage
$\mathbf{U}$	upper triangular matrix; matrix resulting from singular value decomposition
$\mathbf{V}$	matrix resulting from singular value decomposition
w_i	weight assigned to observation y_i, $i = 1, 2, \ldots, N$
$\mathbf{W} = (w_{ij})$	matrix of weights
$w_{ij}^{(k)}$	weight of a neural network that connects the jth neuron of the kth layer with the ith neuron of the preceding layer

x	condition, independent variable
$\mathbf{x}$	vector of conditions $(x_1 \ \ldots \ x_k \ \ldots)$
$\mathbf{x}_i$	vector of conditions under which a particular value y_i was observed
$\hat{x}[n]$	estimated value of $x[n]$
y	dependent variable
$\overline{y}$	mean value of y
y_i	value of a measurement, observation
$\hat{y}_i$	estimate of y_i based on the model function and estimated parameters $\hat{\mathbf{a}}$
$\mathbf{y}$	vector of observations $(y_1 \ldots y_N)^{\mathrm{T}}$
$\mathbb{Z}$	set of integers

Bibliography

[Bar94] Barnett, V.; Lewis, T.: *Outliers in Statistical Data.* 3rd edition, WILEY, 1994, ISBN 0-471-93094-6

[Bec83] Beckman, R.J.; Cook, R.D.: Outliers. *Technometrics*, Vol.25, No.2, May 1983, 119–149

[Bev92] Bevington, P.R.; Robinson, D.K.: *Data reduction and error analysis for the physical science.*, 2nd edition, WCB/McGraw-Hill, 1992, ISBN 0-07-911243-9

[Bir32] Birge, R.T.: The calculation of errors by the method of least squares. *Physical Review* Vol. 40, April 15, 1932, 207–227

[Bis95] Bishop, C.M.: *Neural Networks for Pattern Recognition.* Oxford University Press, 1995, ISBN 0-19-853864-2

[Bjo96] Björk, Å.: *Numerical methods for least squares problems.* SIAM, Philadelphia, 1996, ISBN 0-89871-360-9

[Bog87] Boggs, P.T.; Byrd, R.H.; Schnabel, R.B.: A stable and efficient algorithm for nonlinear orthogonal distance regression. *SIAM J. Sci. Stat. Comput.*, Vol.8, No.6, November 1987, 1052–1078

[Bre00] Breuning, M.M; Kriegel, H.-P.; NG, R.T.; Sander, J, "LOF: Identifying density-based local outliers", *Proc. of the ACM SIGMOD Int. Conf. on Management of Data*, Dallas, Texas, 2000

[Car88] Carroll, R.J.; Ruppert, D.: *Transformation and Weighted Regression.* Chapman and Hall, New York, 1988

[Cha71] Chauvenet, W.: Method of Least Squares. Appendix to *Manual of Spherical and Practical Astronomy*, Vol.2, 4th edition, J.B. Lippincott & CO., Philadelphia, 1871, 469–566

[Cho09] Choi, S.; Kim, T.; Yu, W.: Performance Evaluation of RANSAC Family. *Proc. of the British Machine Vision Conference (BMVC)*, 2009

[Coo93] Coope, I.D.: Circle Fitting by Linear and Nonlinear Least Squares. *Journal of Optimization and Applications*, Vol.6, No.2, February 1993, 381–388

[Cox06a] Cox, M.G.; Eiø, C.; Mana, G,; Pennecchi, F.: The generalized weighted mean of correlated quantities. *Metrologia*, Vol.43, No.4, 2006, 268–275

[Cox06b] Cox, M.G.; Harris, P.M.: *Software Support for Metrology. Best Practice Guide No,6*, NPL Report DEM-ES-011, 2006

[DIN96] DIN 1319-3: *Fundamentals of metrology - Part 3: Evaluation of measurements of a single measurand, measurement uncertainty.* Beuth Verlag, May 1996

[DIN99] DIN 1319-4: *Fundamentals of metrology - Part 4: Evaluation of measurements; uncertainty of measurement.* Beuth Verlag, February 1999 (D)

[ESH] *Engineering Statistics Handbook.* www.itl.nist.gov, accessed 31 July 2015

[Fis81] Fischler, M.A.; Bolles, R.C.: Random Sample Consensus: A Paradigm for Model Fitting with Applications to Image Analysis and Automated Cartography. *Comm. of the ACM*, Vol.24, No.6, June 1981, 381–395

[Gel68] Gellert, W.; Küstner, H.; Hellwich, M.; Kästner, H. (eds.): *Mathematik – Kleine Enzyklopedie.* VEB Bibliographisches Institut Leipzig, 3. Auflage, 1968

[Goe89] Göhler: *Höhere Mathematik – Formeln und Hinweise.* VEB Deutscher Verlag für Grundstoffindustrie, Leipzig, 10. Auflage, 1989

[Gnupl] www.gnuplot.info, accessed 31 July 2015

[GUM] http://www.bipm.org/en/publications/guides/gum.html, accessed 31 July 2015

[Ham71] Hampel, F.R.: A general qualitative definition of robustness. *Ann. Math. Stat.*, Vol.42, 1971, 1887–1896

[Ham86] Hampel, F.R.; Ronchetti, E.M; Rousseeuw, P.J.; Stahel, W.A.: *Robust Statistics: The Approach Based on Influence Functions.* Wiley, New York, 1986

[Hod04] Hodge, V.J; Austin, J., "A survey of Outlier Detection Methodologies", *Artificial Intelligence Review*, 2004, 22, 85–126

[Hol77] Holland, P.W.; Welsch, R.E.: Robust regression using iteratively reweighted least-squares. *Comm. Statist.-Theor. Meth.* Vol.A6(9), 1977, 813–827

[Igl93] Iglewicz, B.; Hoaglin, D.C.: *Volume 16: How to Detect and Handle Outliers*. The ASQC Basic References in Quality Control: Statistical Techniques, 1993, Edward F. Mykytka, Ph.D., Editor.

[ISO00] ISO/IEC 14495-1, *Information technology – Lossless and near-lossless compression of continuous-tone still images: Baseline (JPEG–LS)*. International Standard, corrected and reprinted version, 15 September 2000

[Jai88] Jain, A.K. and Dubes, R.C.: *Algorithms for Clustering Data*. Prentice Hall, Englewood Cliffs, New Jersey 07632, 1988, ISBN 013022278X

[Kas76] Kása, I.: A Circle Fitting Procedure and its Error Analysis. *IEEE Transactions on Instrumentation and Measurement* March 1976, 8–14

[Law95] Lawson, C.L.; Hanson, R.J.: *Solving Least Squares Problems*. SIAM Classics in Applied Mathematics, 1995, ISBN 0-89871-356-0

[Lev44] Levenberg, K.: A Method for the Solution of Certain Problems in Least Squares. *Quart. Appl. Math.* 2, 1944, 164–168

[Mad04] Madsen, K.; Nielsen, H.B.; Tingleff, O: *Methods for Non-Lineaer Least Squares Problems*. 2nd edition, published on the WWW, April 2004

[Mar03] Markou, M.; Singh, S.: Novelty Detection: A Review. Part I + II, *Signal Processing*, 2003, Vol.83, 2481–2521

[Mar63] Marquardt, D.: An Algorithm for Least-Squares Estimation of Nonlinear Parameters. *SIAM J. Appl. Math.* 11, 1963, 431–441

[Met53] Metropolis, N.; Rosenbluth, A.; Rosenbluth, M.; Teller, A.; Teller, E.: Equation of State Calculations by Fast Computing Machines. *Journal of Chemical Physics*, Vol.21, 1953, 1087 – 1092, doi:10.1063/1.1699114

[Mey84] Meyer, E.F.: *A Practical Guide to Curve-Fitting for Chemists*. Chicago, 1984, ASIN B0007B77NK

[Mey01] Meyer, B.; Tischer, P.: Glicbawls — Grey Level Image Compression By Adaptive Weighted Least Squares. *Proc. of Data Compression Conference, DCC'01*, 27–29 March 2001, Snowbird, Utah, USA, 2001

[Pap01] Papula, L.: *Mathematik für Ingenieure und Naturwissenschaftler*. Bd.3, Vieweg, 4th edition, 2001, ISBN 3-528-34937-9

[Pot06] Pothineni, S.B.; Strutz, T.; Lamzin, S.: Automated detection and centring of cryocooled protein crystals. *Acta Crystallographica* D62, November 2006, 1358–1368

[Pre92] Press, W.H.; Teukolsky, S.A.; Vetterling, W.T.; Flannery, B.P.: *Numerical Recipes in C.* 2nd edition, Cambridge University Press, 1992, ISBN 0-521-43108-5

[Pri92] Prince, E.; Boggs, P.T.: Least squares. in *International Tables for Crystallography.* Vol.C, Mathematical, physical and chemical tables, 1992, ISBN 0-7923-1638-X, 672–682

[Rou11] Rousseeuw, P.J.; Hubert, M.: Robust Statistics for Outlier Detection. *WIREs Data Mining and Knowledge Discovery*, Wiley, Vol.1, 2011, 73 – 79

[Rya97] Ryan, T.R.: *Modern Regression Methods.* John Wiley & Sons, Inc., 1997, ISBN 0-471-52912-5

[Sch89] Schwarzenbach, D., et al: Statistical Descriptors in Crystallography. *Acta Cryst.* A45, 1989, 63–75

[Strd] http://www.itl.nist.gov/div898/strd/index.html, accessed 31 July 2015

[Str88] Strang, G.: *Linear Algebra and Its Applications.* 3nd edition, Brooks Cole, 1988, ISBN 0155510053

[Str09a] Strutz, T.: *Bilddatenkompression.* 4th edition, Vieweg+Teubner, Wiesbaden, 2009, ISBN 978-3-8348-0472-3

[Str09b] Strutz, T.: Adaptive Cluster-Based Outlier Detection. *EUSIPCO'09*, Glasgow, UK, 24-28 Aug 2009

[Str14] Strutz, T.: Adaptive context formation for linear prediction of image data. *IEEE International Conference on Image Processing 2014*, Paris, France, 27-30 Oct 2014, 5631 – 5635

[Str15] Strutz, T.: Context-based Predictor Blending for Lossless Colour Image Compression. *IEEE Transactions on Circuits and Systems for Video Technology*, Vol.x, No.99, 2015

[SVD] Singular Value Decomposition (SVD): `http://www.pdas.com/programs/fmm.f90`, accessed 31 July 2015

[Tau91] Taubin, G.: Estimation Of Planar Curves, Surfaces And Nonplanar Space Curves Defined By Implicit Equations, With Applications To Edge And Range Image Segmentation. *IEEE Trans. PAMI*, Vol.13, 1991, 1115-1138

[Tay97] Taylor, J.R.: *An Introduction to Error Analysis.* 2nd edition, University Science Books, Sausalito, California, 1997, ISBN 0-935702-42-3

[Wei98] Weisstein, E.W.: *CRC Concise Encyclopedia of Mathematics.*, CRC Press, 1998, ISBN 0849396409

[Wri84] Wright, T.W.: *A Treatise on the Adjustment of Observations by the Method of Least Squares.* Van Nostrand, New York, 1884

[Zim12] Zimek, A.; Schubert, E.; Kriegel, H.-P.: A Survey on Unsupervised Outlier Detection in High-Dimensional Numerical Data. *Statistical Analysis and Data Mining* 2012, Vol.5, No.5, 363–387

[zPWA] http://www.science20.com/, search items: puzzling weighted average, accessed 31 July 2015

Index

Made in the USA
Monee, IL
12 September 2026

62973772R00181